Microbiology and Biotechnology

Microbiology and Biotechnology

Edited by **Dean Watson**

SYRAWOOD
PUBLISHING HOUSE

New York

Published by Syrawood Publishing House,
750 Third Avenue, 9th Floor,
New York, NY 10017, USA
www.syrawoodpublishinghouse.com

Microbiology and Biotechnology
Edited by Dean Watson

International Standard Book Number: 978-1-68286-179-0 (Hardback)

Printed in the United States of America.

Contents

Preface

Over the recent decade, advancements and applications have progressed exponentially. This has led to the increased interest in this field and projects are being conducted to enhance knowledge. The main objective of this book is to present some of the critical challenges and provide insights into possible solutions. This book will answer the varied questions that arise in the field and also provide an increased scope for furthering studies.

Biotechnology plays a vital role across various industries to develop and manufacture products that involve biological systems and organisms. Microbial organisms are widely used in food processing and pharmaceutical industries. They are also utilized in distinct industrial processes like fermentation and waste water treatment. Some of the diverse topics encompassed within this book are microbial cell biology, environmental microbiology and engineering, genetics, enzymology, applied genetics, etc. The topics in this book are compiled by internationally renowned panel of authors and industry experts. It will serve as a reference to a broad spectrum of readers.

I hope that this book, with its visionary approach, will be a valuable addition and will promote interest among readers. Each of the authors has provided their extraordinary competence in their specific fields by providing different perspectives as they come from diverse nations and regions. I thank them for their contributions.

Editor

Study on combined antimicrobial activity of some biologically active constituents from wild *Moringa peregrina* Forssk

M. A. Abdel-Rahman Tahany[1]*, A. K. Hegazy[1], A. Mohsen Sayed[1], H. F. Kabiel[1], T. El-Alfy[2] and S. M. El-Komy[2]

[1]Cairo University, Faculty of Science, Cairo, Egypt.
[2]Cairo University, Faculty of Pharmacy, Cairo, Egypt.

Three plants were used in this study: *Moringa pregrina*, *Achillea fragrantissima* and *Coleome droserifolia*. Six active constituents were separated from n-hexane and ethyl acetate fractions of *Moringa pregrina*. These active constituents were lupeol acetate, α-amyrin, β-amyrin, β-sitosterol, β-sitosterol-3-O- β-D-glucoside and apignin were assayed individually and in combination against pathogenic bacteria and fungi. All constituents were proved to be more antibacterial than antifungal agents. *Aspergillus flavus* and *Fusarium solani* were completely resistant to all constituents. α-amyrin was the most effective antibacterial compound. The least relative activity was achieved by β-sitosterol against *Bacillus subtilis* compared to ampicillin. Reasonable antifungal activity was recorded in case of lupeol acetate, α-amyrin and β-amyrin, while β-sitosterol and β-sitosterol-3-O- β-D-glucoside, revealed no antifungal activity. Apignin missed both antifungal and antibacterial activities. Low MICs were detected by α-amyrin, β-amyrin and β-sitosterol-3-O-β-D-glucoside against all tested bacteria. Concerning fungi, β-sitosterol and β-sitosterol-3-O-β-D-glucoside showed no antimycotic activity. Lupeol acetate, α-amyrin and β-amyrin, however, have slightly high MICs for all tested dermatophytic fungi compared to that of fluconazole. Among ninety nine assayed combination mixtures, thirty seven synergistic combination mixtures were detected which exerts 37 synergisms against different pathogens with FICI less than 0.5, which indicates high efficacy of combination mixtures over monotherapy treatments.

Key words: Combined antimicrobial therapy, active constituents, *Moringa peregrina*, *Achillea fragrantissima*, *Coleome droserifolia*.

INTRODUCTION

In the field of ethno medicine it has been recorded that therapeutic efficacy was more pronounced when active compound was left in a particular combination with other principles (Obute, 2005). Combination therapy has many advantages and disadvantages. The advantages are additive or synergistic effect, increased spectrum of activity and decreased resistance, while the disadvantages are antagonistic effects, increased risk of drug interactions, increased toxicity and increased cost (Baddely and Pappass, 2005). Although the use of combination

therapy is an appealing alternative to monotherapy to improve the treatment outcome of invasive microbes, it remain controversial, this controversy is based on the specific mode of action of the agent used. When the combined agents act on the same molecule differently, therefore theoretically, using these two agents may lead to antagonism (Sanati et al., 1997). There are several arguments that justify the strategy of combining anti-fungal drugs to optimize therapy such as the *in vitro* data showing a potential for a synergistic effect, broader spectrum of activity and decreased risk of emergence of resistant strains and absence of a negative or harmful effects of monotherapy (Kontoyian-nis and Lewis, 2004 and Marr, 2004; Ramesh Putheti, Okigbo, R. N 2008; Chandraseker et al., 2004; Steimbach, 2005; Baddley

---

*Corresponding author. E-mail: mohsenab2005@yahoo.com.

and Pappas, 2005;).

The terminology of combination therapy used to place results in interpretive categories is often the subject of debate and confusion. Synergism and antagonism have clear and intuitive meanings. The terms (additive) and (summation) may refer to positive interaction and lead to misinterpreted. Mathematically the terms (indifferent) and "no interaction" can be used to describe results with precise interpretation (Johnson et al., 2004).

This work was conducted to evaluate the antimicrobial activity of active constituents of *Moringa pregrina, Achillea fragrantissima* and *Coleome droserifolia* in different combinations on some pathogenic bacteria and fungi.

MATERIALS AND METHODS

Study species

***Moringa peregrine* Forssk. Fiori (Family Moringaceae):** The plant grows on steep slopes of the mountains and in gravel or coarse sand sides of the wadi in Egypt. It becomes threatened by habitat distribution, over collection as fuel and seed collection. Traditionally it is used to treat slimness, constipation, headache, fever, burns, back and muscle pains. Seeds are used as coagulants in developing countries.

***Achillea fragrantissima* Forssk. Sch. Bip. (Family: Compositae):** The plant grows in the limestone wadis in Egypt and occupies various desert soil types. It becomes threatened by drought conditions and over collection for herbal medicine. Traditionally, plant is used for hyperglycemic treatments, cough and cold flu.

***Coleome droserifolia* Forssk. Delile (Family: Cleomaceae):** The species distribution in Egypt covers a wide range of phytogeographical regions. It becomes threatened by over-collection for herbal medicine. It is used traditionally for wound healing, treatment of hyperglycemia, diabetes mellitus, skin allergy and dermatitis (Abd El Ghani, 1988; Boulos, 1995 and 1999)

Extraction

UV spectra were measured using a Shimadzu UV 240 (P/N 204-58000) spectrophotometer, Mass spectra were measured using Finningan Mat SSQ 7000, 70 eV. NMR spectra were recorded at 300 (^{1}H) and 75 MHz (^{13}C) on a Varian Mercury-300 instrument. NMR spectra were recorded in DMSO-d_6, and chemical shifts were given in δ (ppm) relative to TMS as internal standard. Electrothermal 9100 for determination of melting points (uncorrected) (U.K.).For column chromatography, Sephadex LH-20 (pharmacia, Uppsala, Sweden), Silica gel: precoated plates G$_{60}$ F$_{254}$ for thin layer chromatography (TLC) and for column chromatography (Merck).

Plant material

The aerial parts of *Moringa peregrina* (Forssk.) Fiori (Family Moringaceae) were collected and identified by Prof. Hegazy, (Department of Botany, Faculty of Science, Cairo University). The collected material was air-dried reduced to powder and kept for extraction.

Extraction and Isolation

The air-dried aerial parts (650 g) were powdered then extracted by percolation with 95% ethanol (4 x 7 L) to yield (250 g) ethanolic extract residue. The residue (200 g) was suspended in distilled water and partitioned between n-hexane, chloroform, ethyl acetate and *n*-butanol (saturated with water). The solvents were separately evaporated under reduced pressure to yield 6, 3, 4.7 and 5 g respectively.

n-Hexane fraction (HF)

4 g was chromatographed over a vacuum liquid chromatography column (VLC) (Si gel H, 30 g, 5 x 3 cm). Gradient elution was carried out using n-hexane-chloroform mixtures, and chloroform-ethyl acetate mixtures. Fractions 100 ml each were collected to yield four main fractions (A-D). Fraction A (15 - 20% chloroform/n-hexane, 0.5 g) was rechromatographed over a Si gel 60 column (25 x 2 cm, 50 g), using *n*-hexane as an eluent to give 18 mg of compound 1 [R$_f$ 0.51 in system (*n*-hexane-ethyl acetate (9.5 : 0.5 v/v)] it gives a reddish purple colour with *p*-anisaldehyde-sulphuric acid spray reagent. Fraction B (25 - 30% chloroform/n-hexane, 0.6 g) was rechromatographed over a Si gel 60 column (25 x 2 cm, 50 g), using *n*-hexane-ethyl acetate (9.9: 0.1 v/v) as an eluent to give 15 mg of white needle crystals of compound 2 [R$_f$ 0.56 in system *n*-hexane-ethyl acetate (9 : 1 v/v)]. It gives a purple colour with *p*-anisaldehyde-sulphuric acid spray reagent. Fraction C (40% chloroform/n-hexane, 1.2 g) was purified by passing several times over sephadex LH-20 columns (40 × 2 cm) using chloroform-methanol (1:1 v/v) as an eluent. The purified fraction was rechromatographed over a Si gel 60 column (25 x 2 cm, 50 g), using *n*-hexane-ethyl acetate (9.5: 0.5 v/v) as an eluent to yield compound 3 (20 mg) and compound 4 (33 mg) [R$_f$ 0.4 and 0.25, respectively, in system *n*-hexane-ethyl acetate (9:1 v/v)] they give a purple and violet colours respectively, with *p*-anisaldehyde-sulphuric acid spray reagent. Fraction D (100% ethyl acetate, 0.9 g) was rechromatographed over a Si gel 60 column (25 x 2 cm, 50 g), using chloroform-methanol (9.6: 0.4 v/v) as an eluent to give white powder of compound 5 [0.37 in system Chloroform-methanol (9.5 : 0.5 v/v), it gives a violet colour with *p*-anisaldehyde-sulphuric acid spray reagent].

Chloroform fraction (CF)

2 g was chromatographed over VLC column as mentioned under the n-hexane extract. Similar fractions were pooled together to yield two main fractions. Purification of these fractions over several silica columns as under fraction C yielded compounds 4 and 5.

Ethyl acetate fraction (EF)

2 g was fractionated over a sephadex LH-20 column (25 x 3 cm) using 20, 40, 60 and 80% methanol in water mixtures as an eluent. Fractions 200 ml were collected to yield two main fractions (E and F). These fractions were purified by passing several times over sephadex LH-20 columns, using methanol as an eluent to yield compounds 6 (11 mg) and 7 (10 mg) [R$_f$ 0.45 and 0.42, respectively, in system Chloroform-methanol (9:1 v/v)].

n-Butanol fraction (BF)

4 g) was fractionated over a sephadex LH-20 column (25 x 3 cm) using 20, 40, 60 and 80% methanol in water mixtures as an eluent. Fractions 200 ml were collected to yield three main fractions (G-I).

Fractions G, H and I were purified by passing several times over sephadex LH-20 columns, using methanol and methanol-water mixtures (1:1 v/v) as an eluent to yield compounds 8 (20 mg), 9 (45mg) and 10 (30 mg), respectively [Rf 0.23, 0.34 and 0.5 respectively, in system ethyl acetate-methanol-water-formic acid (100: 16: 12:1: 0.1 v/v/v/v).

Compound 1: White microcrystalline powder, m.p. 222 - 224°C, EIMS (70 ev, rel. int.), m/z at 468 [M]$^+$ (7.9 %), 408 (40 %) [M-CH$_3$COO]$^+$, 218 (56%), 203 (77%), 189 (100%).

Compound 2: White needle crystals from *n*-hexane, m.p. 195 - 197°C, EIMS (70 ev, rel. int.), m/z at 426 [M]$^+$ (10%), 218 (100%), 203 (79%) and 189 (60%).

Compound 3: White needle crystals from *n*-hexane, m.p. 185 - 186°C. EIMS (70 ev, rel. int.), m/z at 426 [M]$^+$ (12.3 %), 218 (100%), 203 (38.46%) and 189 (34.61%).

Compound 4: White needle crystals from *n*-hexane, m.p. 140-141°C, **EIMS:** (70 ev, rel. int.), m/z at 414 [M]$^+$ (100%), 396 (51%), 329 (42%), 303 (44%), 273 (60 %) and 255 (80%).

Compound 5: White microcrystalline powder, m.p. 290°C, ^{1}H-NMR: δ (300 MHz, DMSO) 0.66 (3 H, d, J = 5.5 Hz, Me-21), 0.78 (3H, t, J = 6.3,Me-29), 0.83 (3H, d, J = 6.2 Hz, Me-26), 0.90 (3 H, d, J = 6.3 Hz, Me-27), 0.92 (3 H, s, Me-18), 0.96 (3H, s, Me-19), 3.03 (1 H, m, H-3), 4.21 (1H, d, J = 7.5, H-1`), 5.33 (H, br.s, H-6) ppm.

Compound 6: Yellow microcrystalline powder, soluble in methanol, m.p. 348-350°C. UV spectral data (Table 1).

Compound 7: Yellow microcrystalline powder, soluble in methanol, m.p.294-296°C. UV spectral data (Table 1).

Compound 8: Yellowish-white amorphous powder, soluble in methanol, UV spectral data (Table 1). ^{1}H-NMR δ ppm [300 MHz, DMSO]: 1.59 (1H,dd,J = 15,4,H-6 ax), 1.78 (2 H,m, H-2 ax,eq), 1.94 (1 H,dd,J = 13,9,H-6 eq), 3.79 (1 H,br.s, H-4), 3.94(1 H,br.s, H-5), 5.14(1 H,m,H-3), 6.18(1 H,d,J = 15.9 Hz, H-8`), 6.73 (1 H,d,J = 6.6Hz, H-5`), 6.94 (1H,dd,J = 8.1,2Hz, H-6`), 7.04 (1 H,br.s, H-2`), 7.40 (1 H,d,J = 15.9Hz , H-7`).

Compound 9: Yellow amorphous powder, soluble in methanol, UV spectral data (Table 1). ^{1}H-NMR δ ppm [300 MHz, DMSO]: 0.97 (3 H,d,J = 5.1, CH$_3$-6```), 3.84(3 H,s,OCH$_3$), 4.39(1 H, d,J = 2.1,H-1````),5.42(1 H,d,J = 7.2,H-1``), 6.18 (1 H, d, J = 2.4 Hz, H-6), 6.37 (1 H, d, J = 2.1 Hz, H-8), 6.90 (1 H,d, J = 8.7 Hz, H-5`), 7.50 (1 H,dd, J = 1.2, 6.6 Hz, H-6`), 7.53(1 H,s,H-2`). $^{13}$C-NMR: δ [75MHz, DMSO] 66.84 (C-6``), 70.09(C-4``), 74.27(C-2``),75.91(C-5``),76.39(C-3``), 101.22 (C-1``), 17.68(C-6```), 68.26(C-5```) 70.29(C-2```), 70.58(C-3```), 71.79 (C-4```), 100.86(C-1```), 93.60(C-8), 98.78 (C-6), 103.88(C-10), 115.07(C-2`), 116.22(C-5`), 121.03(C-6`), 122.26 (C-1`), 133.00(C-3), 144.73(C-3`), 146.87 (C-2), 148.43(C-4`), 156.37 (C-9), 161.14(C-5), 164.44 (C-7), 178.07(C-4).

Compound 10: Yellow amorphous powder, soluble in methanol, UV spectral data (Table 1). ^{1}H-NMR δ ppm [300 MHz, DMSO]: 3.70 (3H,s,OCH$_3$), 3.74 (3 H,s,OCH$_3$), 4.66 (1 H,d,J = 9.6,H-1``), 6.75 (1 H, s, H-3), 6.89 (2 H,d,J = 8.2Hz, H-3`,5`), 7.98 (2 H,d,J = 8.2Hz, H-2`,6`). $^{13}$C-NMR : δ [75 MHz, DMSO] 61.96 (C-6``), 70.85(C-4``), 70.86 (C-2``), 73.45(C-1``), 78.83(C-3``),82.01(C-5``), 56.61(4`-OCH$_3$), 61.7(6-OCH$_3$),102.41(C-3), 103.96(C-8), 104.60(C-10), 116.14(C-3`,5`), 121.95 (C-1`), 129.36 (C-2`,6`), 133.07 (C-6), 156.05 (C-9), 160.43(C-5,4`), 162.88 (C-2), 164.11 (C-7), 181.99(C-4).

Active constituents

Five compounds were isolated from the n-hexane fraction and were identified as lupeol acetate (1), β-amyrin (2), α-amyrin (3), β-sitosterol (4) and β-sitosterol -3-O- β-D-glucoside (5). One compound was isolated from the ethyl acetate was identified as apigenin (6) Table 1 (El-Alfy et al., 2009).

Antimicrobial activity

The antimicrobial activity was assayed against three Gram +ve: *Bacillus subtilis* (ATCC 6051), *Staphylococcus aureus* (ATCC 12600) and *Streptococcus faecalis* ATCC 19433) and three Gram--ve: *Escherichia coli* (ATCC 11775) *Neisseria gonorrhoeae* (ATCC 19424) and *Pseudomonas aeruginosa* (ATCC 10145), and seven fungal species: two filamentous moulds: *Aspergillus flavus* L. and *Fusarium oxysporum* L. and four dermatophytic fungi: *Candida albicans* (ATCC 26555), *C. krusei* (ATCC 6258), *C. parapsilosis* (ATCC 22019), *C. tropicalis* (ATCC 750) and *Saccharomyces cerevisiae* (2180-1A).

Bacteria and dematophytic fungi were grown and maintained on nutrient agar slants and Sabouraud dextrose agar slants, respectively. They were then stored under aerobic conditions. The dermatophytic fungi were cultured overnight, while moulds for 3 days at 30°C in Sabouraud dextrose broth and bacteria were cultured overnight at 35°C in nutrient broth. The minimum inhibitory concentration (MIC) was determined according to National Committee for Clinical laboratory Standards (NCCLS) M38-A microdilution method (Ibrahim et al., 2009), using (12 x 8 wells) microtitre plates. Aliquots (50 µl) of the sample (single or combined 1:1 v/v) stock solution and 200 µl of the inoculum were pipetted to the well labeled as (A). Only 100 µl of each inoculum were added to the wells labeled (B-H). The inoculum and sample in the well (A) were mixed thoroughly before transferring 100 µl of the resultant mixture to well B. the same procedure was repeated for inoculum mixture in well (B) to (C) and repeated from wells (C-H) thus creating a serial dilution of the test materials.

Ampicillin was used as a standard reference antibiotic for comparison with the antibacterial activities, while fluconazol was used as the standard in the antifungal activity test. After an incubation period at 30°C for 24 h. Turbidity was taken as indication of growth. Thus the lowest was taken as the minimum inhibitory concentration (MIC). The MIC was recorded as the mean of triplicates.

Quantitative mathematical analysis of combination therapy

The fractional inhibitory concentration index (FICI), calculated by use of checkerboard method, has long been the most commonly used way to characterize the activity of antimicrobial combinations in the laboratory (Mukherjee et al. 2005). FICI is determined by dividing the MIC of each drug when used in combination by the MIC of each drug when used alone.

FICI = MIC$_a$ in combination / MIC$_a$ tested alone + MIC$_b$ in combination / MIC$_b$ tested alone, Where MIC$_a$ and MIC$_b$ are the MICs of drugs a, b. FICI > 4 defines antagonism FICI 0.5-4 defines no interaction (indifference), FICI < 0.5 defines synergism

Statistical analysis

Chi-square test and one way ANOVA test were used. All results are expressed as mean ± St. Error. F values < 0.05 were considered significant (Lewis, 1984).

Table (1). FICI of the synergistic combination mixtures to the tested microorganisms.

Synergistic combinations	B. subtilis FICI	E. coli FICI	N. gonorrhoroea FICI	P. aeruginosa FICI	S. aecalis FICI	C. lbicans FICI	C. krusei FICI	C. arapsilesis FICI	C. tropicalis FICI	S. cereviseae FICI	S. aureus FICI	Total no. of synergy
Lupeol acetate + β-amyrin			0.37									1
Lupeol acetate + β-sitosterol		0.49										1
Lupeol acetate + β-sitosterol -3-O- β-D-glucoside		0.35									0.39	2
β-amyrin + α-amyrin			0.47									1
β-amyrin + β-sitosterol	0.35			0.32								2
α-amyrin + β-sitosterol -3-O- β-D-glucoside			0.41		0.46							2
β-sitosterol + β-sitosterol -3-O- β-D-glucoside				0.40								1
Lupeol acetate + A. vol. oil				0.48		0.41	0.28	0.39	0.35			5
Lupeol acetate + C. vol. oil							0.38					1
α-amyrin + A. vol. oil						0.30	0.40					2
α-amyrin + C. vol. oil			0.40			0.30	0.40					3
β-amyrin + A. vol. oil						0.40		0.40				2
β-amyrin + C. vol. oil			0.40			0.30						2
β-sitosterol + A. vol. oil		0.40	0.40									2
β-sitosterol + C. vol. oil	0.40		0.30									2
Lupeol acetate + Flaconazole						0.27	0.30					2
α-amyrin + Fluconazole							0.30			0.40		2
β-amyrin + Fluconazole						0.30	0.40					2
β-amyrin + ampicillin	0.35											1
β-sitosterol -3-O- β-D-glucoside + ampicillin				0.40								1
Total no. of synergy	3	3	7	4	1	7	7	2	1	1	1	37

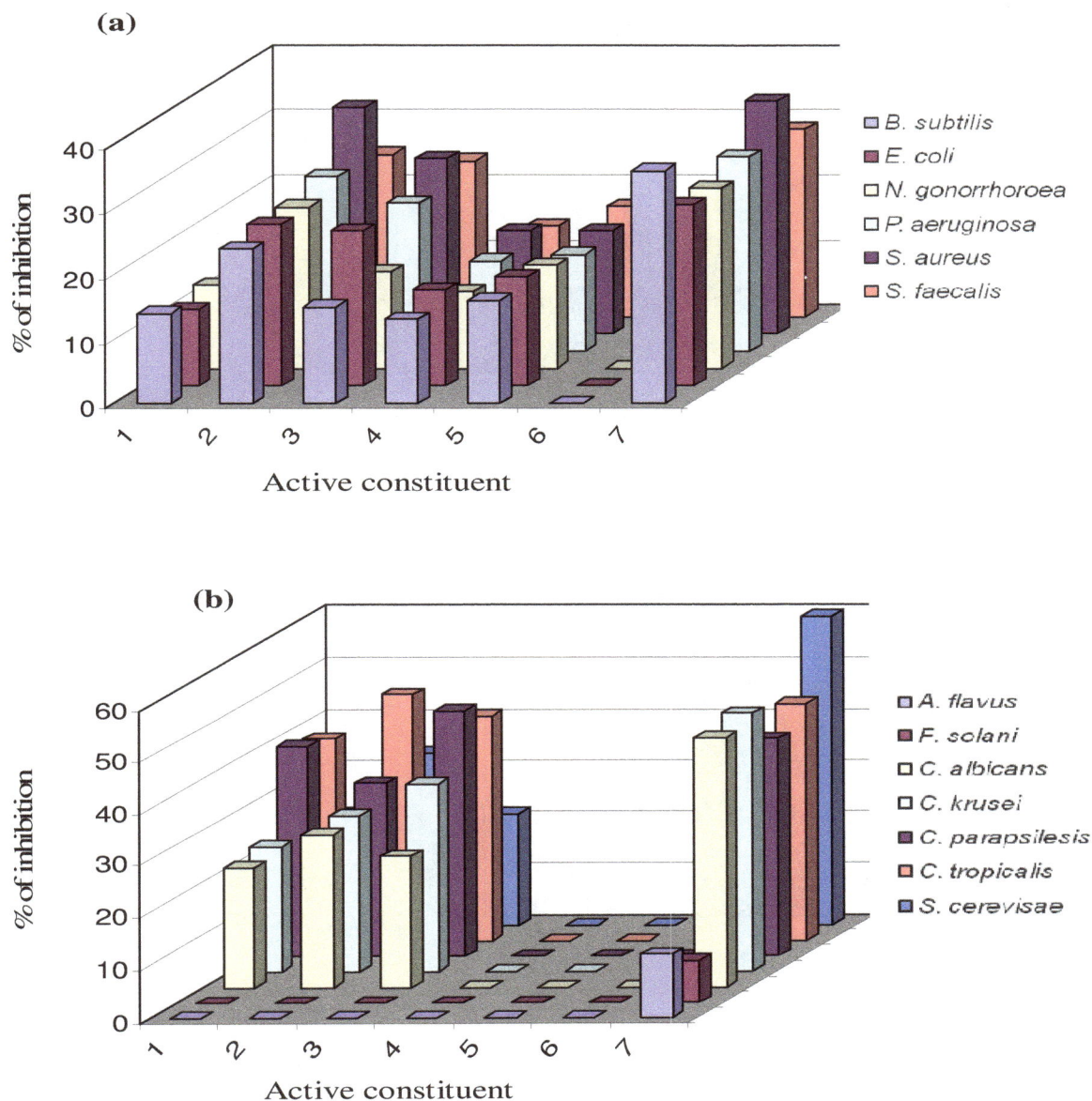

Figure 1. Antimicrobial susceptibility of the active constituents separated from *Moringa Peregrina* for (a) Bacteria: $P < 0.01$, $F = 0.5$ and (b) Fungi $P < 0.01$, $F = 0.6$)). 1: Lupeol acetate, 2: β-amyrin, 3: α-amyrin, 4: β-sitosterol, 5: β-sitosterol-3-O-glucoside, 6: apigenin, 7: ampicillin or fluconazol (standard drugs).

RESULTS AND DISCUSSION

The data presented in Figure (1) showed that all compounds separated from n-hexane and ethyl acetate fractions of *M. peregrina* proved to be more antibacterial than antifungal agents. The filamentous moulds *A. flavus* and *F. solani* were completely resistant to all compounds. β-amyrin was the most effective antibacterial compound and achieved high relative activity with the reference antibiotic apmicillin reaching 97.2% in case of *S. aureus*. Least relative activity (36.1%) was achieved by β-sitosterol against *B. subtilis*.

Reasonable antifungal activity was recorded in case of lupeol acetate, α-amyrin and β-amyrin with relative activity ranging from 21.6 - 47.6% compared to fluconazole activity. B-sitosterol and B-sitosterol-3-o-glucoside however, revealed no antifungal activity. On the other hand, Apignin missed both antifungal and antibacterial activities, so it was excluded from the next experiments. Relatively low MICs were detected by α-amyrin, β-amyrin and B-sitosterol-3-o-glucoside against all tested bacterial species while lupeol acetate recorded higher MIC (Figure 2). Concerning fungi, β-sitosterol and β-sitosterol-3-o-glucoside showed no antimycotic activity, lupeol acetate,

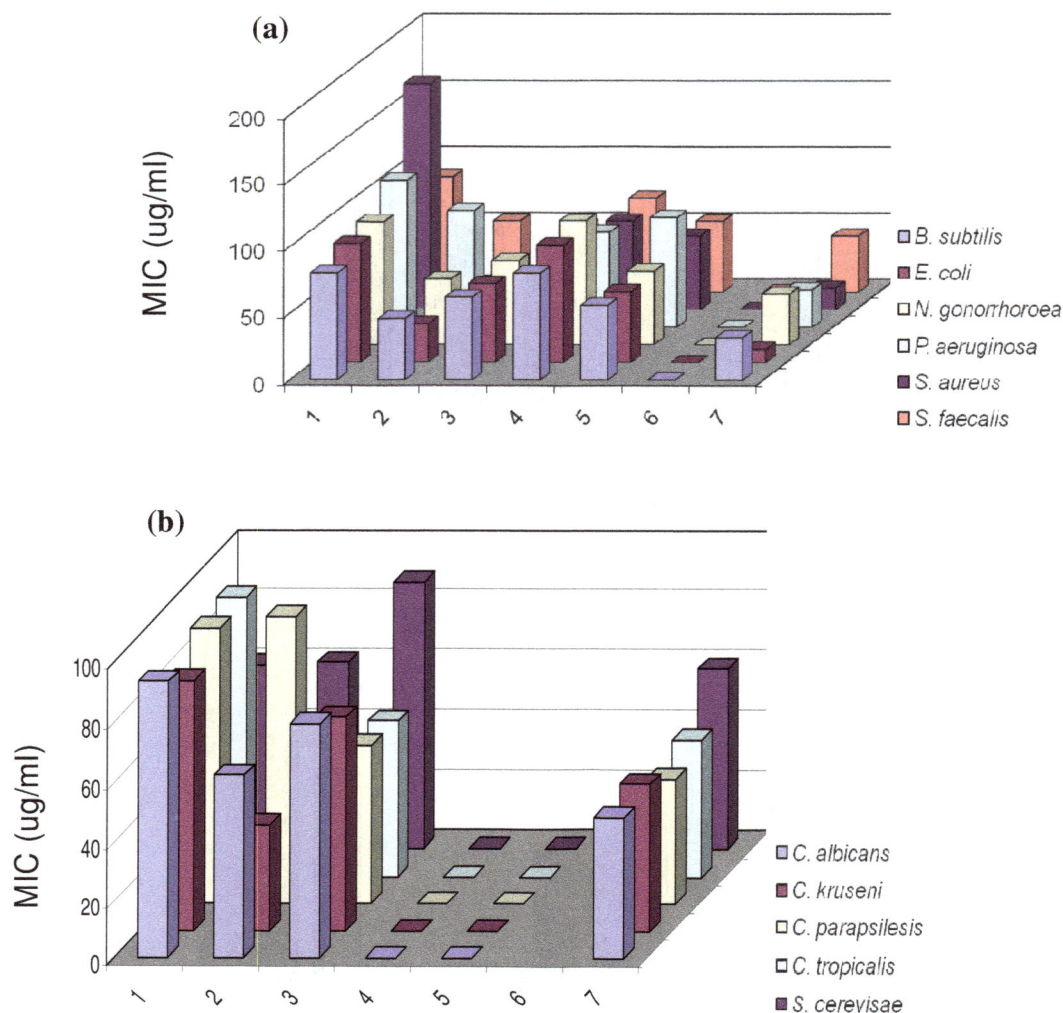

Figure 2. MIC for the active constituents separated from *Moringa Peregrina* for (a) Bacteria (P < 0.02, F = 2.1) and (b) Fungi (P < 0.03, F = 1.2)). 1: Lupeol acetate, 2: β-amyrin, 3: α-amyrin, 4: β-sitosterol, 5: β-sitosterol-3-O-glucoside, 6: apigenin, 7: ampicillin or fluconazol (standard drugs).

β-amyrin and α-amyrin, however, have slightly high MICs for all tested dermatophytic fungi compared to that of fluconazole.

Previous surveys demonstrated the wide occurrence of active antimicrobial substances in higher plants. For centuries plant and herbs have been used in various parts of the world for the treatment of certain diseases. Yet a scientific study of plants to determine their content of antimicrobial materials is comparatively new. Our finding that *M. peregrina* compounds have wide spectrum antimicrobial activity was also recorded by Eilert et al. (1981) who screened the antimicrobial activity of *M. peregrina* against six Gram +ve and seven Gram –ve bacteria and found that water extract, ethanol and petroleum ether extracts possess high antimicrobial activity. *Moringa oleifera* provides a rich and rare combination of zeatin, quercetin, β-sitsterol, caffeoylquinic acid and

kaempferol which have antifungal and antibacterial activities (Anwar et al., 2006; Ramesh Putheti, Okigbo RN, 2008).

The high mortality of microbial infections and the relatively limited efficacy of current agents have produced significant interest in combination therapy for these agents to treat infections so in the present study, ten combination mixtures of dual active constituents were investigated for MIC and FICI determination against bacteria and fungi.

The fractional inhibitory concentration index (FICI) recorded "no interaction" (FICI 0.5-4) in relatively large number of combinations between the active constituents of *M. peregrina* reaching 100% in combination of α-amyrin plus β-sitosterol (mixture 8) (Figure 3). Antagonism, however, rarely occurred as it recorded once in case of lupeol acetate plus β-amyrin (mixture 1) against

(a)

(b)

Combination mixture

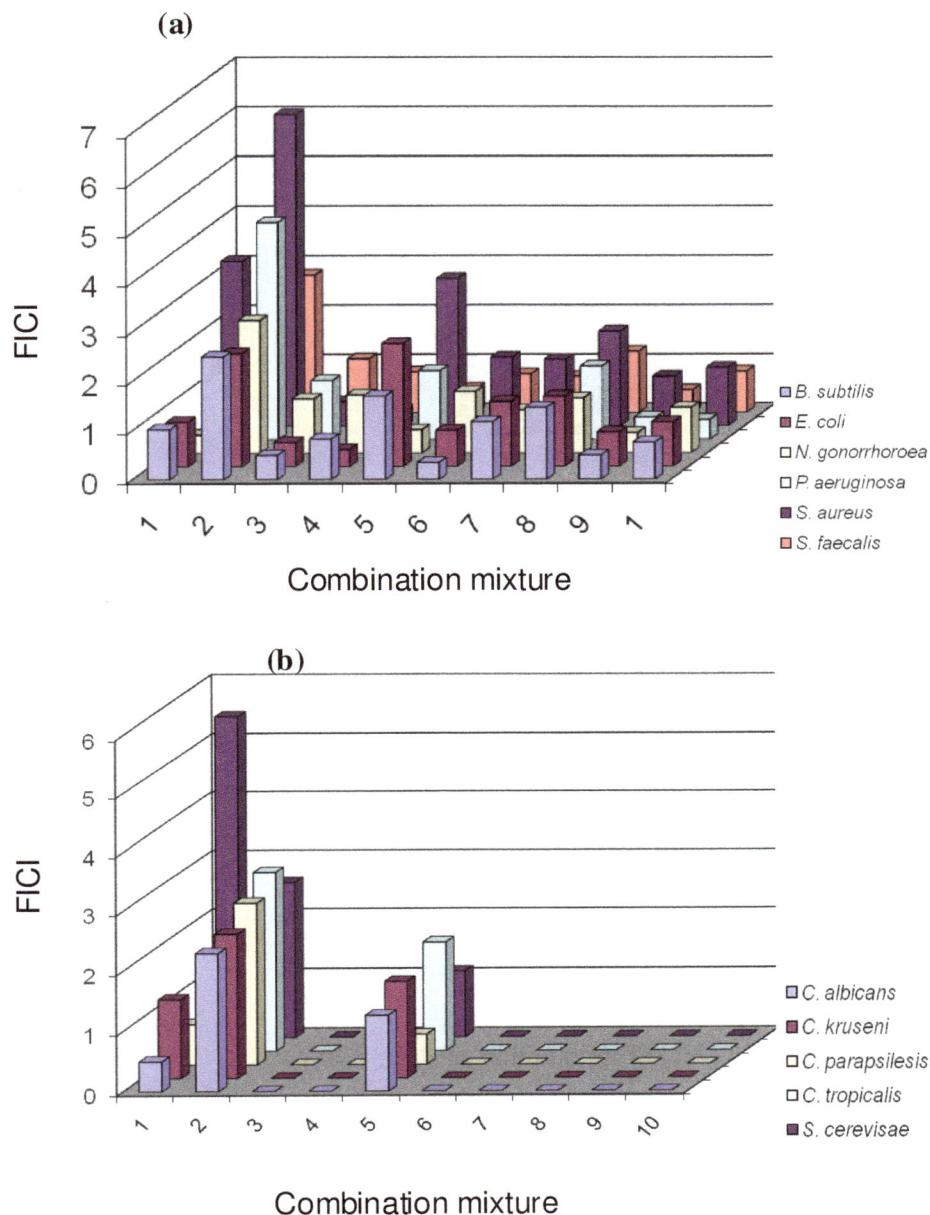

Figure 3. Combination antimicrobial of dual active constituents separated from of *M. peregrine* (Bacteria (P < 0.03, F = 2.1), Fungi (P < 0.03, F = 1.9)). Combination mixtures: 1: Lupeol acetate + β-amyrin, 2: Lupeol acetate + α-amyrin, 3: Lupeol acetate + β-sitosterol, 4: Lupeol acetate + β-sitosterol-3-O-glucoside, 5: β-amyrin + α-amyrin, 6: β-amyrin + β-sitosterol, 7: β-amyrin + β-sitosterol-3-O-glucoside, 8: α-amyrin + β-sitosterol, 9: α-amyrin + β-sitosterol-3-O-glucoside, 10: β-sitosterol + β-sitosterol-3-O-glucoside

S. cerevisiae (FICI, 5.44) and twice in case of lupeol acetate plus α-amyrin (mixture 2) against *P. aeruginosa* and *S. aureus* (FICI, 4.4 and 6.3 respectively). It was worthy noting that eleven cases of synergy between *M. peregrina* active constituents against all tested bacterial species (Table 1). Synergy means that the combined therapy is more effective in killing pathogen than mono-therapy.

High synergistic mixtures were detected for fungi, but no interaction mixtures were detected for fungi and no antagonistic interactions were recorded except mixture 2 was antagonistic to *S. cerevisiae*. As the focus of this study is on the efficacy of combination antimicrobial drugs especially pathogens, another combination experiment

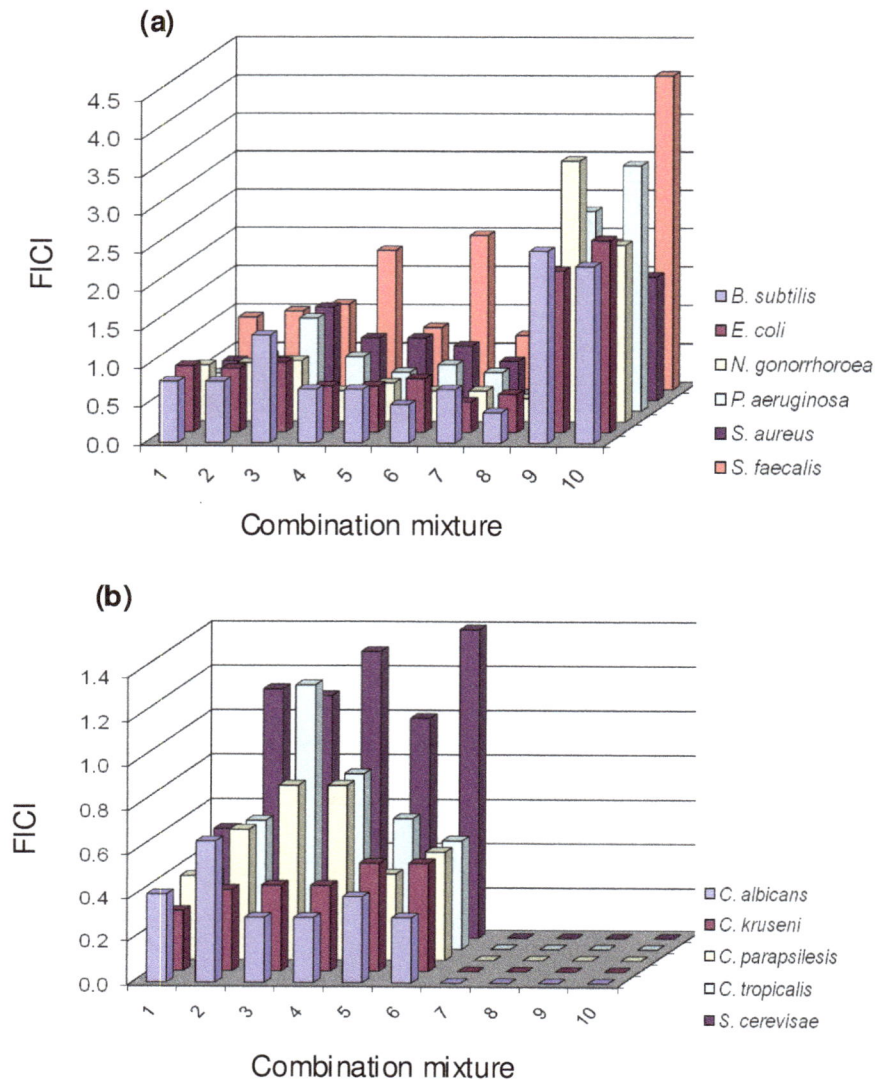

Figure 4. Combination antimicrobial therapy of the active constituents separated from *M. peregrine* plus volatile oils extracted from *Achillea fragrantissima* and *Coleome droserifolia* (Bacteria ($P < 0.03$, $F = 1.2$), Fungi ($P < 0.01$, $F = 0.8$)). Combination mixtures: 1: Lupeol acetate + A. vol. oil, 2: Lupeol acetate + C. vol. oil, 3: α-amyrin + A. vol. oil, 4: α-amyrin + C. vol. oil, 5: β-amyrin + A. vol. oil, 6: β-amyrin + C. vol. oil, 7: β-sitosterol + A. vol. oil, 8: β-sitosterol + C. vol. oil, 9: β-sitosterol-3-O-glucoside + A. vol. oil, 10: β-sitosterol-3-O-glucoside + C. vol. oil.

using *M. peregrina* active constituents plus volatile oils from *Achillea fragrantissima* or *Cleome droserifolia* was carried out.

Data in (Figure 4) revealed "no interaction" against bacteria in almost all combinations (FICI 0.5-4) except seven cases of synergism and one case of antagonism. The antagonistic interaction was between the B-sitosterol-3-o-glucoside and volatile oil of *C. droserifolia* (mixture 10) against *S. faecalis*. Four of the seven synergistic mixtures acted against the serious pathogen *N. gonorrhoeae* one acted against each of *P. aerugnosa*,

E. coli and *B. subtilis* (Table 1).

Regarding fungi, high synergy was recorded in the combination between volatile oil of *A. fragrantissima* plus lupeol acetate covering 4 of the 5 tested dermatophytes (table 1) which represent promising finding in antimycosis therapy. Five of 6 mixtures acted synergistically against *C. albicans* while 4 of 6 mixtures were synergistic against *C. krusei* where the rate of killing increased with reduction of MIC (Table 1). Mixtures from 7 to 10 have lost the antifungal activity totally. The antimicrobial activity of *A. fragiantissima* and *C. viscose* was also reported by Barel

Intergeneric protoplast fusion of yeast for high ethanol production from cheese industry waste – Whey

Krishnamoorthy, R. Narayanan[*], K. Vijila and K. Kumutha

Department of Agricultural Microbiology, Centre for Plant Molecular Biology and Biotechnology, Tamil Nadu Agricultural University, Coimbatore, Tamil Nadu, India, Pin: 641 003.

In this study, Intergeneric protoplast fusion of the yeast cultures *Viz., Saccharomyces cerevisiae* and *Kluyveromyces marxianus* was carried out for enhancing production to ethanol temperature tolerance and lactose utilizing characters in a single stain. *S. cerevisiae* (Parent 1) has ethanol tolerance, whereas, *K. marxianus* (Parent 2) is temperature tolerant and also has lactose utilizing capacity. Twelve fused cultures were obtained by protoplast fusion. Fused cultures recorded higher DNA content than the parent strains, which showed complementary banding pattern of two parental Strains. SDS-PAGE confirms the presence of HSP 70 in the fused culture, which is responsible for temperature tolerance. Fermentation of cheese whey was carried out with two parental and fused cultures. The results revealed that the ethanol production was higher with fused culture (12.5%, with 18.09 g/l of biomass) after 72 h of fermentation. Parent 1 showed poor growth on the cheese whey medium, but growth of the Parent 2 was inhibited when the ethanol production reached 6%.

Key words: Marker selection, protoplast fusion, RAPD, SDS-PAGE, whey.

INTRODUCTION

Natural energy resources such as petroleum and coal have been consumed at high rate over the last decades. The heavy reliance of the modern economy on these fuels is bound to end, due to their environmental impact and to the fact that they might eventually run out. Therefore, alternative resources such as ethanol are becoming more important. Bio-ethanol is one of the most important renewable fuels contributing to the reduction of negative environmental impacts generated by the worldwide utilization of the fossil fuels.

Some biological processes have rendered possible route to produce ethanol in large volume using the cheap resources (Gunasekaran and Raj, 1999). The massive need of fuel ethanol in the world requires that, its production technology should be cost-effective and environmentally sustainable. For current technologies employed at commercial level, the main share in the cost structure corresponds to the feedstocks (above 60%)

followed by the cooling cost of the fermentor (Sing and Lindquist., 1998). High energy input is required to maintain tempe-rature between 25 and 35°C in fermentation process to maximize ethanol production and prevent irreversible heat inactivation of yeast cells.

Many opportunities may be explored using different cheaper renewable waste materials with a lot of usable substrate (for example, whey, agricultural food waste, wood chips, molasses and newspaper waste) for microorganisms to grow upon and then produce useful products for society.

The production of cheese has reached $11 - 12 \times 10^6$ tones per year and liquid whey pro-duced in these processes has reached around 10^8 tones per year. This whey contains 5 - 6% lactose, 0.8 - 1% protein and 0.06% fat (Kosikowsk, 1979). However, dis-posal of whey without expensive sewage treatment can represent a source of water pollution because it's high BOD which is 50,000 ppm. This is one of the least expensive Carbon source for ethanol production.

Although, earlier attempts were made to use lactose for alcohol production, the major problem was the inability of *Saccharomyces cerevisiae* to ferment lactose. *Kluyveromyces marxianus* is known to ferment lactose (Ferrari et al., 1994).

*Corresponding author. E-mail: krnarayanan_2001@Yahoo.com.

Abbreviation: MTCC, Microbial type culture collection.

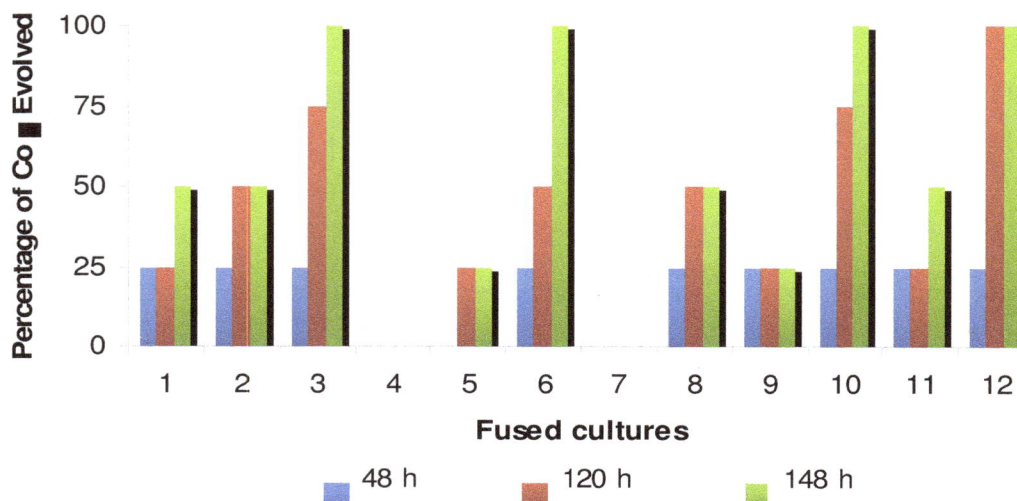

Figure 1. Percentage of CO_2 production by the recombinant yeast strains at 42°C.
*1-12 - Fused culture 1 to 12

However, it has been shown that only a fraction of the available lactose is converted to ethanol, possibly due to ethanol inhibition. *S. cerevisiae* cells can grow in high ethanol concentration, but it does not have capacity to utilize lactose as carbon source and high temperature tolerance like *K. marxianus*. To overcome the above-mentioned problems, a genetically engineered strain of *S. cerevisiae* that expresses β-galac-tosidase activity was reconstructed with the capabilities for the bioconversion of lactose in whey into ethanol and temperature tolerance. Successful attempts in this respect yielded the recombinant yeast strains that were developed by intergeneric protoplast fusion of *S. cerevisiae* and *K. marxianus* (Figure 1).

METHODOLOGY

Collection of standard cultures and genotypic difference

S. cerevisiae (MTCC 181) and *K. marxianus* (MTCC 188) were obtained from Microbial Type Culture Collection (MTCC), Chandigarh, India. Five differentiating characters *viz*, antibiotic tolerance (Barnett et al., 1986), crystal violet tolerance (Lin et al., 1985), ethanol tolerance, temperature tolerance (Banat et al., 1992) and carbon source preference (Stvitree et al., 2000) were used to differentiate two parental cultures.

Protoplast formation and fusion (Farahnak et al., 1986)

The two parental cultures were inoculated in 100 ml YEPD (Yeast extract- 10.0 g, Peptone – 20 g, Glucose - 20.0 g, Distill water – 1000 ml) broth and incubated at 35°C for 24 h. Five ml of the 24 h old parent cultures ($\sim 4 \times 10^7$ cfu per ml) were taken (three set) in centrifuge tube and centrifuged at 6000 rpm for 5 min. The supernatant was discarded and pellet was resuspended in 1 ml of protoplasting solution with different concentration (2, 5 and 10 ppm) of cell wall lysing enzyme (Glucanex) (Petit et al., 1994). This suspension was kept at 35°C for 1 h with occasional shaking. After 1 h the protoplasts were harvested by centrifugation (6000 rpm for 10 min), washed three times with protoplast buffer and resuspended in 1 ml of same buffer.

Protoplasts of each parental culture suspended in 5 ml of protoplast buffer was taken and added into Eppendorf tube containing 1 ml of PEG solution. This mixture was incubated in room temperature for 20 min, followed by centrifugation at 6000 rpm for 5 min and pellet was resuspended on 1 ml of protoplast buffer. Fused protoplasts in one ml of this buffer were transferred to 10 ml of regeneration medium taken in conical flask and incubated at 30°C for 72 h.

Screening of the fused cultures

The fermentation ability of the fused cultures was assessed (Van der walt, 1970) by growing the cultures at 42°C in the test tubes containing 10 ml of YEPD medium with lactose and 10% ethanol. The percentage of CO_2 evolved was calculated by the amount of CO_2 filled in the Durham's tube after 48, 72 and 120 h of fermentation. Based on the Durham's tube, amount of CO_2 filled in the percentage was recorded as 25, 50, 75 and 100%. Genotypic stability of the recombinant cultures was tested by streaking the slant cultures on the YEPD medium with markers. This was assessed on 0, 2, 4, 6, 8, 10, 15 and 20 weeks after fusion (Dziuba and Chmielewska, 2002).

Confirmation of protoplast fusion

The genomic DNA of the parental and fused culture was extracted by the method given by Melody (1997) with slight modifications. The Nanodrop ® ND-1000 was used to quantify the DNA content and random amplified polymorphic DNA (RAPD) was carried out as per the method given by Martins et al. (1999).

Table 1. Differentiating characters of parental cultures.

Yeast cultures	Cycloheximide tolerance (ppm)	Crystal violet tolerance (ppm)	Temperature tolerance (°C)	Carbon source	Ethanol tolerance (%)
Parent 1	< 10	>100	39	Glucose	>14
Parent 2	>100	<10	42	Lactose	<10

Parent 1 – *S. cerevisiae*
Parent 2 – *K. marxianus*

SDS -PAGE (Sodium Dodecyl Sulphate Polyacrylamide Gel Electrophoresis)

The protein extraction was done by the method described by Saumya and Hemchick, 1983 and it was estimated by Lowry's method (Lowry et al., 1951). The extracted protein samples were used in SDS -PAGE.

Whey fermentation

Substrate preparation (Mahoney et al., 1975)

Cheese industry waste was collected from Acres wild organic cheese making farm, Coonoor, Nilgris. The whey was heated at 90°C for 15 min, cooled and centrifuged at 1000 rpm for 10 min and removed the coagulated proteins. The whey medium containing 4.8% lactose was supplemented with 0.03 g K_2HPO_4, 0.05 g yeast extract per 100 ml of whey. The medium was adjusted to pH 5.5 with 0.01 N sulphuric acid, autoclaved for 15 min at 121°C, cooled to 30°C and used for the fermentation.

Fermentation and analysis

Fermentation using the parental and fused cultures was conducted in 250 ml of Erlenmeyer flasks, which contained 100 ml of cheese whey medium (CWM) (Mahoney et al., 1975). In the medium 1.0% of inoculum, which contains approximately 40×10^6 cfu per ml, was added and incubated in shaker at 42°C, 100 rpm. Biomass, lactose content and ethanol was estimated calorimetrically.

RESULT AND DISCUSSION

In order to differentiate the hybrids from parent cultures, five different set of parameters were taken into account. The first one is that anti-fungal antibiotic resistance characters of the both parental cultures were assessed. It has been found that parent 2 has tolerance to high level of cycloheximide concentration compared to parent 1 (Table 1). Tahoun et al. (1999) used cycloheximide, imazilil to differentiate the parental cultures *S. cerevisiae* ATTC 4129 and *K. fragilis* CBS 683. The results of the *S. cerevisiae* showed susceptibility to cycloheximide even at low concentration and *K. fragilis* was susceptible to imazilil.

As the second marker, the temperature tolerance was taken in to account. Table 1 represents a comparison of parental strains to different temperatures which indicated that, parent 2 grown well in 42°C compared to parent 1.

Dziuba and Chmielewska (2002) differentiated the parental strains *S. cerevisiae*, *Pachysolen sp* and *Candida shehatae* based on the temperature tolerance. They observed that *S. cerevisiae*, *Pachysolen sp* showed tolerance upto 42°C and *C. shehatae* was sensitive to high temperature. The ethanol tolerance capacity of the parental strains were tested in the parent study and it was found that parent 1 showed tolerance to high ethanol concentration (12%) compared to parent 2 (6%). Ethanol has long been known to inhibit the growth of yeast at high concentration. Holzberg et al. (1967) observed that a threshold concentration of ethanol (about 0.6 M) below which there was no inhibition and above that level inhibition followed a linear pattern. The above finding supported the result of the present investigation; the parent 1 was able to grow upto 12% ethanol concentration. However, the decreased growth was observed above 12% concentration.

Screening of fused culture

Twelve fused cultures were taken from fusion studies, in which the cultures FC3, FC6 and FC10 with good fermenting ability were selected and there strains took 120 h (Figure 1) to fill the Durham's tube. The fused culture FC12 was the best lactose fermenting culture, as Durham's tube was filled within 72 hr. FC1, FC2, FC5, FC8, FC9 and FC11 were found to be poor in fermenting the lactose. The key issue in the assessing the usefulness of recombined strains are the ability to consume lactose in the medium, high ethanol and high temperature tolerance. Based on the growth and CO_2 evolved, the fused culture FC12 was selected as the best culture. Similarly Dziuba and Chmielewska (2002) found that 6 out of 18 hybrids of *S. cerevisiae* CD 43 and *C. shehatae* ATTC 58779 have not shown the ability to convert the xylose to ethanol. The hybrids CD43-7 and CD 43-9 were characterized as the best converting ability, because they finished the fermentation after 120 h.

The stability was assessed on 2, 4, 6, 8, 10, 15 and 20 weeks after fusion, almost 40% of the fused cultures lost their features received from parental cultures in the 4th week after fusion (Table 2). Pasha et al. (2007) tested stability of 3 fused cultures for 12 month and found that only one fused culture Viz., CP11 was stable after 12 month; another 2 cultures lost their stability during that

Table 2. Genotypic stability of the recombinant strains tested for a period of 20 weeks after protoplast fusion in YEPD slant for 20 weeks.

Stability of the fused culture	Weeks after fusion Weeks							
	0	2	4	6	8	10	15	20
FC 1	+	+	-	-	-	-	-	-
FC 2	+	+	-	-	-	-	-	-
FC 3	+	+	+	+	+	+	+	+
FC 5	+	-	-	-	-	-	-	-
FC 6	+	+	+	+	+	-	-	-
FC 8	+	+	+	+	+	+	+	-
FC 9	+	+	+	+	+	-	-	-
FC 10	+	+	+	+	+	+	+	-
FC 11	+	-	-	-	-	-	-	-
FC 12	+	+	+	+	+	+	+	+

*FC - Fused culture; (+) - Stable ; (-) - Not stable.

Table 3. DNA quantification of the parental and the selected recombinant strains using Nanodrop.

Yeast strain	*S cerevisiae*	*K. marxianus*	Fused culture 12
DNA content (ng/µl)	456.4 ± 19.02	571.4 ± 12.52	892.5 ± 15.68

period. The result of present investigation is similar to above findings. Out of the ten fused culture (except FC4 and FC7) studied for the biomass and ethanol production at 42 C using lactose as carbon source, fused cultures FC3 and FC12 were found to be stable for growth and ethanol production during the study period (20 weeks). Remaining 8 cultures lost the ability to utilize the lactose after 15 weeks (Table 2). Based on the fermentation ability and stability of the culture, FC 12 was taken for the further studies to confirm their stability and fermentation efficiency.

Recombinant characters of the fused cells

Genetic level confirmations of fusion were done by DNA quantification and RAPD. DNA content of the parent and fused culture were estimated using Nanodrop, the DNA content of parent 1, parent 2, fused culture were 456.4, 571.4 and 892.5 ng/µl of genomic DNA solution, respectively. The DNA content of the fused culture was higher than the both parents, because of the nuclear fusion that occurred between both the parents (Table 3). Tahoun et al. (1999) measured the DNA content of the parental and fused culture by using diphenylamine reagent method. Result showed that DNA content of the fused culture was higher than the parental cultures. The result of the present investigation was in accordance with the above finding.

RAPD was done for the further confirmation of fusion;

using 3 primers (OPQ2, OPQ4 and OPQ6) and the amplification pattern of the DNA fragments for the strains under investigation were obtained, as shown in Figure 2. The fused culture FC 12 was confirmed as nuclear fusion of both parental cultures, because of the complementary banding pattern of the parental strains. This is conformed with the findings of Francis and Clair (1993) who demonstrate that F1 hybrids showed bands characteristic of both parental strains.

Polypeptide profile of temperature tolerant strains by SDS-PAGE analysis

The polypeptide profile was seen with reference to protein marker ranging from 14.3 - 97 KDa. Lane 1 of 10% polyacrylamide shows the medium molecular weight marker, lane 2, 3, 4 were represent the protein profile of parent 1 exposed to different temperatures (35, 40 and 45 C. The parent 2 protein profiles and fused culture protein profile are lanes shown in 5, 6, 7, 8, 9 and 10. HSP 70 was seen in all the lanes of the parent 2 and fused culture (Figure 3). In the present investigation, the higher thermotolerance of parent 2 and fused cultures were positively correlated with the synthesis of HSP 70 KD. It has already been reported that HSP 70 KD was involved in a variety of functions, such as helping newly synthesized proteins to assemble and fold correctly, denaturation of proteins (or prevention of denaturation) during a heat shock, facilitation of protein transport and

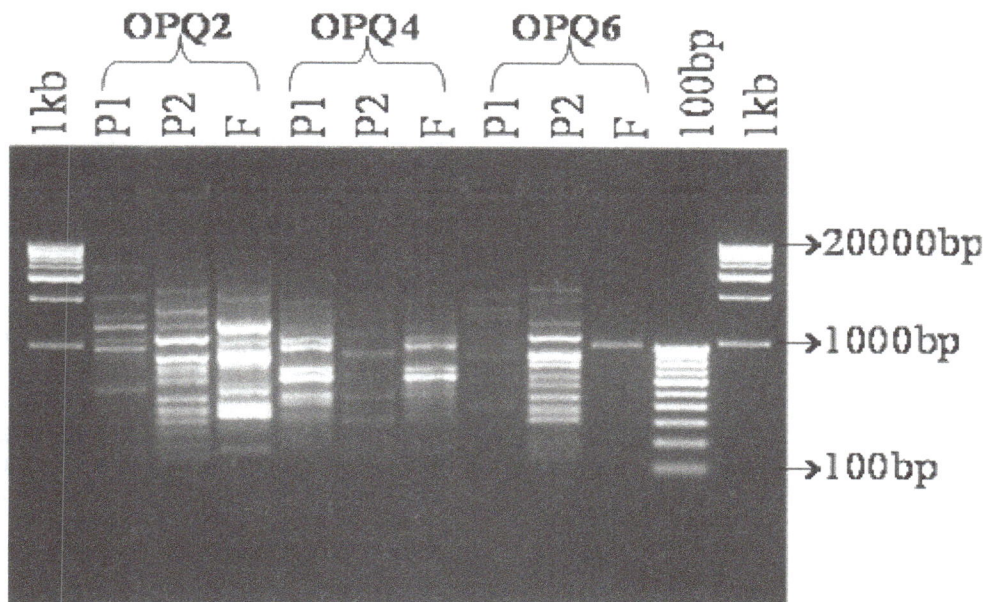

Figure 2. Random amplified polymorphic DNA of the parental and fused cultures.
*OPQ2, OPQ4, OPQ6 – Primer, P1 - Parent 1 – *S. cerevisiae*, P2 - Parent 2 – *K. marxianus*, F - Fused culture 12.

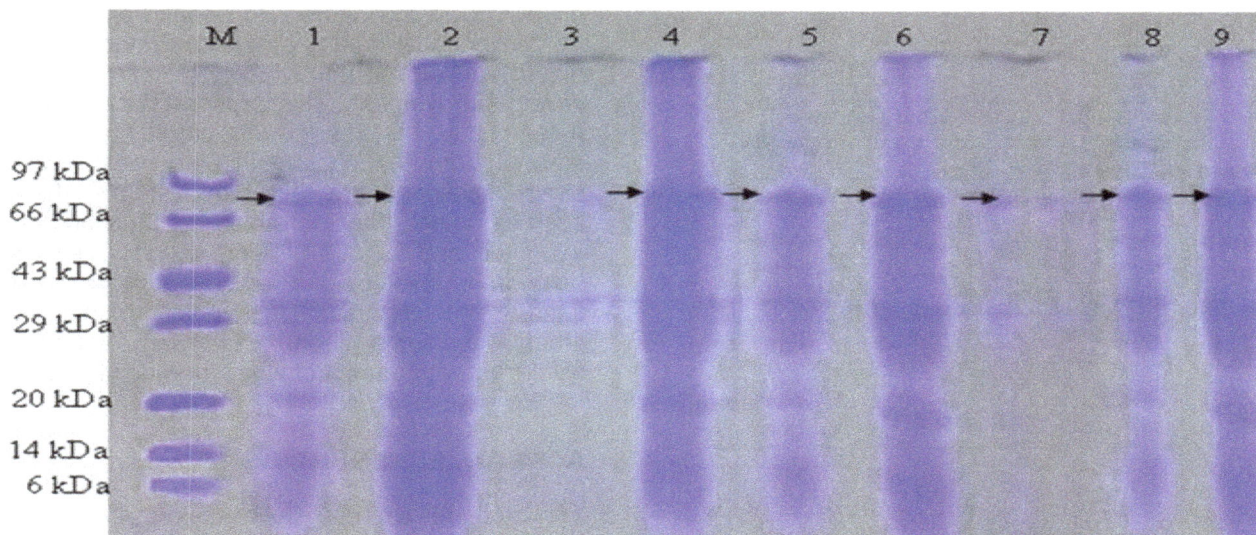

Figure 3. SDS-PAGE for parental and fused culture at different temperature.
1- Parent 1 at 35°C, 2- Parent 1 at 40°C, 3 - Parent 1 at 45°C, 4- Parent 2 at 35°C, 5- Parent 2 at 40°C, 6 - Parent 2 at 45°C, 7- Fused culture at 35°C, 8- Fused culture at 40°C, 9- Fused culture at 45°C.

delivery in cell membrane system and promoting disassembly, dissociation and elimination of proteins (Hayes and Dice, 1996).

Substrate fermentation

Fermentation was performed in 250 ml Erlenmeyer flask containing 100 ml of CWM. The flasks were inoculated with 2% of inoculum obtained from 24 h old cultures of parental and fused culture and incubated at 42°C with 100 rpm/min. Samples were taken periodically from the CWM and analyzed for live cell population, biomass, lactose and ethanol concentration during fermentation and the results were shown in Figures 4a, b and c. The biomass yield and ethanol production was higher in the

4a. Parent 1

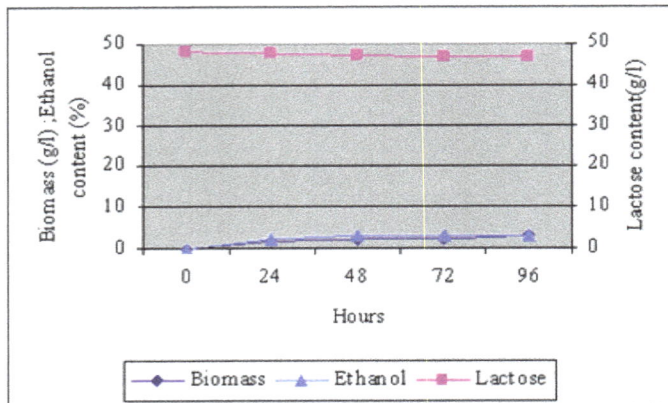

4b. Parent 2

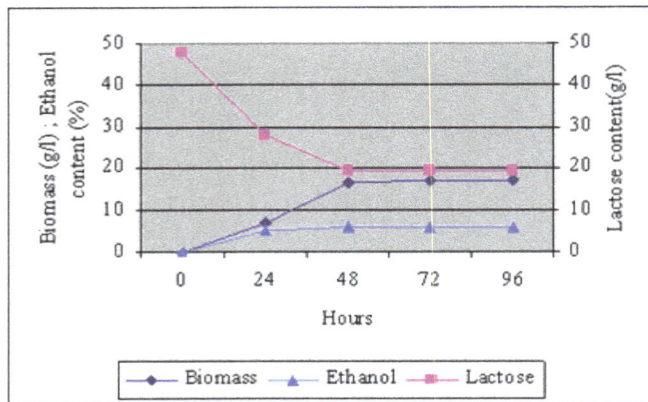

4c. Fused culture 12

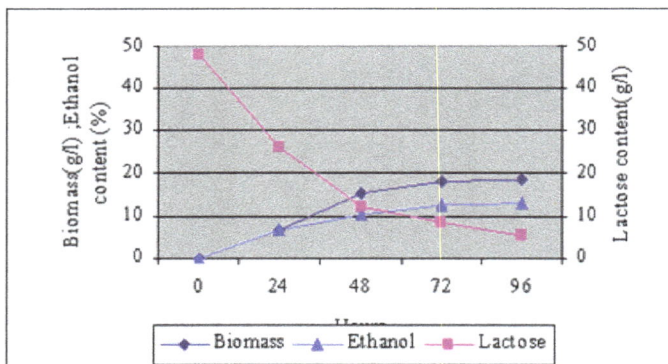

Figure 4. Growth and ethanol production of the parental and fused culture 4a Parent 1.

fused culture compared to the parental cultures. In 72 h of fermentation period, the fused culture produced the biomass of 18.09 g/lit with an ethanol yield of 12.5%, which was higher than the parental cultures (Figure 4c).

Kargi and Ozmihci (2006) found that ethanol concentration increased with time and reached a constant final concentration at the end of 72 h of incubation in CWP (Cheese whey powder) concentration between 50 and

150 g/l. Similar to sugar utilization, ethanol fermentation was slow for the first 72 h for sugar concentration above 100 g/l, probably due to osmotic pressure caused by high sugar concentrations. Ethanol production increased considerably after the first 72 h of adaptation period for sugar concentration above 100 g/l. The maximum ethanol concentration was 10.5% (V/V). Ethanol was obtained at the end of 216 h with initial sugar concentrations above 100 g/l slowed down ethanol formation; however, improved the final ethanol concentration considerably. Similarly, Pasha et al. (2007) showed the fused culture (CP 11) produced more biomass and ethanol than any of its parental cultures using xylose at 42°C. More ethanol was produced by fused mutant CP 11 at 42°C with xylose 150 g/l, Compared with parental strains. The amount of ethanol produced by parent VS3 strain was nil, whereas CP11 strain produced 26.1 g/l ethanol at 42°C. The fused yeast was found to be superior to the parent VS3 for ethanol production using lignocellulosic substrates.

Conclusion

Bioethanol production from waste is one of the preferable areas of alternate energy production. Protoplast fusion is an important tool for *gene* manipulation, because, it breaks down the barriers to genetic exchange imposed by conventional mating systems. Protoplast fusion technique has a great potential for genetic analysis and for strain improvement. This present study found that the fused culture FC12 is the efficient strain to produce high ethanol from the cheese industry waste (Whey). The strain developed has got the desirable characteristics such as having the high ethanol production, temperature tolerance and lactose utilizing capacity. This study has helped in developing a recombinant yeast strains by intergeneric protoplast fusion for, higher amount of ethanol production from waste with reduced cost for waste water treatment of whey for reduction of BOD

ACKNOWLEDGEMENT

The authors express their sincere thanks to Dr. D. Balachandar Associate Professor, Department of Agricultural Microbiology, TNAU and Dr. T. K. Raja, Retired Professor, VIT for their encouragement, spontaneous help and interest in the execution and refinement of this study.

REFERENCES

Banat IM, Nigam P, Marchant R (1992). Isolation of thermo tolerant, fermentative yeasts growing at 52°C and producing ethanol at 45°C and 50°C. World J. Microbiol. Biotechnol., 8: 259-263.
Barnett JA, Payne W, Yarrow D (1986). "Yeasts: characteristics and identification", Cambridge University Press, Cambridge, New York, New Rochelle, Melbourne, Sydney.
Dziuba E, Chmielewska J (2002). Fermentative activity of somatic

hybrids of *saccharomyces cerevisiae* and *candida shehatae* or *pachysolen tannophilus*. Elec. J. Pol. Agri. Univ. Biotechnol., 5: 1-12.

Farahnak F, Seki T, Ryu DY, Ogrydziak D (1986). Construction of lactose-assimilating and high ethanol producing yeasts by protoplast fusion. Appl. Environ. Microbiol., 51: 362-367.

Ferrari MD, Loperena L, Varela H (1994). Ethanol production from concentrated whey permeates using a fed-batch culture of *Kluyveromyces fragilis*. Biotechnol. Lett., 16: 205-210.

Francis DM, Clair (1993). Outcrossing in the homothallic oomycete, *Pythium ultimatum,* detected with molecular markers. Curr. Genet., 24: 100-106.

Gunasekaran P, Raj KC (1999). Ethanol fermentation technology – *Zymomonas mobilis*. Curr. Sci., 77: 56-68.

Hayes SA, Dice (1996) J F Roles of molecular chaperones in protein degradation. J. Cell Biol., 132: 255-258.

Holzberg I, Finn RK, Steinkraus KH (1967). A kinetic study of the alcoholic fermentation of grape juice. Biotechnol. Bioeng., 9: 413-427.

Kargi F, Ozmihci S (2006). Utilization of cheese whey powder for ethanol fermentation: Effects of operating parameters. Enzyme Microbial technol., 38: 711-718.

Kosikowsk (1979). Whey utilization and whey products. J. Dairy Sci., 62: 1149-1160.

Lin CCL, Fung DYC (1985). Effect of dyes on the growth of food yeasts. J. Food Sci., 50: 241-244.

Lowry OH, Rose NJ, Farr AL (1951). Protein measurement with folin phenol reagent. J. Biol. Chem., 193: 265-275.

Mahoney RR, Nickerson TA, Whitaker JR (1975). Application of acid lactase to wine making from cottage cheese whey concentrates. J. Dairy Sci., 58: 1620-1629.

Martins C, Horii J, Pizzirani-Kleiner A (1999). Characterization of fusion products from protoplasts of yeasts and their segregants by electrophoretic karyotyping and RAPD. Revista De Microbiologia. 30: 71-76.

Pasha CR, Kuhad C, Rao CV (2007). Strain improvement of thermo tolerant *Saccharomyces cerevisiae* VS3 strain for better utilization of lignocellulosic substrates. J. Appl. Microbiol., 103: 1480-1489.

Petit J, Boisseau P, Arveiler B (1994). Glucanex: a cost - effective yeast lytic enzyme. Trends genet., 10: 4-5.

Saumya B, Hemchick PH (1983). Simple and rapid method for disruption of bacteria for protein studies. Appl. Environ. Microbiol., 46: 941-943.

Sing D, Nigam P, Banat IM, Marchant R, Mchale AP (1998). Review: Ethanol production at elevated temperatures and alcohol concentrations: Part II – Use of *Kluyveromyces marxianus* IMB3. World J. Microbiol. Biotechnol., 14: 823-834.

Sing MA, S Lindquist (1998). Thermo tolerance in *Saccharomyces cerevisiae*: the Yin and Yang of trehalose. Trends Biotechnol., 16: 460-468.

Stvitree L, Sumpradit T, Kitpreechavanich V, Tuntirungkij M, Seki T and Yoshida T (2000). Effect of Acetic Acid on Growth and Ethanol Fermentation of Xylose Fermenting Yeast and *Saccharomyces cerevisiae*. Kasetsart J. (Nat. Sci.), 34 : 64-73.

Tahoun MK, Nemr T M, Shata OH (1999). Ethanol from lactose in salted cheese whey by recombinant *Saccharomyces cerevisiae*. Z. Lebensm. Unters. Forsch. a. 208: 60-64.

Van der Walt JP (1970). "The yeasts. A taxonomic study". 2nd edition, North-Holland Publishing Company, Amsterdam. pp. 34-113.

A specific inhibitory protein to a restriction enzyme from *Saccharomyces cerevisiae*

Mukaram Shikara*

Biotechnology Division, Department of Applied Sciences, University of Technology, Baghdad, Iraq.
E-mail: mukaramshikara2010@yahoo.com.

A specific protein inhibitor for the restriction enzyme (SacC1) has been purified from *Saccharomyces cerevisiae* approximately 21,000 fold and its inhibitory properties have been characterized. The isoelectric points (pI) of SacCI and its inhibitor are 9.0 and 5.22, respectively. The molecular weight of SacC1, the inhibitor and SacC1-inhibitor complex were estimated by gel filtration on a Sephadex G-100 column to be 64,000, 32,000 and 85,000, respectively. The inhibitor protein inhibits SacC1 catalytic activities efficiently, but has no effect on other restriction enzymes tested. Inhibition does not occur unless SacC1 enzyme is exposed to the inhibitor protein prior to the reaction of the enzyme with DNA. The inhibitory activity is independent of temperature. The inhibition increased linearly with the addition of inhibitor to various amounts of SacC1, up to 85% inhibition. The slope of inhibition was constant irrespective of the initial amount of SacC1 and Ki value of 3.45×10^{-12} was obtained. The inhibitor interacts strongly with SacC1 and this interaction could increase the stability of the complex, possibly manifesting itself as SacC1 decreases in the dissociation rate due to the electrostatic attraction between the two groups or the stability may increase by potentially stronger electrostatic interaction. The conformational specificity between SacC1 and its inhibitor seems to be essential for their interaction. The extremely strong affinity of the inhibitor to SacC1 is remarkable and stronger than the affinity of several restriction enzymes.

Key words: *Saccharomyces cerevisiae*, inhibitor, protein, restriction enzyme, yeast, purification, Ki.

INTRODUCTION

The yeast, *Saccharyomyces cerevisiae*, has become an important organism in molecular, biochemical and genetic analysis. The organism has specific requirements for growth under a variety of conditions to produce specific or non-specific inhibitors against their restriction enzymes. The characterization of the interaction between restriction enzyme and its inhibitor is of interest for at least three reasons. Firstly, the restriction enzyme has an important genetic roles and understanding inhibition could result in new strategies for the genetic engineering. Secondly, the activity of both restriction enzyme and its inhibitor will form a high affinity complex. Finally, the mechanism of interaction between the enzyme and its inhibitor is of great interest (Richard et al., 2003).

The wide range of restriction enzymes enables the researcher to manipulate DNA in such a simple yet specific ways. Natural inhibitors are better equipped than chemical artificial inhibitors in order to play an important role in genetic engineering. Several natural inhibitors were discovered during the past years in bacteria and seem to protect the bacterium from adventitious proteolysis, probably, during secretion (Braciak et al., 1988; Muralidhara et al., 2006; Richter and Conti, 2004).

The researcher purified a restriction endonuclease (SacC1) from *S. cerevisiae* (Shikara, 2010) and in due course an inhibitor protein was discovered, inhibiting the purified enzyme. Many natural DNases and RNases inhibitors were purified, but very few (to the author's knowledge) have been purified and the goal of this study is to purify the inhibitor and investigate its mechanism

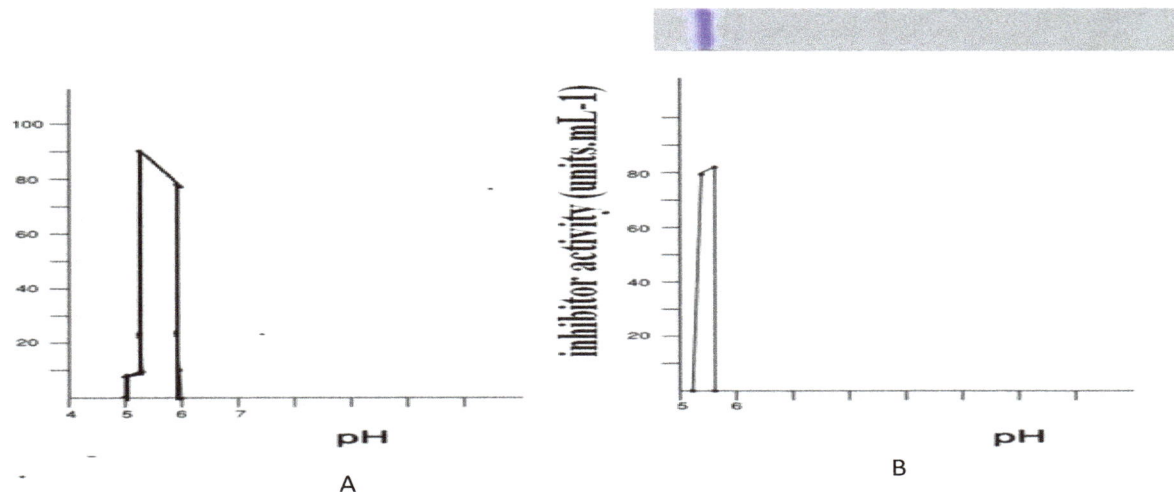

Figure 1. Isoelectricfocusing chromatography of the inhibitor A) Fraction IV was charged on an isoelectricfocusing column at a narrower pH range (4.0 to 6.0). The inhibitor was recovered with pH values of 5.15 to 5.40 and pooled (Fraction V). B) Fraction V was electrofocused to achieve at a narrower pH range (pH 5 to 6). The inhibitor was found to focus at pH 5.2 to 5.3 with the peak value being 5.22. The fractions of maximum specific activity were pooled, concentrated dialyzed against 0.02M Tris-HCl containing 0.1% glycerine, pH 7.5 with three changes for 5 h and 10 ml of dialysate (Fraction VI).

and properties.

MATERIALS AND METHODS

Sources of media and analytical chemicals

All chemicals used were of analytical grade. Media and chemicals used in this study were purchased from Sigma. Lambda DNA 32300 KD, size 48502bp, concentration 250 mg/ml and standard restriction enzymes (EcoRI and SacI) were obtained from Sigma Chemical Company. Molecular weight markers were obtained from Boehringer, Mannheim, Germany.

Isolation of the inhibitor

Isolation of yeast

S. cerevisiae strain R-Z128 was used throughout the study. Yeast was grown in YPD medium (1% yeast extract, 2% Bacto-peptone and 2% dextrose) at 30°C with constant shaking for 3 d and then the cells were lysed by sonication in ultrasonic bath (Sonicator Branson 5210) for 20 x 10 s and broken cells were removed by centrifugation (2 min, 15,600 g). The supernatant was centrifuged at 3000 g for 30 min at 4°C and the supernatant has been used as the "crude extract", source of the enzyme and inhibitor.

Purification of the inhibitor

Solid ammonium sulphate was added to the "crude extract" to form 0 to 30%, 30 to 50% and 50 to 80% saturation fractions, respectively. After centrifugation at 4000 g for 15 min, the pellet (of each fraction) was suspended in 40 mM potassium phosphate (pH 7.5) containing 5 mM 2-mercaptoethanol and 10% glycerol (buffer A) and then dialyzed with two changes against 4 L of the above buffer for 24 h and measured for endonuclease activity.

The 50 to 80% saturation fraction was found to have a high inhibitor activity, so, it has been purified further by layering onto a 1.5 x 40cm phosphocellulose column that was previously equilibrated with 4 L of the same buffer. The inhibitor activity was eluted from the column with a linear gradient of 0-0.6N NaCl in buffer A. The peaked fractions were pooled together and dialyzed for 5 h against 4 L of 40 mM Tris-HCl, pH 8.0 containing 5 mM 2-mercaptoethanl and 10% glycerol (buffer B). The pooled fractions (Fraction II) were loaded onto a Sephadex G-100 column (1.5 x 18cm) that was previously equilibrated with buffer B. The enzyme was eluted with 120 ml linear gradient of 0-1N NaCl in buffer B and one major peak was observed.

The active fractions from Sephadex G-100 were pooled and dialyzed against 4 L of buffer B for 5 h with one change. The pooled fractions (Fraction III) were concentrated by filtration through collodion bag (Sartorious, Germany) almost to dryness (Fraction IV). This fraction was charged on to an isoelectricfocusing column at a narrower pH range from pH 4.0 to 6.0. The inhibitor was recovered with pH values of 5.15 to 5.40, pooled and then concentrated (Fraction V) and was electrofocused without the addition of an ampholyte to achieve at a narrower pH range (pH 5 to 6). The inhibitor was found to focus at pH 5.2-5.3 with the peak value being 5.22. The fractions of maximum specific activity were pooled, concentrated with collodion bag and dialyzed against 0.02M Tris-HCl containing 0.1% glycerine, pH 7.5 with three changes for 5 h and 10 ml of dialysate (Fraction VI) (Figure 1).

All operations were carried out at 4°C. Endonuclease activity, inhibitor activity, protein and carbohydrate concentrations were determined for all fractions.

Determination of endonuclease and inhibitor activities

Endonuclease activity determined by measuring the amount of acid-soluble nucleotide liberated from DNA according to Brown and Smith method (1980) by incubation of incubation of 0.3 pmol lambda DNA with 3.0 pmol of the purified endonuclease enzyme in

Table 1. Purification steps of the inhibitor.

Step	Total volume (ml)	Total activity (units)	Total protein (mg)	Specific activity (unit/mg)	Purification	Recovery %
Fraction I	30	1656.93	162.44	10.2	1	100
Fraction II	65	1326.5	3.31625	400	39.21	80
Fraction III	80	1124	0.7025	1600	156.86	68
Fraction IV	20	900.5	0.225125	4,000	392.156	54
Fraction V	22	634	0.048	13,000	1274.50	38
Fraction VI	12	345.44	0.0164	21,000	2058.8	21

a final volume of 20 µl containing 40 mM Tris–HCl (pH 7.5), 10 mM MgCl$_2$ and 0.1 mg/ml bovine serum albumin for 1h at 37°C. The reaction mixture was stopped by cooling at 0ºC and with the addition of 20mM EDTA. The cleavage products were analyzed using 0.7 to 1.2% TBE agarose, and 5% neutral polyacrylamide gels. DNA was visualized by staining with ethidium bromide.

One unit of the enzyme was defined as the amount of the enzyme that can digest 1 µg of Lamba DNA for 1 h at 37°C.

The inhibitor was incubated briefly with the endonuclease prior to the addition of the substrate (DNA) and one unit of the inhibitor was obtained by subtracting the remaining endonuclease activity measured by the addition of the inhibitor from the original endonuclease activity of the enzyme. One unit of the inhibitor is defined as the amount that inhibits one unit of the enzyme.

Estimation of protein and carbohydrates

Protein contents were determined by the methods of Lowry et al. (1951) using Bovine serum albumin `as a standard. Carbohydrates contents were determined by the method of Dubois et al. (1956) using glucose as a standard.

Determination of Ki

The inhibitor constant (Ki) was determined by the method Lineweaver and Burk (1934) as modified by Oda et al. (2002) and Serap et al. (2006). The Inhibition was measured with lower concentration of the enzyme and inhibitor, where the dissociation of the enzyme-inhibitor complex became evident. Ki was calculated from the curves near the origin by the equation: $Ki = [I_i]/[V_o/V_{-1}]$, where $[I_i]$ is the concentration of the inhibitor. V_o and V_{-1} are the velocities of the enzyme reaction in the absence and presence of the inhibitor, respectively.

Examination of purity and estimation of the molecular weight

The molecular mass of the enzyme was estimated by gel filtration on a Sephadex G-100 column according to the method of Andrews (1964) and by SDS-gel electrophoresis with 12% polyacrylamide gels as described by Laemmli (1970) as modified by Maizel (1971).

SacC1 and inhibitor samples were loaded (separately) onto a Sephadex G-100 column (1 x 50 cm) which was equilibrated with a buffer B. Separation was carried out at a flow rate of 0.2 ml/min. The molecular mass of the endonuclease (and the inhibitor) was estimated by comparing its elution volume with those of calibration standards, such as blue dextran blue (>100 kDa), bovine serum albumin (66.2 kDa), Egg albumin (45 KDa), chymotrypsinogen A (25 kDa), lysozyme (14.4 kDa) and cytochrome C (12.4 kDa). Elution profiles were monitored by measuring absorbance at 280 nm. For the interpolation of unknown molecular mass, a linear

dependence of the logarithm of the molecular mass on the elution time was assumed.

RESULTS

Purification of the inhibitor

A summary of the purification steps (as in methods) is presented in Table 1.

Properties of the inhibitor

Homogeneity

The final preparation of the inhibitor (Fraction VIII) seemed homogenous on SDS-PAGE carried out in 12% acrylamide gels (Figure 1).

Ultraviolet absorption

The inhibitor showed a typical ultraviolet absorption curve of protein with maximum 270 nm and the A260:A280 ration was 1.40

SacC1-inhibitor complex formation

As long as the inhibitor was not obtained in high purified form, the existence of the specific complex between SacC1 and its inhibitor could not be ascertained because the association of the molecules could be mediated by other proteins.

When nearly equal volumes or amounts (in units) of the purified SacC1 and the inhibitor protein were mixed in buffer B and immediately applied to Sephadex G-100 column, SacC1-inhibitor complex was eluted at a position earlier than either of that of SacC1 or the inhibitor, whereas no SacC1 activity or an inhibitor activity were observed at their original position (Figure 2). The complex had little endonuclease activity by itself.

Molecular mass estimation

Molecular weights of SacC1, the inhibitor and the SacC1-

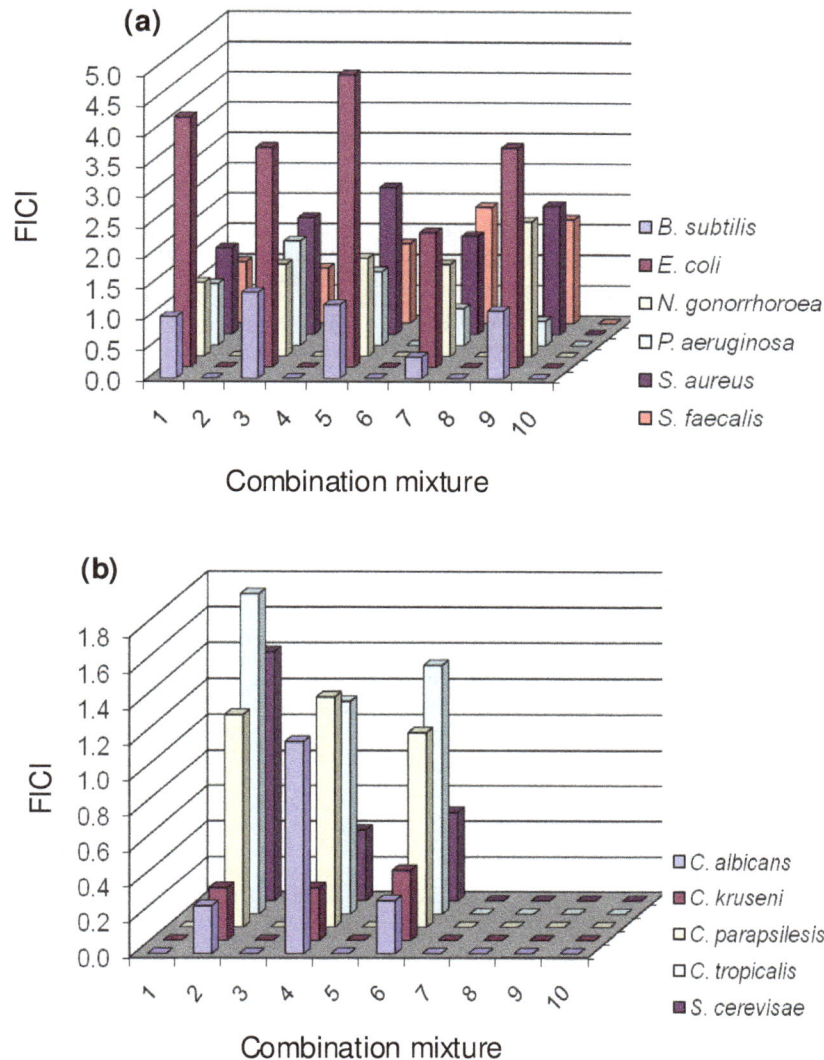

Figure 5. Combination antimicrobial therapy between active constituents separated from *M. peregrina* plus reference antibiotics from local market (Bacteria ($P < 0.01$, F = 1.9), Fungi ($P < 0.02$, F = 0.9)). Combination mixtures: 1: Lupeol acetate + ampicillin, 2: Lupeol acetate + flaconozote, 3: α-amyrin + ampicillin, 4: α-amyrin + fluconazole, 5: β-amyrin + ampicillin, 6: β-amyrin + fluconazole, 7: β-sitosterol-3-O-B-d glucoside + ampicillin, 8: β-sitosterol + fluconazole, 9: β-sitosterol + ampicillin, 10:6- β-sitosterol-3-O-glucoside + fluconazole

et al. (1991), Hashem and Wahba, (1999) and Sudhakar et al. (2005).

The efficacy of the active constituents separated from *M. peregrina* was finally assayed in combination with ampicillin and fluconazol from the local market (Figure 5). The data revealed "no interaction" in most combination for mixtures in inhibiting bacterial growth (FC1C 0.5-4). Antagonistic effects recorded in two were lupeol acetate plus ampicillin (mixt. 1) and □-amyrin plus ampicillin (mixture 5) against *E. coli* (FICI 4.1 and 4.8) where the antibacterial activity decreased in the individually tested compound. Synergy was recorded also in two are β-

sitostrol plus ampicillin (mixture 1) against *S. aureus* and B-sitosterol-3-o-glucoside plus ampicillin (mixture 9) against *P. aeruginosa* (Table 1).

High increase in antifungal activity was observed in these combination mixtures. In case of fungi 6 of 15 synergistic mixtures were recorded against dermatophytes (Table 1) were mixture (1) against *C. albicans* and *C. krusei* mixture (4) against *C. krusei* and *S. cerevisiae* and mixture (6) against *C. albicans* and *S. cerevisiae*. This observation indicated the more efficacy of combined treatment for candidiasis than monotherapy. Polak (1989) reported that antifungal combination may increase the magnitude and rate of microbial killing *in vitro*, shorten the total duration of therapy, prevent the emergence of drug resistance, expand the spectrum of activity and decrease the drug related toxicities by allowing the use of lower doses of antifungal. Sugar (1995) found that multi-compound therapies along the disease pathway may need to be manipulated simultaneously from an effective treatment. When one drug is used, the required high dosage for efficacy often produce bioavailability problems and unwanted side effects and drug resistance problems may also emerge (Zhang et al., 2007). Perhaps if we can focus on multiple targets in the microbial pathway through the use of co-drugs, high dosage of single drug will not be necessary (Cowen and Lindquist, 2005 and Heitman, 2005).

There are several mechanisms produced from synergy in combined antimicrobial therapy: (i) inhibition of different stages of some biochemical pathways represent one type of interaction; (ii) increased penetration of the antimicrobial agent as a result of cell wall or cell membrane antimicrobial activity from another agent which facilitate the 2nd agent to reach their target site; (iii) a transport interaction where one antimicrobial degrade cell wall allowing the other to remain at the site of its action within the wall; and (iv) Simultaneous inhibition of different microbial cell wall targets, such as cell wall and membrane targets. Antagonism among antimicrobial agents might occur in one of several ways including: (i) direct act at the same site which render the second agent inactive; (ii) adsorption to the surface by one agent inhibits binding of another antimicrobial agent; and (iii) modification of the pathogen upon exposure to one antimicrobial agent renders it less susceptible to the effect of the second agent (Johnson et al., 2004).

In conclusion twenty synergistic combined mixtures were detected in this *in vitro* study which is pathogen dependent. These mixtures need further *in vivo* studies to evaluate their actual effect. Table (1) summarizes these important conclusions. We want to pay attention to the higher efficacy of combined therapy against *Candida* species and some bacterial species than monotherapy. When one of the combined agents is of natural source, the concentration of chemical decreased, and the side effect and appearance of resistant pathogens delayed.

ACKNOWLEDGEMENT

This work was funded by a research grant from Cairo University titled "Production of pharmacologically active materials from medicinal plant resources in Egypt"

REFERENCES

Abd El Ghani M (1988). Moringaceae. In: Hadidi, M. N. (ed.) Flora of Egypt. Taeckolmia add. Ser. 3:21-40.

Anwar F, Latif S, Ashraf M, Gilani A H (2006). *Moringa oleifera*: a food plant with multiple medicinal uses. Phytother. Res. 21(1): 17-25.

Baddely JW, Pappas PG (2005). Antifungal combination therapy: Clinical potential. Drugs 65: 1461-1480.

Barel S, Segal R, Yashphe J (1991). The antimicrobial activity of the essential oil from *Achillea fragrantissima* J. Ethnopharmacol. 33(1-2), 187-191.

Boulos L (1995). Flora of Egypt, Checklist. Al Hadara publishing, Cairo, Egypt. 283p.

Boulos L (1999). Flora of Egypt, 1 (Azollaceae-Oxalidaceae) Al Hadara publishing, Cairo, Egypt. 283, plate 42 (1) cl. Plate 390p.

Chandraseku PH, Cutright JL, Manavathu EK (2002). Efficacy of vericonazole in the treatment of acute invasive aspergillosis. Clin. Infect. Dis. 34: 563-571.

Cowen LE, Linduquist S (2005). Science 309: 2185-2189.

Eilert U, Wolters B, Nahrstedt A (1981). The antibiotic principle of seeds of *Moringa oleifera* and *Moringa stenopetala*. Plant Med. 42: 55-61.

El-Alfy T, Hegazy AK, Mahrous A, El-Komy SM (2009). Isolation of biologically active constituents from Moringa peregrine (Forssk.) Fiori (Family: Moringaceae). Natural product communication (under puplication).

Hashem F, Wahba HE (1999). Isothiocyanates in myrosinase treated herb extract of *Coleome chrysantha*. Decne and their antimicrobial activity. Phytother. Res. 14(4) 284-287.

Heitman J (2005). Science. 309: 2175-2176.

Ibrahim H, Aziz AN, Syamsir DR, Ali NAM, Mokhtar DM and Awang K (2009). Essential oils of *Abpinia conchigera* Griff. And their antimicrobial activities. Food Chem. 113: 575-577.

Johanson MD, Mac Dongall C, Ostrosky-Zeichner L, Perfect JR, Rex JA (2004). Combination antifungal therapy. Antimicrob. Agents Chemother. 48(3): 693-715.

Kontoyiannis DP, Lewis RE (2004). Toward more effective antifungal therapy: the prospects of combination therapy. Br. J. Haematol 126(2): 165 - 175.

Lewis AE (1984). Biostatistics. New York: Van Nostrand Reinholc Company Inc. pp. 84-94.

Marr k (2004). Combination antifungal therapy: where are we now and where are we going? Oncology 18(S7): 24-29.

Mukherjee PK, Danniel JS, Christopher AH (2005). Comb nation treatment of invasive fungal infections. Clin. Microbiol. Rev. 18(1), 163-194.

Obute GC (2005). Ethnomedicinal plant resouces of south Eastern Nigeria, http/www.sin.edu/~ebl/leaflets/obute.html.

Ramesh Putheti, Okigbo, RN (2008). Effects of plants and medicinal plant combinations as anti-infective. Afr. J. Pharm. Pharmacol. 2(7): 130-135.

Polak A (1989). A combination therapy of experimental candidiasis, cryptococcosis, aspergilosis systemic mycosis. Infection 17: 203-209.

Sanat II, Ramos CF, Bayer AS, Channoum SA (1997). Combination therapy with amphotericin β and Fluconazol against invasive candidiasis and neutropenic – mouse and infective – endocarditis rabbit models. Antimicrob. Agents Chemother. 41: 1345-1348.

Steinbach W (2005). Combination antifungal therapy for invasive aspergillosis utilizing new targeting strategies. Curr. Drug Targets Infect. Disord. 5: 203-210.

Sudhakar CH, Rao V, Raija DB (2005). Evaluation of antimicrobial activity of *Coleome viscose* and *Gmelina asiatica* (2005) Fitoterrapia. 16325351 (P.S.G.E.B.D.).

Sugar AM (1995). Use of amphotericin B with azole antifungal drugs: what are we doing? Antimicrob. Agents Chemother. 39: 1907-1912.

Zhang L, Yan K, Zhang Y, Huang R, Bian J, Zheng C, Sun H, Chen Z, Sun N, An R, Min F, Zhao W, Zhuo Y, You J, Song Y, Yu Z, Liu Z, Yang K, Gao H, Dai H, Zhang X, Wang J, Fu C, Pei G, Liu J, Zhang S, Goodfellow M, Jiang Y, Kuai J, Zhou G, Chen X (2007). High-throughout synergy screening identifies microbial metabolites as combination agents for the treatment of fungal infections, Proc. Natl. Acad. Sci. USA. 104(11): 4606-4611.

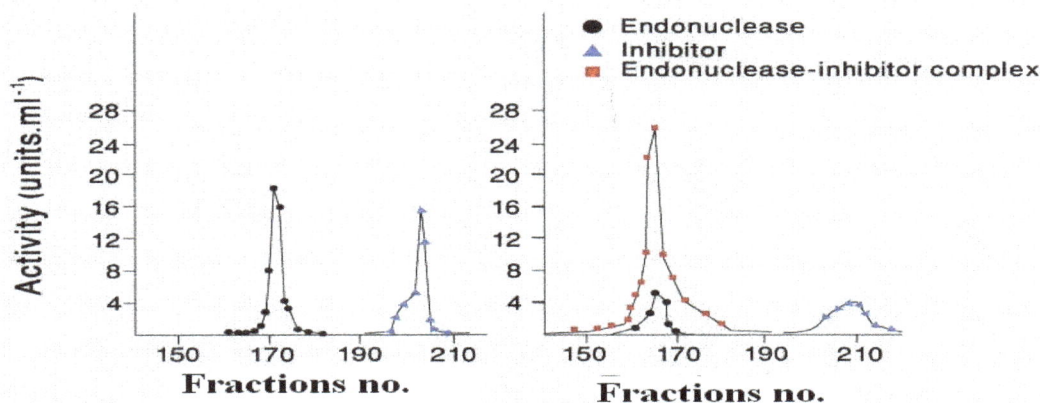

Figure 2. Sephadex G-100 column chromatography of SacC1, the inhibitor and SacC1-inhibitor complex. A) Each SacC1 and the inhibitor were chromatographed separately on the column; B) A mixture of SacC1 and the inhibitor was chromatographed on the column.

Table 2. Effect of the pH on the inhibitor.

pH	Activity (units)
4	20
5	40
6	45
7	76
8	100
9	70
10	12

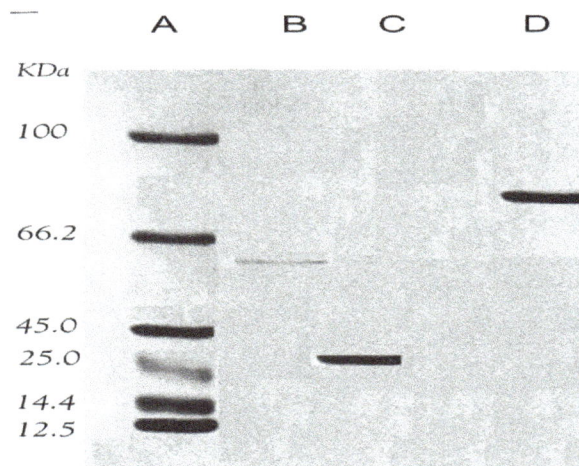

Figure 3. Determination of the molecular weight of SacC1 (B) its inhibitor (C) and SacC1-inhibitor complex(D), while (A) is the protein ladder (10-100 kDa) that was used as the protein molecular weight marker.

inhibitor complex were estimated from their elution from Sephadex G-100 column to be 64,000, 32,000 and 85,000, respectively using standards (as in methods). By using SDS-PAGE, the molecular weight of the enzyme, the inhibitor and enzyme-inhibitor complex were 68,000, 35,000 and 87,000, respectively.

pH optimum and effect of different cations

The inhibitor has a maximum activity in buffer B, pH 8.0 (Table. 2). Mg^{+2} or Mn^{+2} reduced the inhibition rate, while all other cations have no effect on the inhibitor.

Isoelectric point

The isoelectric points (pI) of SacC1 and its inhibitor are 9.5 and 5.22, respectively at 4 °C.

Properties of the inhibition reaction

Optimal conditions for inhibition

The primary requirement of the inhibition is a brief incubation of the SacC1 with the purified inhibitor protein

prior to the addition of substrate DNA inhibition, is nearly maximal after 5 to 10 min. This level of maximum inhibition, as well as the rate at which it achieves, increased with the increasing amount of the inhibitor protein (Figure 3 and 4).

By contrast, if the inhibitor protein is added to the DNA before SacC1, or if the inhibitor protein added after the reaction has been initiated, no inhibitor is observed (Figure 5). It is clear that the inhibitor protein does not interact with the DNA, but rather with SacC1 before it initiates DNA degradation.

The mode of inhibition

The inhibitor inhibits SacC1 noncompetitively with Km of

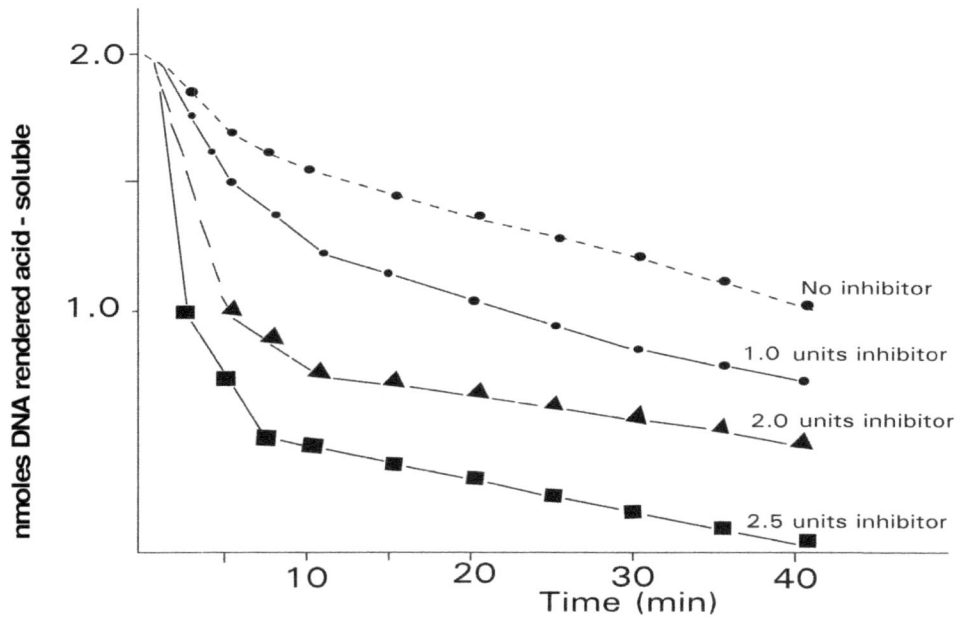

Figure 4. The rate and the limit of inhibition as a function of time of the incubation of SacC1 with the inhibitor protein, SacC1 (2 units) was incubated at 37°C with various amounts of the inhibitor protein in a standard mixture as described under methods.

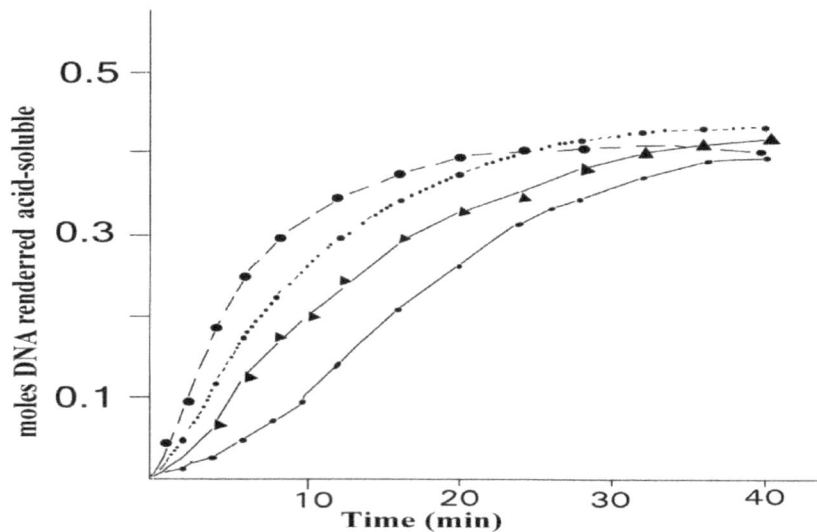

Figure 5. The rate and the limit of inhibition as a function of time of the incubation of SacC1 with the inhibitor protein.
The inhibitor protein SacC1 (2 units) was incubated at 37°C with various amounts of the inhibitor protein were added after the reaction has been initiated in a standard mixture as described under Methods;•- 2.5 units inhibitor; •......2.0 units inhibitor; ◄-1.0 units inhibitor ◄...... no inhibitor.

the enzyme for DNA as a substrate is 2.5×10^{-8} M according to Lineweaver-Burk plot (Figure 6). When the increasing amounts of the inhibitor were added to a constant amount of SacC1, the inhibition increased linearly with the addition of the inhibitor up to 85% inhibition. The

slope of inhibition was constant irrespective of the initial amount of SacC1. Further inhibition required disproportionably larger amounts of the inhibitor and the residual SacC1 activity cannot be entirely demolished (Figure 7).
When increasing amounts of SacC1 were added to a

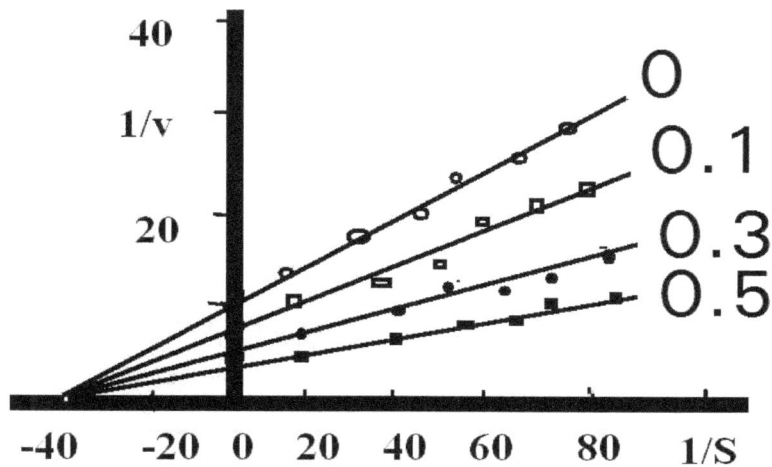

Figure 6. Lineweaver–Burk plot of SacC1 and the substrate with various amounts of the inhibitor. Each reaction mixture (1 ml) contains 0.4milliunits/ml of endonuclease, 0.05 milliunits.mL^{-1} of inhibitor and 0.01 mg/ml of substrate in buffer B. Reaction carried out at 30°C for 10 min. The reaction was stopped at zero and 10 min time by 0.2 ml uranyl reagent, cooled for 10 min and centrifuged at 10,000 g for 5 min. A260 was measured with zero time supernatant as a control. The reaction velocity (V) was expressed directly by A260. Substrate concentration was expressed as millimolar nucleotides. Inhibitor concentrations are in milliunits/ml.

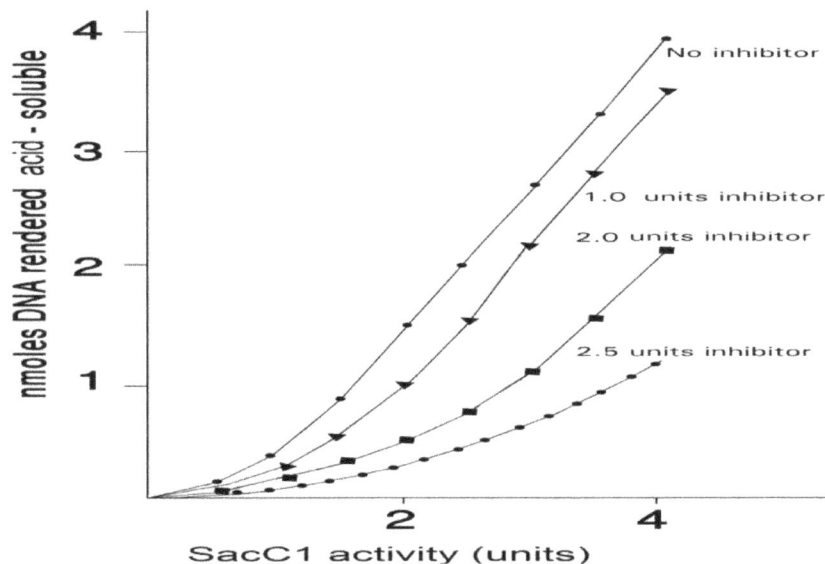

Figure 7. The effect of the inhibitor on various amounts of SacC1. SacC1 and the inhibitor protein were mixed in the amounts indicated and incubated at 37°C for 10min as described under "methods".

fixed amount of the inhibitor protein, no or very little SacC1 activity was observed until the amount of SacC1 exceeded the amount of the inhibitor. After that, SacC1 activity increased linearly with the same slope as observed in the absence of the inhibitor (Figure 8).

The amount of inhibition is independent of temperatures and independent of the pH during the pre-assey

incubation.

Determination of Ki

The inhibitor constant Ki was determined by measuring the inhibition with lower concentrations of SacC1 and the

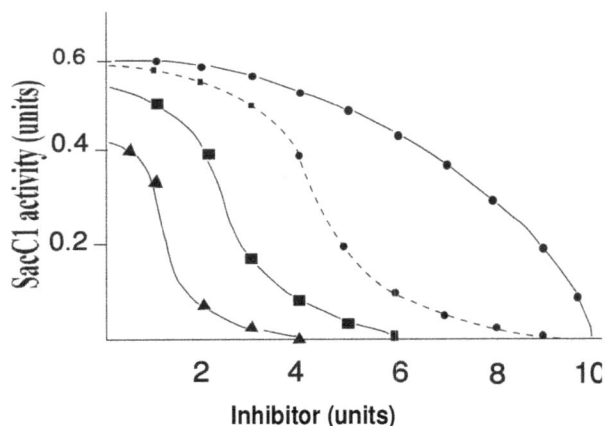

Figure 8. The effect of SacC1 on various amounts of the inhibitor. SacC1 and the inhibitor protein were mixed in the amounts indicated and incubated at 37°C for 10min as described under "Methods".SacC1 concentration 4 units.
● no inhibitor; ●...... 1.0 units inhibitor; ■- 2.0 units inhibitor and ◄...... 2.5 units inhibitor.

inhibitor, where the dissociation of SacC1–inhibitor complex becomes evident (Figure 9). From the curves near the origin, Ki was calculated with the equation $Ki=[It]/[v_o/v_{-1}]$, where [It] is the concentration of the total inhibitor and v_o and v_{-1} are the velocities of SacC1 reaction in the absence and presence of the inhibitor, respectively. As shown in Figure 9, a Ki value of 3.45×10^{-12} was obtained as the mean Table 3.

Specificity

The inhibitor nearly inhibited SacC1, but has no effect on EcoRI or SacI. This indicates the specificity of this inhibitor.

DISCUSSION

A protein inhibitor, which specifically inhibits SacC1 (Shikara, 2010) was purified by ammonium sulphate precipitation, dialysis and gel filtration using phosphorcellulose, Sephadex G-100 and isoelectricfocusing columns. The affinity chromatography method was very effective and allowed the isolation of the inhibitor in homogenous state and the removal of all interacting proteins

The purity of the enzyme was judged by the appearance of one band in SDS-gel electrophoresis. SacC1, its inhibitor and SacC1-inhibitor complex have a molecular weight of 64,000, 32,000 and 85,000 by using gel filtration and 68,000, 35,000 and 80,000 by using SDS-PAGE, respectively. This deviation is expected since SDS-PAGE has its limitations and most proteins will give estimates within a few percentage of their actual weight

(Sallantin et al., 1990).

Sac I (pl=6.2) and EcoRI (pl = 5.0)(Sallantin et al., 1990; Kim et al., 1990) have no interaction with the inhibitor of SacC1 (pl=5.22) since they have the same electric charge, while the inhibitor interacts with SacCI (pl=9.0) and this interaction could increase the stability of the complex, possibly manifesting itself as a decrease in the dissociation rate due to the electrostatic attraction between the two groups (Sallantin et al., 1990; Zhuravleva et al., 1987) or the stability may increase by potentially stronger electrostatic interaction (Spector et al., 2000; Strickler et al., 2006).

This may facilitate the electrostatic force between SacC1 and the inhibitor, which lead to a complex formation since they charged oppositely at pH 8.0

The inhibition increased linearly with the addition of the inhibitor up to 85% inhibition. Further, inhibition required disproportionably larger amounts of the inhibitor and the residual SacC1 activity cannot be entirely demolished.

It is unclear why total inhibition (100%) is not achieved. It is unlikely that the uninhibited activity results from the presence of small amounts of SacC1, which is resistant to inhibition since the purified inhibitor protein inhibits various preparations of SacC1.

Alternatively, the uninhibited activity may be a consequence of an equilibrium established by a reversible interaction between the inhibitor and the enzyme in the form of $E + I \leftrightarrow EI$ between the free and complexed (inhibited) forms of the enzyme. Other possibilities cannot be ruled out.

The stoichiometric measurements conclude that the SacC1 and the inhibitor have an extremely high affinity and constitute a mutual depletion system. The enzyme and the inhibitor combined stoichiometrically with each other form enzyme-inhibitor complex, which consists of one molecule of SacC1 and one molecule of the inhibitor until the concentration of one of them becomes near zero. This indicates that the inhibitor exerts a direct titration effect upon any given quantity of SacC1.

On the other hand, the determination of interaction parameters in vitro indicates that the SacC1 possesses a higher affinity for the substrate than for the inhibitor. That means one molecule of SacC1 will bind to one molecule of the inhibitor. These results, as well as the kinetic data are consistent with the model of enzyme-inhibitor complex, which composed of catalytic and regulatory subunits. The values obtained from molecular mass estimation strengthened the hypothesis that the complex is formed from one molecule of SacC1 and one molecule of the inhibitor. It has been shown that the hydrophobic effect, hydrogen bonding and packing interactions between residues in the protein interior are dominant factors that define protein stability. These results suggest that surface charge–charge interactions are important for protein stability and that rational optimization of charge–charge interactions on the protein surface can be a viable strategy for enhancing protein stability. Charge-charge interactions on the surface of native proteins

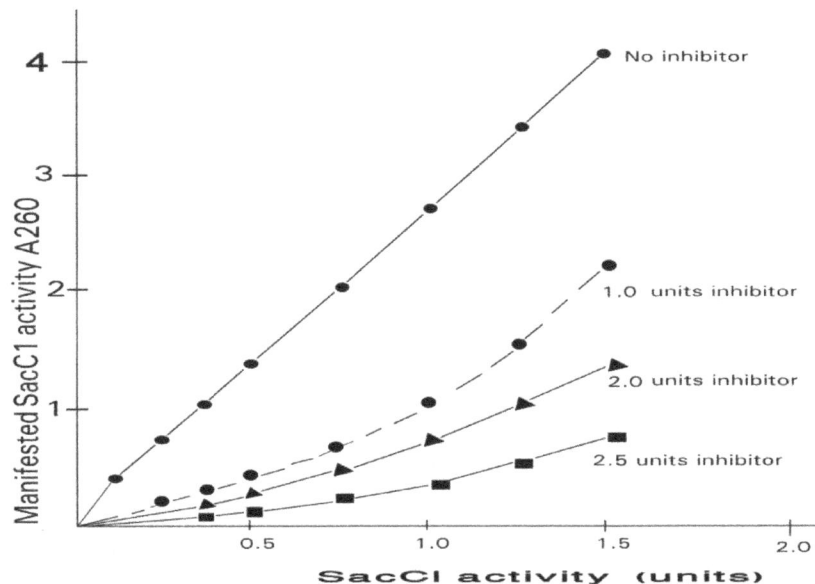

Figure 9. Residual SacC1 activity in the presence of different concentrations of the inhibitor. The inhibitor constant Ki was calculated as described under methods.

Table 3. Determination of Ki.

Total inhibitor		Ki	
Milliunits/ml	M	V_0/V_{-1}	M
0.1	3.6×10^{-12}	9.8	3.67×10^{-12}
0.3	4.2×10^{-12}	12.7	3.30×10^{-12}
0.5	4.9×10^{-12}	14.5	3.37×10^{-12}
0.7	5.7×10^{-12}	15.1	3.77×10^{-12}
1.0	6.6×10^{-12}	21.1	3.12×10^{-12}
Mean value			3.45×10^{-12}

are important for protein stability (Gribenko and Makhatadze, 2007).

The conformational specificity between SacC1 and its inhibitor seems to be essential for their interaction. The extremely strong affinity of the inhibitor to SacC1 ($Ki= 3.45 \times 10^{-12}$) is remarkable and weaker than the affinity of several restriction enzymes, such as EcoRI ($Ki=1.38 \times 10^{-11}$) (Malin et al., 1999), Serine Protease ($Ki= 2.3 \times 10^{-10}$) (Milstone et al., 2000), λ Integrase ($Ki= 3.8 \times 10^{-10}$) (Boldt et al., 2004) and Astacin Metalloproteinase ($Ki= 1.9 \times 10^{-11}$) (Tsai et al., 2004), so, it is safe to say that the enzyme activity is regulated by the content of the inhibitor. The inhibitor does not affect other restriction enzymes.

The development of inhibition required 3 to 4 h in order to reach completion at the mM concentrations of the proteins. The mechanism of inhibition is consistent with a noncompetitive, single-step bimolecular reaction between enzyme and inhibitor. The rate constants for formation and breakdown of the enzyme-complex along with values for the apparent dissociation constant and change in

Gibbs free energy of binding derived from these constants are decreased affinity for the inhibitor. The decrease in binding affinity resulted predominantly and definitely from an increase in the rate of dissociation, with an increase in association rate of ~2- and ~5-fold, respectively, whereas the rate of association of the enzyme-complex was about the same as that for the weight protein.

It seems that SacC1 and its inhibitor is a good system for the study of physiological role inside yeast cell and is of interest to compare the effects with similar studies.

ACKNOWLEDGMENTS

The author is grateful for the skillful technical assistance of Mr. Jalal Ahmad Odeh, Faculty of Pharmacy, University of Al-Zaytoonah, Amman, Jordan. The author express his profound gratitude to professor Julia Bejar Alvarado, Department of Cellular Biology, Genetics and

Physiology, University of Malaga and professor Manchikatla Venkat Rajam, Department of Genetics, University of New Delhi India for their valuable suggestions and criticism in kinetics studies.

REFERENCES

Andrews P (1964). Estimation of the molecular weights of proteins by Sephadex gel-filtration. Biochem. J., 91: 222-233.

Boldt JL, Pinilla C, Segall AM (2004). Reversible Inhibitors of λ Integrase-mediated Recombination Efficiently Trap Holliday Junction Intermediates and Form the Basis of a Novel Assay for Junction Resolution. J. Biol. Chem., 279: 3472-3483.

Braciak TA, Northemann W, Hudson GO, Shiels BR, Gehring MR, Fey GH (1988). Sequence and acute phase regulation of rat alpha 1-inhibitor III messenger RNA. J. Biol. Chem., 263: 3999-4012.

Brown NL, Smith M (1980). A general method for defining restriction enzyme cleavage and recognition site. Methods Enzymol., 65: 391-404.

Dubois M, Gilles KA, Hamilton JK, Rebers, RA, Smith F (1956). Colorimetric method for determination of sugars and related substances, Anal. Chem., 28: 350.

Gribenko AV, Makhatadze GI (2007). Role of the Charge-Charge Interactions in Defining Stability and Halophilicity of the CspB Proteins. J. Mol. Bio., 366(3): 842-856.

Kim YC, Grable JC, Love R, Greene PJ, Rosenberg JM (1990). Refinement of EcoRI endonuclease crystal structure. A revised protein chain tracing. Science, 249(4974): 1307-1309.

Lammeli UK (1970). Cleavage of structural proteins during the assembly of the head of Bacteriophage T4. Nature, 227: 680-685.

Lineweaver H, Burk D (1934). The Determination of Enzyme Dissociation Constants. J. Am. Chem., Soc. 56: 658-666.

Lowry OH, Rosebrough NJ, Furr AL, Randall RJ (1951). Protein measurement with the Folin phenol reagent. J. Biol. Chem., 193: 1209-1220.

Maizel JV (1971). Polyacrylamide gel electrophoresis of viral protein. In: Maramorosch, K, Koprowshi H (Eds.). Meth. Virol., 5: 179-246.

Malin G, Lakobashvili R, Lapidot A (1999). Effect of tetrahydropyrimidine Derivatives on Protein-Nucleic Acids Interaction: Type II restriction endonuclease as a model system. J. Biol. Chem., 274: 6920-6929.

Milstone AM, Harrison LM, Bungiro RD, Kuzmic P, Cappello M (2000). A Broad Spectrum Kunitz Type Serine Protease Inhibitor Secreted by the Hookworm Ancylostoma ceylanicum J. Biol. Chem. 275: 29391-29399.

Muralidhara BK, Negi S, Chin CC, Braun W, Halpert JR (2006). Conformational Flexibility of Mammalian Cytochrome P450 2B4 in Binding Imidazole Inhibitors with Different Ring Chemistry and Side Chains: Solution thermodynamics and molecular modeling. J. Biol. Chem. 281: 8051-8061.

Oda M, Tamura A, Kanaori, K, Kojima S, Miura KI, Momma K, Tonomura B, Akasaka K (2002). Functional tolerance of Streptomyces subtilism inhibitor toward conformational and stability changes caused by single-point mutations in the hydrophobic core. J. Biochem. 132: 991-995.

Richard JR, Belfort M, Bestor T, Bhagwat AS, Bickle TA, Bitinaite J, Blumenthal RM, Degtyarev SKH, Dryden DTF, Dybvig K, Firman K, Gromova ES, Gumport RI, Halford SE, Hattman S, Heitman J, Hornby DP, Janulaitis AJ, Jeltsch A, Josephsen J, Kiss A, Klaenhammer, Kobayashi I, Kong H, Krüger DH, Lacks S, Marinus MG, Miyahara M, Morgan RD, Murray NE, Nagaraja V, Piekarowicz A, Pingoud A, Raleigh E, Rao DN, Reich N, Repin VE, Selker EU, Shaw PC, Stein DC, Stoddard BL, Szybalski W, Trautner TA, Van Etten, JL, Vitor, JMB, Wilson GG, Xu SY (2003). Survey and Summary: A nomenclature for restriction enzymes, DNA methyltransferases, homing endonucleases and their genes. Nucleic Acids Res. 31(7): 1805-1812.

Richter W, Conti M (2004). The Oligomerization State Determines Regulatory Properties and Inhibitor Sensitivity of Type 4 cAMP-specific Phosphodiesterases. J. Biol. Chem. 279: 30338-30348

Sallantin M, Huet JC, Demarteau C, Pernollet JC (1990). Reassessment of commercially available molecular weight standards for peptide SDS polyacrylamide gel electrophoresis using electroblotting and microsequencing. Electrophoresis, 11(1): 34-36

Serap Doğan S, Turan P, Doğan M (2006). Some kinetic properties of polyphenol oxidase from Thymbra spicata L. var. spicata". Process Biochem. 41(12): 2379-2385.

Shikara M (2010). Purification of a restriction enzyme from Saccharomyces cerevisiae. J. Yeast and Fungal Res. 1(7): 127-135.

Spector S, Wang M, Carp SA, Robblee J, Hendsch ZS, Fairman R, Tidor B, Raleigh DP (2000). Rational Modification of Protein Stability by the Mutation of Charged Surface Residues. Biochemistry, 39(5): 872-879.

Strickler SS, Gribenko AV, Gribenko AV, Keiffer TR, Tomlinson J, Reihle T, Loladze VV, Makhatadze GI (2006). Protein Stability and Surface Electrostatics: A Charged Relationship, Biochemistry, 45(9): 2761-2766.

Tsai PL, Chen CH, Huang CH, Chou CM, Chang GD (2004). Purification and Cloning of an Endogenous Protein Inhibitor of Carp Nephrosin, an Astacin Metalloproteinase J. Bio. Chem. 279: 11146-11155.

Zhuravleva LI, Oreshkin EN, Bezborodov AM (1987). Isolation and purification of restriction endonuclease Sac I from Streptomyces achromogenes ATCC 12767. Prikl. Biokhim. Mikrobiol. 23(2): 208-215.

Solid state fermentation of *Jatropha curcas* kernel cake: Proximate composition and antinutritional components

M. A. Belewu[1]* and R. Sam[2]

[1]Department of Animal Production, Microbial Biotechnology and Dairy Science Laboratory, University of Ilorin, Ilorin, Kwara State, Nigeria.
[2]School of Medical Sciences, University of Cape Coast, Ghana.

Five fungi (*Aspergillus niger, Penicillium chrysogenum, Rhizopus oligosporus, Rhizopus nigricans and Trichoderma longibrachitum*) were used in the fermentation of *Jatropha curcas* kernel cake for a 7 days period in a completely randomized design model. The results revealed significant increase in the crude protein content of all the fungi treated samples with *Aspergillus niger and T. longbrachitum* treated cake recorded higher value compared to other treated samples. There was no significant difference in the crude fiber content among all the samples. With the exception of ether extract content of sample treated with *Rhizopus nigricans* which was similar to that of the control (untreated sample) other samples showed lower significant values than the untreated (control) sample. Contrarily, the ash content was significantly lower in the control sample compared to other fungi treated samples. The content of the trypsin inhibitor was highest (18.6 mg/kg) in the control but reduced significantly in the fungi treated samples (6.50 - 8.23). The lectin, saponins, phytate and phorbolester contents followed similar trend. It could be concluded from this study that solid state fermentation of *Jatropha* kernel cake detoxified and inactivate almost 100% of the antinutrient contents expect phorbolester to a tolerable level in the *A. niger* treated sample.

Key words: *Jatropha curcas* kernel cake, proximate composition, trypsin inhibitor, lectin, phytic acid, saponins, phorbolester.

INTRODUCTION

Jatropha curcas L. Linnaeus 1753 is a small shrub plant which grows wildly in the tropics and sub-tropics but it is used as fencing in Nigeria. Apart from the University of Ilorin that has the plantation of *Jatropha*, its economic importance is still at the infancy stage in Nigeria (www.unilorin.edu.ng). The plant could adapt to marginal areas with poor soils and low rainfall (480 mm per annum and 28.5°C) where growth is not in competition with annual food crops. The cake which is obtained after the extraction of the oil contains a crude protein content of between 58 and 64 percent. Hence, it has high potential to complement and substitute soybean meal as a protein source in livestock diets (Makkar and Becher, 1997b).

The percentage of essential amino acids and mineral contents can be compared to those of other seed and press cakes used as a fodder (Trabi et al., 1997). Additionally, *J. curcas* contains various antinutrients like trypsin inhibitor, lectin, tannins, saponins, phytate and phorbolester. The above named toxins can be removed either by chemical or physical methods while phorbo-lester is the most difficult toxin to be detoxified by these methods. Phorbolester is a natural plant derived from organic compound which is a member of the tigliance family of diterpene.

Symptoms of phorbolester toxicity include dehydration, sunken eyes, skin irritation, loss of appetite, loss of condition and finally death (Belewu, 2008). Makkar and Becker (1997a) reported on the chemical treatment of *Jatropha* seed cake with no encouraging result .Hence, the thrust of this study was to evaluate the efficacy of fungal fermentation of *J. curcas* kernel cake on proximate composition and antinutrient content of spent substrates.

---

*Corresponding author. E-mail: milkyinka@yahoo.com, mabel@unilori.edu.ng.

Table 1. Proximate composition of the fungi treated and untreated *J. curcas cake*.

Parameters (%)	Untreated sample (control)	*Penicillium* treated sample	*R. oligosporus* treated sample	*R. nigricans* treated sample	*A. niger* treated sample	*T. longibrachitum* treated sample
Dry matter	80.70	77.40	76.35	75.65	90.20	88.05
Crude Protein	37.82[c]	48.18[b]	49..22[b]	52.49[b]	65.75[a]	63.06[a]
Crude fibre	6.50[b]	6.70[b]	5.96[b]	6.87[b]	5.70[b]	11.70[a]
Ether extract	13.18[b]	12.00[b]	12.55[b]	16.20[a]	10.70[b]	11.75[b]
Ash	4.68[b]	5.80[b]	6.00[a]	5.17[b]	5.26[b]	5.17[b]

Means along the same row with similar superscripts are not significantly different (p > 0.05).

MATERIALS AND METHODS

Collection and processing of *Jatropha curcas* seed

Mature seeds of *J. curcas* were collected around the University of Ilorin as well as Ilorin metropolis, Nigeria. The seeds were hand picked to get rid of stones and other debris. The cleaned seeds were weighed and later cracked individually to remove the kernel which was equally weighed and expressed as the percentage of the seed. The kernel was later milled using grinder and then defatted using mechanical hydraulic press. The defatted kernel cake was kept in a polythene bags for autoclaving at 121°C for 30 min, so as to get rid of any microbes that could be present in the cake (Belewu, 2008).

Fungi used

The fungi used which was collected from International Institute of Tropical Agriculture Ibadan (IITTA), Nigeria include *Rhizopus oligosporus, Rhizopus nigericans, Aspergillus niger, Trichoderma longibrachiatum* and *Penicillium chrysogenum*. The fungi were maintained on potato dextrose agar (PDA) and later used in inoculating the cooled autoclaved dried substrate.

Inoculation and Incubation of the substrate

The cooled autoclaved *J. curcas* kernel cake containing in Petri dishes were inoculated each with the fungi (*R. oligosporus, R. nigericans, A. niger, T. longibrachiatum,* and *P. chrysogenum*), while the control experiment was not inoculated with any fungus. Each of the fungi was replicated five times throughout the experimental period. The spores of each fungus was harvested with Tween 80 solution and adjusted to 10^7 - 10^8 spores per ml with sterile water (serial dilution and haemocytometer methods). After inoculation the samples were incubated for each inoculum to colonize the substrate in about 7 days. The fermented substrates were ovendried at 70°C to terminate the fungi growth.

Chemical analysis

Samples of the fungi treated and untreated were collected for proximate analysis and antinutrient determinations (AOAC, 2000).

Statistical analysis

All data collected were subjected to analysis of variance of a completely randomized design model while the means were separated using Duncan (1955) multiple range test.

RESULTS AND DISCUSSION

The average weight of the Jatropha seed found in Nigeria was 0.60g while the weight of the kernel expressed as the weight of the seed was 0.63%. These values agreed with the reported values of Makkar et al. (1998). The dry matter (Table 1) content of the *Aspergillus niger* treated *Jatropha curcas* kernel cake was significantly higher than that of the control and other samples. The higher content was consistent with the report of Belewu (2008) for similar substrate. The improvement in the crude protein content (Table 1) of the fungi treated samples was in agreement with the work of Jacqueline and Visser (1996), Belewu (2008) and Belewu et al. (2009) who used similar fungi in the treatment of Jatropha seed cake and discarded cell phone recharged cards. The increment in the protein content could be due probably to the addition of microbial protein during the process of fermentation.

With the exception of highly significant ether extract content of *R. oligosporus* treated sample (Table 1) other samples had similar values of ether extract (p>0.05). The higher content of the ether extract in the *R. oligosporus* treated sample could be due to ability of the fungus in producing lipase. The ash content and the crude fibre content showed no significant difference in all the samples.

The reduction of trypsin inhibitor activity in the fungi treated samples (Table 2) showed the potential of the various fungi in reducing the toxins. There are few reports on the reduction of such compounds. Other toxins decreased appreciably in all the samples as compared to the untreated sample (control). The decreased in the various toxins levels could be due to the production of various enzymes during the vegetative and reproductive phases of the fungi (Jacqueline and Visser (1996).The various enzymes secreted during incubation period include cellulose, xylanase, xylosidases, hemicellulase, amylases beta glycosidase, proteinases, pectinases, alpha-galactosidase etc and these could have contributed to the detoxification of the kernel cake.

through sanitary engineering, as a solution to the problems generated by the wastes released on the environment. The outstanding capacity of degradation of microorganisms is the consequence of the evolution of the enzymatic systems of prokaryote and eukaryote cells, which have co-existed with a wide variety of natural substances of diverse origins (Vasconcelos et al., 2003). This diversity of potential substrates for microbial growth resulted, then, in the appearance of enzymes capable of transforming organic molecules of rather distinct structures. This response of the metabolism of certain microorganisms undoubtedly confers some additional advantages to the microbial cells, such as exploration of new ecologic niches and energy sources. In addition, despite the importance given to the *amylase*s, few are the information concerning these enzymes from endophytic fungi, especially their function, molecular and kinetic properties (Spier, 2005).

Despite that traditionally enzymes are produced through submerse fermentation because of the easier control of the process, the fermentation in semi-solid medium has some advantages over the former (Santos et al., 2006), for instance: the culture medium is simpler and less expensive, often made up of non-refined agricultural residues, like the residues from the processing of rice, which are frequently disposed of by the industries; less production of residues by the fermentation; less use of water, preventing the contamination through bacteria that generally demand larger amounts of liquid in the medium; less energy input; easy aeration with greater diffusion of gases; possibility of production of spores and mushrooms, which cannot be obtained through submerse fermentation (Hölker et al., 2004; Kwiatkowski et al., 2006).

Fungi from the genus *Colletotrichum* are ascomycetes found in every environment, especially in association with plants, either as pathogens, symbionts or endophitic. Tavares (2004) described that the conidia of the fungus *Colletotrichum gloesporioides* are salmon-colored, straight, cylindrical, with obtuse apex and truncated base.

Assis et al. (2010) report that *C. gloeosporioides* has a large potential for producing the α-*amylase* and *glucoamylase* enzymes. The *amylase*s are enzymes that hydrolyze starch releasing several products, including dextrines and small polymers of glucose units.

The use of residues of rice processing is desirable because they can represent up to 22% of the grain weight. The inadequate destination of these residues can cause large passive environmental issues for the industries. One of the destinations of these residues can be the feeding of domestic animals, when associated to other nutrients. In this way they become a low-cost substrate for fungal growth and the obtention of enzymes of biotechnological interest.

The present work evaluated the capacity of the endophytic fungus *C. gloeosporioides* of producing the amylolytic enzymes α-*amylase* and *glucoamylase*

through fermentation in semi-solid medium composed of residues of the processing of rice.

MATERIALS AND METHODS

Microorganism under study

For this investigation it was used the endophytic fungus *C. gloesporioides*, lineage D4-FB, isolated from *Baccharis dracunculifolia* D.C. (Asteraceae) between 2008 and 2009 and kept in the fungus library of the Laboratory of Microbiology of the Paranaense University– UNIPAR – Unit Campus of Francisco Beltrão – Paraná – Brazil.

Fermentation process

In the production of the enzymes α-*amylase* and *amyloglucosidase*, it was used as the solid fermentation in 250 ml Erlenmeyer flasks. The following culture media was employed for the enzymatic production: 100 ml distilled water for each 50 g of rice residue. The pH was adjusted to 6.8 and this medium was sterilized at 121°C for 15 min. The medium was inoculated with the spore suspension at a ratio of 10^7 spores per gram of rice. After being homogenized and mixed in the Erlenmeyer, the medium was incubated at 28°C for 144 h (Pandey et al., 2005).

Analysis of the fermented substrate

Each 24 h, samples of five grams were mixed with 50 ml of distilled water. This suspension remained under agitation during 30 min. Next, it was filtered to remove the solid debris, yielding a clear extract used for pH determination. The extract was centrifuged at 3000 rpm for 15 min and the supernatant was used to determine the enzymatic activity.

pH

The pH was measured on a suspension obtained after homogenization of five grams of fermented material in 50 ml of distilled water, which was agitated continually for 30 min.

Dosage of the amylolytic activity of the α-*amylase*

The activity of the α-*amylase* was determined measuring the concentration of starch through iodine dosage. One unit of α-amylase is defined as the amount of enzyme capable of hydrolyzing 10 mg of starch in 30 min under the conditions described by Soccol (1992). The calculations for the determination of the α-amylase activity were carried out according to the methodology described by Pandey et al. (2005).

Glucoamylase

The *glucoamylase* activity was determined through the release of reducing sugars, dosed through the DNS method (Miller, 1959) cited by Costa (1996), described by Soccol (1992) and Pandey et al. (2005). The sugars were expressed as glucose equivalents. One unit of *glucoamylase* was defined as the amount of enzyme releasing 1 μmol of reducing sugar (expressed as glucose) per minute under the conditions of the assay (Alazard and Raimbault, 1981), cited by Soccol (1992). The calculations for the

Table 1. Units of enzyme produced during the periods from 24 to 144 h of fermentation, at 28°C.

Time (h)	α-Amylase (U/g)*	Glucoamylase (U/g)	pH
24	145.95±28.95D#	28.80±10.23E	6.29
48	392.90±23.98B	78.75±18.43D	6.57
72	510.67±35.76A	342.90±15.87A	6.05
96	563.32±45.54A	345.30±14.98A	5.49
120	498.54±43.65B	278.98±36.45B	5.35
144	328.23±28.56C	178.65±30.23C	5.30

*Means followed by capital letters on the column do not differ at the level of 5% by Tukey's test. #Mean of the activities obtained from triplicates. Mean from the enzymatic activities ± standard deviation.

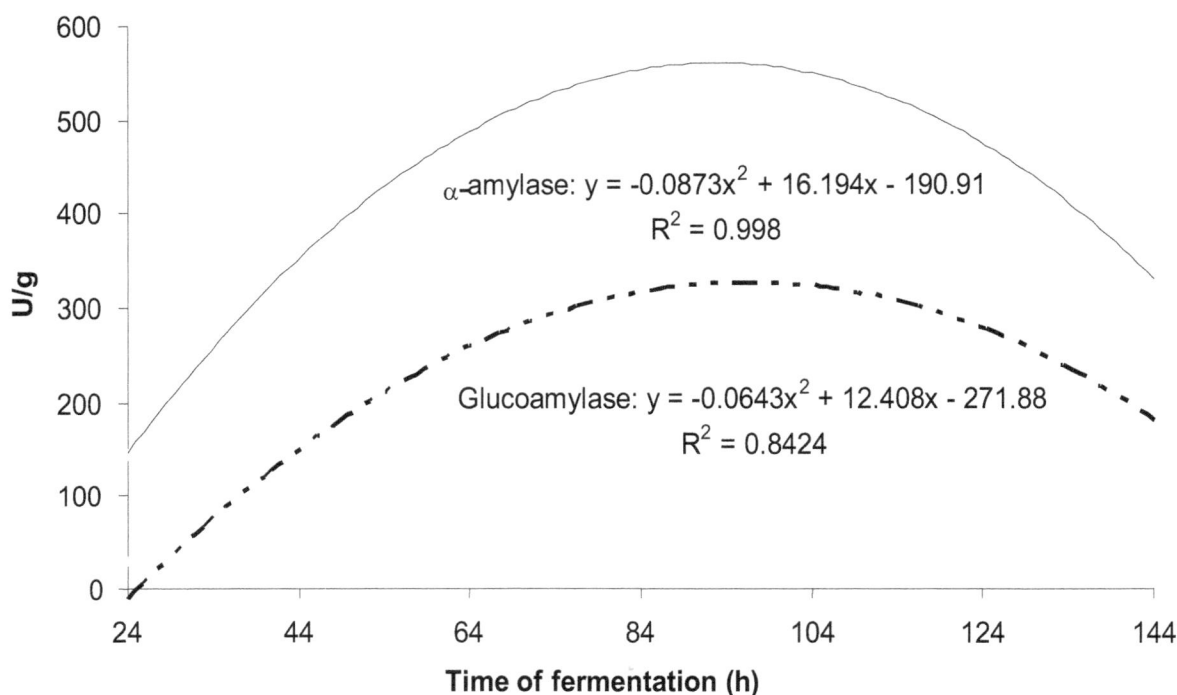

Figure 1. Kinetic of the production of the enzymes α-*amylase and amyloglucosidase* at 28°C, initial pH 6.8, rice-based substrate without supplementation relative to time (hours) and the production of enzymes U/g of substrate.

determination of the α-☐amylase activity were carried out according to the methodology described by Pandey et al. (2005).

Statistical analysis

The statistical analysis was carried out with the Statistical software, version 5.0. The analyses of variance were carried out according to the rules of the ANOVA. The significant differences between the means were determined through Tukey's test at the level of 5%. All the activities were made in triplicate.

RESULTS AND DISCUSSION

The determination of the enzymatic activity produced by *C. gloeosporioides* was expressed in units of activity

(UI – International Units) as the amount of enzyme catalyzing the transformation of 1 µmol of product per minute for each gram of fermented substrate. These data are presented in Table 1 and in Figure 1.

With the data obtained, it is observed that the fungus *C. gloeosporioides* inoculated in rice-based culture medium caused hydrolysis of the substrate in response to the action of the enzymes produced. The high level of starch makes rice an excellent source of carbon, in addition of being an excellent source of starch, indispensable for the synthesis of *amylases*. Fellows (1994) states the need for the presence of a source of starch to induce the production of *amylases* by filamentous fungi.

The analysis of the data in Table 1 and Figure 1 reveals that the best time for the production of the two enzymes was from 72 to 96 h, with peak production of 510.67±35.76 and 342.90±15.87 U/g of fermented substrate, respectively for α-☐amylase and glucoamylase. At 96 h the values were higher but statistically equal to those at 72 h of fermentation at the level of 5% by Tukey's test: 563.32±45.54 and 345.30±14.98 U/g of fermented substrate for α-☐amylase and glucoamylase, respectively. After this time, the values became lower at 120 and 144 h, this behavior being observed in the expression of both enzymes. At 24 and 48 h, it was observed that the values for both α-amylase and glucoamylase were increasing, indication a strong gene expression for both enzymes.

As for glucoamylase, it is noticed that the initial values at 24 h are statistically lower by Tukey's test than those obtained at 48 h, because these values were 28.80±10.23 and 78.75±18.43 U/g of fermented substrate for 24 and 48 h, respectively. The results obtained allowed the construction of a curve representing the kinetics of the behavior of the two enzymes during the whole fermentative process. The following variables were considered: rice-based substrate, temperature of 28°C and initial pH 6.8, and the equations representing this behavior were $y = -0.0873x^2 + 16.194x - 190.91$ with $R^2 = 0.998$ for α-amylase and $y = -0.0643x^2 + 12.408x - 271.88$ with $R^2 = 0.8424$ for glucoamylase.

For the mathematical model observed, there is the ratio of the relative expression of the error in the process, which is given by R^2 and was very close to 1.00 in the quantification of α-☐amylase, indicating a small variation in the increasing expression of the gene for this enzyme during the period of observation. This points to a system with stable control of the variables and efficient process of enzyme quantification (Pfaffl, 2001).

The ratio between the values of R^2 of the gene expressing the production of the enzymes under study normalizes the expression of this gene. If the reference gene has a stable behavior, R^2 will be 1.00 and the value of R^2 can be used for the general analysis of the behavior of the whole process (Pfaffl, 2001). The results obtained reveal that rice was an excellent substrate for the production of amylolytic enzymes, because it induced the fungus C. gloeosporioides to express the genes for the two enzymes assayed.

Norouzian et al. (2006) reported that some fungal lineages produce enzymes capable of hydrolyzing starch and releasing glucose. Other authors also state the need of the presence of a source of starch for the induction of amylase production by fungi, yeast and bacteria (Fellows, 1994; Pandey et al., 1999; Gupta et al., 2003). The microorganisms do not directly uptake complex molecules such as starch, a polysaccharide. In the absence of another usable component in the medium, the microorganism synthesizes those specific enzymes that degrade the complex substrate to simpler molecules – in this instance amylases – so that these can convert the starch from the rice into usable sugars, thus guaranteeing

the growth and development of the microorganism (Gupta et al., 2003).

According to Pandey et al. (1999), some factors can interfere in the gene expression of amylases, such as the accumulation of substrate, which can interfere in the microbial respiration as well as in the aeration of the medium. Spier et al. (2004) reported the importance of empty spaces among particles to facilitate the transfer of gases and heat, of the size of the particles and of the assembly of the fermentation medium on solid substrates, leading to a greater production of amylases through fermentation with fungal cultures. When studying several substrates with or without supplementation for the production of extracellular amylases by Aspergillus oryzae through solid fermentation, (Galvez, 2005) obtained amylase activity of 255 U/g of starch-based substrate, under the same conditions of this work, after 72 h of fermentation, indicating values inferior to those reported here with the fungus C. gloeosporioides, which presented values of 510.67 U/g of substrate, demonstrating that it is an excellent amylolytic fungus.

Another important variable to be assessed is temperature, because it interferes with the optimal conditions for the activity of an enzyme (Lehninger, 2006). In this work temperature was initially set at 28°C. According to Pandey et al. (2005), excellent yields of α-amylase were reached with temperatures between 28 and 37°C. This same author obtained the best productions of α-☐amylase by Aspergillus niger at temperatures of 28 to 30°C, coinciding with the conditions of the fermentation process used in this work. The monitoring of the pH is also an important variable concerning enzyme activity, because according to Soccol (1992) and Soccol et al. (2005), the fungus has a limited capacity for growth in extreme conditions of acidity or alkalinity. This characteristic is of extreme importance for the fermentative processes, because they show that under these conditions most of the bacteria responsible for the contamination of the fermentative processes are inhibited.

In this work, initial pH was 6.8. This pH was changed during the process from 6.8 to 6.29 in the first 24 h, reaching 5.49 at 96 h of fermentation and keeping close to 5.30 until the end of the process. It is observed that the best productions of both enzymes were observed at 72 and 96 h – pH 6.05 and 5.49, respectively – indicating that the fungus C. gloeosporioides expresses its genes for these amylolytic enzymes in more acidic environments. These results are in accordance with the data obtained by Costa (1996), who studied the influence of pH in the production of amyloglucosidase by A. niger by solid fermentation in rice bran, and verified an optimal pH of 5.6 for the production of this enzyme. In comparison, the fungus C. gloeosporioides has preference for pH 5.5 for the production of α-☐amylase and glucoamylase.

In face of the results obtained, it is verified the capacity of *C. gloeosporioides* of producing the amylolytic enzymes α-☐*amylase* and *glucoamylase,* using residues of the processing of rice without supplementation as substrate, under the conditions of this work for solid-state fermentation.

Conclusion

After this investigation, it could be concluded that the fungus *C. gloeosporioides* demonstrated capacity for the production of *amylases* (α-*amylase* and *glucoamylase*) through rice-based solid-state fermentation without supplementation. The highest activity of α-*amylase* and *glucoamylase* was at 72 and 96 h, with 510.67±35.76 and 563.32±45.54 U/g of substrate for α-*amylase* and 342.90±15.87 and 345.30±14.98 for *glucoamylase*, and was reached at pH of enzymatic incubation of 6.05 and 5.49 and incubation temperature of 28°C.

REFERENCES

Assis TC, Menezes M, Andrade DEGT, Coelho RSB, Oliveira SMA (2010). Estudo comparativo de isolados de *Colletotrichum gloeosporioides* quanto ao efeito da nutrição de carboidratos no crescimento, esporulação e patogenicidade em frutos de três variedades de mangueira. Summa Phytopathologica, 27: 208-212.

Butzen S, Haefele D (2008) Dry-grind ethanol production from corn. Crop Insights, 18: 01-05.

Costa JAV (1996). Estudo da Produção de Amiloglucosidase por *Aspergillus níger* NRRL 3122 em Fermentação Semi-Sólida de Farelo de Arroz. Tese de Doutorado em Engenharia de Alimentos, Universidade Estadual de Campinas, p. 203.

Fellows P (1994). Tecnologia del Procesado de los Alimentos: Princípios e préticas. Zaragoza: Editorial Acribia, p. 172-177.

Galvez A (2005). Analyzing cold enzyme starch hydrolysis technology in new ethanol plant design. Ethanol Prod. Mag., 11: 58–60.

Gupta R, Gigras P, Mohapatra H, Goswami VK, Chauhan B (2003) Microbial α-*amylases*: A biotechnological perspective. Process Biochem., 38: 11, 1599–1616.

Hölker U, Höfer M, Lenz J (2004). Biotechnological advantages of laboratory-scale solid-state fermentation with fungi. Appl. Microbiol. Biotechnol., 64: 175-186.

Kwiatkowski JR, Mcaloon AJ, Taylor F, Johnston DB (2006). Modeling the process and costs of fuel ethanol production by the corn dry-grind process. Ind. Crops Prod., 23(3): 288–296.

Lehninger AL (2006). Princípios de bioquímica. 4 ed: São Paulo. Sarvier, p. 725.

Norouzian D, Akbarzadeh A, Scharer JM, Moo-Young M (2006). Fungal *glucoamylases*. Biotechnol. Adv., 24(1): 80–85.

Pandey A, Benjamin S, Soccol CR, Nigam P, Krieger N, Thomaz-Soccol V (1999). The realm of microbial lipases in biotechnology. Biotechnol. Appl. Biochem., 29: 119-113.

Pandey A, Webb C, Soccol CR, Larroche C (2005). Enzyme Technology. 1ª ed. New Delhi: Asiatech Publishers Inc., p. 760.

Pfaffl MW (2001). A new mathematical model for relative quantification in real-time RT-PCR. Nucleic Acids Res., 29: 9-12.

Santos DT, Sarrouh BF, Santos JC, Pérez VH, Silva SS (2006). Potencialidades e aplicações da fermentação semi-sólida em biotecnologia. *Janus*. 3(4): 164-183.

Soccol CR, Rojan PJ, Patel AK, Woiciechowski AL, Vandenberghe LPS, Pandey A (2005). Glucoamylase. In: Enzyme Technology. New Delhi: Asiatec Publishers Inc., pp. 221-230.

Soccol CR (1992) Physiologie et Métabolisme de Rhizopus en Culture Solide et Submergée en Relation Avec la Dégradation d'Amidon et la Production d'Acide L(+) Lactique. Thèse de Doctorat. Mention Génie enzymatique, Bioconversion et Microbiologie, Université de Tecnologie de Campiègne. Compiègne-France, p. 218.

Spier MR, Woiciechowski AL, Soccol CR (2004). Produção de α-Amilase por *Aspergillus* em Fermentação no Estado Sólido de Amido de Mandioca e Bagaço de Can☐☐de-Açúcar. VI Seminário Brasileiro de Tecnologia Enzimática. Anais Enzitec 2004. Rio de Janeiro: Enzitec, 1: 116.

Spier MR (2005). Produção de enzimas amilolíticas fúngicas α – amilase e amiloglicosidase por fermentação no estado sólido. Dissertação (Mestrado em Tecnologia de Alimentos) – Universidade Federal do Paraná. Curitiba – PR, p. 157.

Vasconcelos WE, Rios MS, Sousa AH, Medeiros EV, Silva GMC, Maracaja PB (2003) Caracterização bioquímica e enzimática de Cunninghamella isoladas de manguezal. Revista de Biotecnologia e Ciência da Terra, 3: 18.

6

Improved antimicrobial activity of the Tanzanian edible mushroom *Coprinus cinereus* (Schaeff) Gray by chicken manure supplemented solid sisal wastes substrates

Liberata Nyang'oso Mwita, Anthony Manoni Mshandete and Sylvester Leonard Lyantagaye*

Department of Molecular Biology and Biotechnology, College of Natural and Applied Sciences, University of Dar es Salaam, Tanzania.

The Tanzanian edible mushroom species *Coprinus cinereus* was grown on sisal waste substrates supplemented with chicken manure with the aim to evaluate the effects of the chicken manure supplement on the antimicrobial activity of the mushroom's extracts. Crude ethyl acetate extracts were prepared from the mushroom's fruiting bodies harvested at pre-capping, capping and post capping stages, and the extracts were tested for antimicrobial activity, using the agar well method. The antimicrobial activity was observed only in capping and post capping stages of the mushrooms and the activity generally increased with increased percentage of manure supplementation. These findings show that Tanzanian edible *C. cinereus* mushroom contains antimicrobial compounds and chicken manure could be used in the cultivation of the mushroom to increase the production of active secondary metabolites, which could be used as lead compounds for discovery of new and more effective drugs against microbial infections.

Key words: *Coprinus cinereus*, antimicrobial, chicken manure.

INTRODUCTION AND LITERATURE REVIEW

Coprinus cinereus belongs to a small genus of mushrooms, Coprinus, in a black spored family Coprinaceae. All species go through an auto digestion at maturity in which the cap forms black spores that become a soupy black glob. *C. cinereus* has several close relatives, which are *Coprinus sterquilinus*, *Coprinus calyptratus*, *Coprinus spadiceisporus* and *Coprinus comatus* (Philips, 1981). However, the classification of the Coprinus species is still unclear because it was only based on morphological characteristics without molecular analyses. As a result, *C. cinereus* has been given different names by different scientists such as *Coprinosis cinerea* or *Coprinus macrorhizus* (Ohtsuka et al., 1973). Some Tanzanian indigenous strains of *Coprinus* were

reported for the first time by Härkönen et al. (2003), also based on the morphological caharacteristics. Mshandete and Cuff (2008) reported for the first time, the successful domestication of wild edible mushroom beleaved to be *C. cinereus* indigenous to Tanzanian. Recently, we showed that the Tanzanian *C. cinereus* can also grow on dried grasses supplemented with cow dung manure (Ndyetabura et al., 2010).

Mushrooms are important sources of medicines and nutritive proteins and minerals (Bahl, 1994). Many cultures worldwide recognize that extracts from certain mushrooms could have profound health promoting benefits and consequently, became essential components in many traditional medicines. There are at least 270 species of mushrooms that are known to possess various therapeutic properties (Ying et al., 1987) and hence the term "medicinal mushroom". Examples of edible mushroom genera mentioned by Ying et al. (1987), which demonstrate medicinal or functional properties are Lentinula, Hericium, Grifola, Flammulina, Coprinus and

*Corresponding author. E-mail: lyantagaye@amu.udsm.ac.tz , slyantagaye@gmail.com.

Pleurotus.

Whereas the majority of fungi produce very similar primary metabolites, secondary metabolites are more species–specific and products are often unique to a particular species (Isaac, 1997). For this reason it is likely that, although a very large number of fungal secondary metabolites have now been identified, many more will be described in future as the activities of more fungal species are investigated. By manipulation of the growth conditions, it is possible for the production of a secondary metabolite to be prevented or a fungus be induced to over-produce some of these compounds.

Different Coprinus species have been found to have medicinal and bioactive compounds by several researches. Polysaccharides extracted from the mycelia culture of *C. cinerea* (as *C. macrorhizus*) have been shown to contain antitumor effects (Ohtsuka et al., 1973). Polysaccharide solutions extracted from *C. comatus* and given to mice had the ability to increase serum lysozyme activity, which is used as a general indicator of immune system fitness. In addition to breaking down polysaccharides found in bacterial cell walls, lysozyme can also bind to the surface of some invading bacteria and make it easier for white blood cells to engulf them (Li et al., 2001). Water extract of *C. comatus* has been identified as containing potent antitumor compounds for breast cancer (Gu and Leonard, 2006). In the more recent study, it was found that the Tanzanian *C. cinereus* grown on dried grasses supplemented with cow dung manure exhibits activity against *Escherichia coli*, *Aspergilus niger* and *Candida albicans*, with the highest activity observed at 40% of cow dung manure supplementation (Ndyetabura et al., 2010). The potential of the Tanzanian *C. cinereus* as a possible source of bioactive compounds which could be used as lead compounds for drug discovery needs to be researched further. Varying the growth environmental conditions, such as supplementing the mushroom growing substrates with chicken manure, could lead the mushroom to produce secondary metabolites of medicinal or commercial significance.

According to 2002 census, Tanzania had 47 million chickens out of them 27 million were free-range chicken (*Gallus gallus* domesticus). Average manure production per chicken is 139 g per day with dry matter content of 22% (Thomsen, 2004). Based on 27 million free-range chicken populations in Tanzania, about 3,753 tons (826 tons dry matter) chicken manure can be generated per day equivalent to 1,369,845 tons (301,366 tons dry matter) per annum. Chicken manure has a high nitrogen and phosphorus contents; fresh chicken manure has 0.9% while on dry weight basis is 4.5% (Thomsen, 2004).

In the present study, crude ethyl acetate extract of *C. cinereus* grown on sisal wastes supplemented with different amounts of chicken manure was prepared and then antimicrobial activity was determined on the crude extracts by agar well method.

MATERIALS AND METHODS

C. cinereus

C. cinereus inoculum was obtained from the collection at the Department of Molecular Biology and Biotechnology of the University of Dar es Salaam, and cultivation was done according to Mshandete and Cuff (2008).

Spawn preparation

Spawn of *C. cinereus* mushroom were prepared with intact sorghum grains, which were bought from Kariakoo market in Dar es Salaam. The grains were first soaked in water overnight and thereafter parboiled for 10 min. After draining excess water, 1% (w/w) of calcium carbonate ($CaCO_3$) was added and properly mixed into the grains before spreading them out on a clean plastic sheath. After air-drying for about 20 min, 150 g of the grains were packed in 330 ml wide mouth bottles (300 ml jar, Kioo Ltd, Dar es Salaam) and sterilized in an autoclave (Koninklijke ad Linden Jr.Bn-Zwijinderect, Holland) at 121°C and 1 atm for 1 h. Thereafter, each cooled bottle of sterilized grains was aseptically inoculated with 1 cm² pieces of mycelium MEA taken from 4 to 7- day-old cultures of *C. cinereus*. The inoculated bottles, with caps closed, were shaken thoroughly by hand to distribute the mycelia to the grains. Before use, the bottles were incubated with their caps loosely in a ventilated incubator (Memmert GMBH kg, Schwabach frg, Germany) set at 28°C for 10 days.

Preparation of substrates and cultivation of *C. cinerius*

Diferent types of sisal wastes namely: sisal leaf decortication waste (also referred to as sisal leaves) and sisal fibre dust (also called sisal dust) were used each supplemented with chicken manure at 5, 10, 15, 20 and 25%. *C. cinereus* fruit bodies were harvested at capping and post capping developmental stages. Small amount of water was added during mixing to just soak and no a drop of water on squeezing the mixed substrate by hand. The mixture was then put in sterilization container pre-prepared. Each mixture composition was separated from the other by aluminium foil. The container and its content was sterilized at 121°C for 1 h after which it was left to cool to room temperature. Different mixture composition was then transferred into different plastic dishes with small hole on their sides and covers. Spawn-running (mycelia colonizing the substrate) and fructification (fruit body - mushroom development) were done as per Mshandete and Cuff (2008).

Harvesting of the mushroom

C. cinereus fruit bodies were harvested at young or pre-capping stage, firm and fleshy (immature/juvenile stage) or capping stage and when the mushroom caps turned into an inky mass considered matured or post capping stage. Mushrooms were harvested from the substrate, the substrate clinging to the stipe or to the volva was removed and the mushrooms in their entirety were weighed the same day.

Extraction of bioactive compounds

Extraction of bioactive compounds was done according to Ndyetabura et al. (2010). Freshly picked mushrooms were crushed using motor and piston, and transferred into 250-ml conical flasks for extraction. The crushed mushroom material was soaked twice in standard grade ethyl acetate for 12 h and then the ethyl acetate extract was decanted in a clean conical flask. Volumes of each extract in different pre-weighed round bottom flask (x) were concentrated, using rotary evaporation at constant temperature of 40°C. Each time when the solvent had been evaporated more extract was added until the whole volume was concentrated. Each round bottom flask with concentrated crude extract was weighed (y) to get the weight of the crude extract (y - x). The extracts were immediately tested for bioactivity or kept in the refrigerator for subsequent bioactivity test.

Test microorganisms and culture preparation

The test microorganisms used were bacteria *Pseudomonas aeuroginosa* and *E. coli* (Gram negative), *Staphylococcus aureus* (Gram positive) and one fungi *Candida albicans*. The test microorganisms were obtained from the Department of Molecular Biology and Biotechnology, University of Dar es Salaam. Nutrient agar (Lab MTM, Lancashire, UK) and malt extract agar (Pronadisa®, Conda Ltd. Madrid, Spain) were prepared for bacteria and Candida respectively, according to the manufacturers' instructions. Immediately after autoclaving, the media was allowed to cool in a 45 to 50°C water bath. The freshly prepared and cooled media was poured into glass, flat-bottomed Petri dishes (90 mm in diameter) placed on a level, horizontal surface (Faster® Laminar flow, Cornaredo via Merendi, 22 20010, Italy) to give a uniform depth of approximately 4 mm. The agar media was allowed to cool and solidify at room temperature. About 0.2 ml of the test inoculum was evenly spread on the surface of the solidified agar media, using a sterile Grigalsky spatula and antimicrobial test was done as explained shortly.

Antimicrobial activity tests

Antimicrobial activity of the crude extracts was tested according to Ndyetabura et al. (2010) by agar well assay methods as previously described by Rojas et al. (2006), Moshi et al. (2006). The concentrated crude extracts were re-dissolved in dimethylsulfoxide (DMSO) to make 0.1 mg/ml solutions. The prepared agar plates were inoculated with 200 µl bacteria/ fungi culture by spreading evenly over the surface of agar plate, using an ethanol flamed glass Drigalsky spatula (spreader). Un inoculated untreated agar plate was incubated at 37°C for 24 h before use, to ensure sterility. Wells of 5 mm in diameter and 4 mm in depth were made on the agar, using a sterile cork borer.

For each test microoganism, 25 µl of each extract and of control were pipetted into different wells (Rojas et al., 2006). The wells were then labeled to correspond with the code numbers of the test crude extracts and controls. The treated plates were stored in a refrigerator (Daewoo®, Daewoo electronics, Europe GMBH, Germany) at 4°C for at least six hours to allow diffusion of the extracts into the agar while arresting the growth of the test microbes. The plates were then incubated for 24 h at 37°C for bacteria and for 48 h at 30°C for fungi. The test was carried out in triplicates. Antimicrobial activity was determined by measuring the diameters of zones of inhibition in mm. The means of the diameters of zones of growth inhibitions for the treatments are shown in Table 1.

RESULTS

Antimicrobial activity was observed only in capping and post capping stages of *C. cinereus* mushrooms extracts and the mocrobial growth inhibition zones (in mm) are presented in Table 1. Figure 1 shows that the sisal leaves and sisal dusts without chicken manure supplementation produced *C. cinereus* mushroms whose extracts exhibited activity only against *E. coli* (Figure 1a to d) and *C. albicans* (Figure 1b). *P. aeuroginosa* and *S. aureus* growth was inhibited only after supplemention with chicken manure *P. aeuroginosa* being sensitive to broader levels of the chicken manure. The activity generally was increased with increased percent of supplementation (Figure 1a to d).

Figure 2 shows that the capping stage of *C. cinereus* mushrooms grown on sisal dust exhibited the highest antibacterial activity 17 mm for *E. coli* at 20% chicken manure supplimentation (Figure 2a), 16 mm for *S. aureus* (Figure 2c) and 17 mm for *P. aeuroginosa* at 25% chicken manure (Figure 2d). For *C. albicans* (fungi) the highest growth inhibition zone was 19 mm observed from the post capping stage of the mushrooms grown on sisal leaves (Figure 2b). *E. coli* was the most sensitive inhibited to grow by mushroom extracts from all levels of chicken manure supplements (Figure 2a) whereas *S. aureus* was the least sensitive inhibited to grow only by mushroom extracts from higher levels of chicken manure supplements (Figure 2c). At 10% chicken manure supplementation a reduced activity for *E. coli* (Figure 2a) or no activity for all other test microbes was observed (Figure 2b to d). Both Figures 1 and 2 show that, the mushrooms grown on sisal dust had broader spectra and higher antimicrobial activity than those grown on sisal leaves.

DISCUSSION AND CONCLUSION

Crude ethyl acetate extracts from mature stages of *C. cinereus* exhibited antimicrobial activity (Table 1) whereas younger stages did not. This observation agrees with the previous study by Ndyetabura et al. (2010), which showed that only extracts from capping stage and post capping stage of *C. cinereus* development exhibited antimicrobial activities against *E. coli*, *C. albicans* and *A. niger*. This observation also agrees with some older literature (Isaac, 1997; Abraham, 2001) which reported that most active secondary metabolites produced at the end of active growth are derived from primary products which were synthesized earlirer, and their formation may accompany differentiation and sporulation in the fungus. It is also true that when growth becomes restricted by some factors, products and intermediate compounds will then accumulate in their bodies or culture medium. This can be brought about by manipulation of the particular nutrients provided for growth or the provision of specific environmental conditions (Isaac, 1997). It is this attribute which is exploited for the commercial production of useful primary and secondary fungal products.

Table 1. The growth inhibition zones (in mm) formed after treatment of the test micro-organisms with the extracts from *C. cinereus* grown on sisal wastes supplemented with different percentages of chicken manure.

Sisal waste/developmental stages of *C. cinereus*	Test micro-organism	Mean of growth inhibition zone (mm)						
		Control	Chicken manure supplements (%)					
			0	5	10	15	20	25
Dust / Capping	*S. aureus*	0	0	0	0	0	0	16
	P. aeuroginosa	0	0	12	0	9	1	17
	E. coli	0	4	12	11	1	17	16
	C. albicans	0	0	0	0	0	0	9
Dust/ Post capping	*S. aureus*	0	0	0	0	7	12	0
	P. aeuroginosa	0	0	14	0	15	12	11
	E. coli	0	5	9	3	10	9	7
	C. albicans	0	8	3	0	8	13	0
Leaves/ Capping	*S. aureus*	0	0	0	0	0	13	0
	P. aeuroginosa	0	0	0	0	5	15	0
	E. coli	0	5	1	0	10	10	0
	C. albicans	0	0	0	0	0	5	19
Leaves/ Post capping	*S. aureus*	0	0	0	0	0	0	0
	P. aeuroginosa	0	0	8	0	0	16	0
	E. coli	0	5	9	4	11	10	9
	C. albicans	0	0	0	0	0	5	0

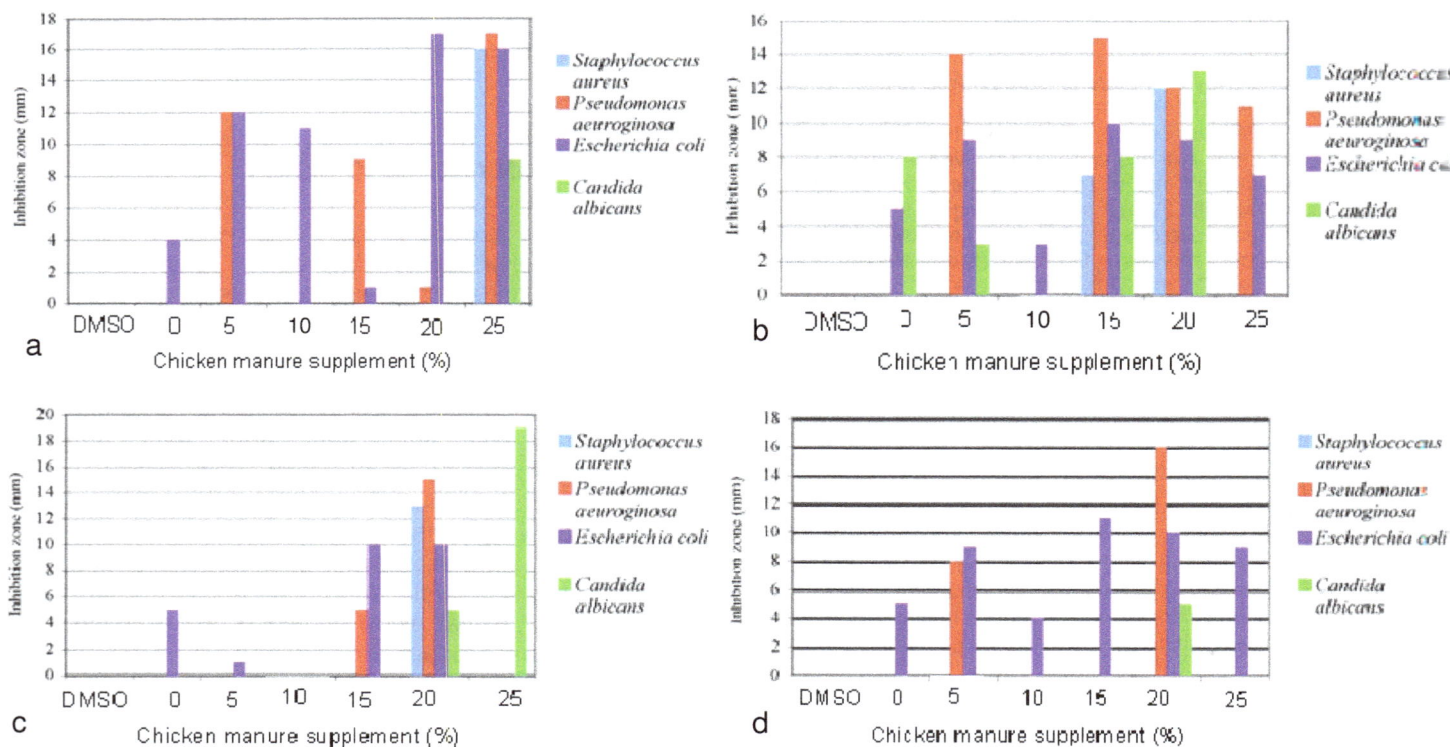

Figure 1. Effects of substrate types and developmental stages of *C. cinereus* on the antimicrobial activity: (a) Activity of *C. cinereus* capping stage grown on sisal dust, (b) activity of *C. cinereus* postcapping stage grown on sisal dust, (c) activity of *C. cinereus* capping stage grown on sisal leaves and (d) activity of *C. cinereus* postcapping stage grown on sisal leaves.

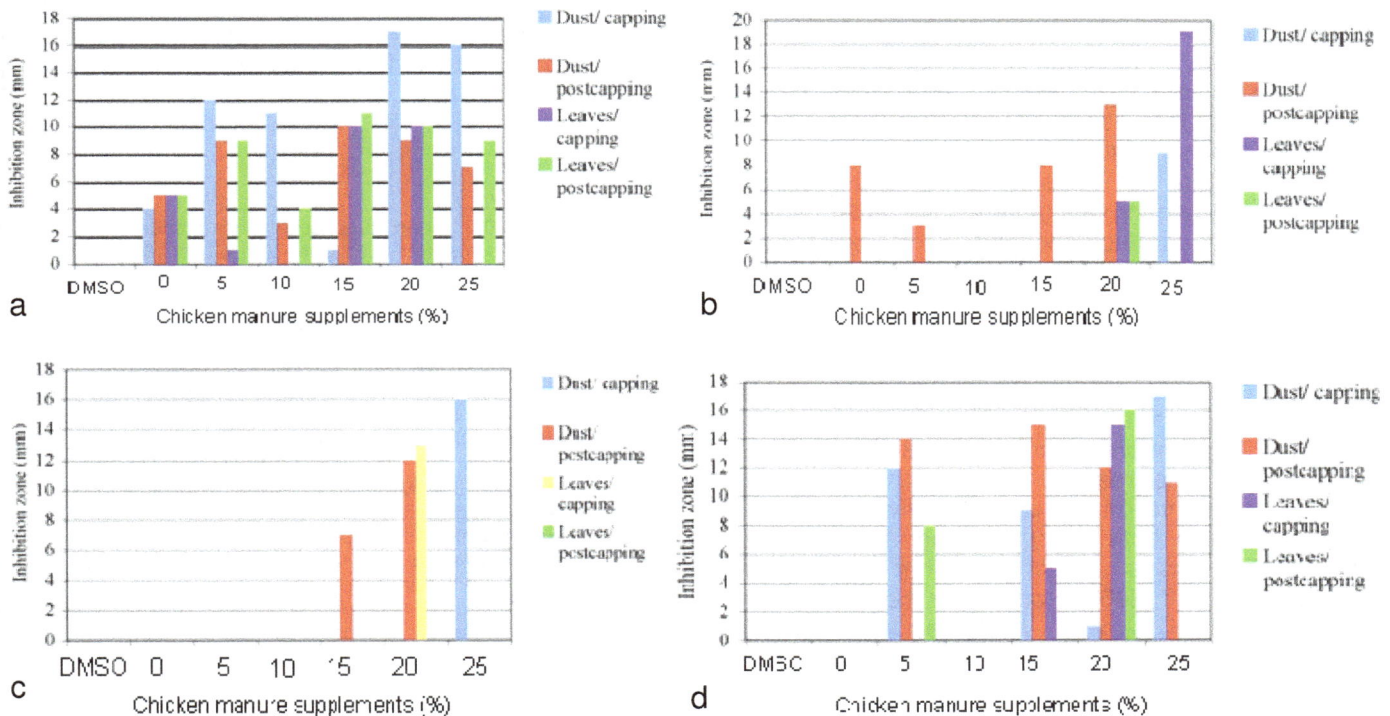

Figure 2. Comparing sensitivity of different microorganisms to *C. cinereus* extracts: (a) sensitivity of *E. coli* to *C. cinereus* extracts, (b) sensitivity of *C. albicans* to *C. cinereus* extracts, (c) sensitivity of *S. aureus* to *C. cinereus* extracts, and (d) sensitivity of *P.aeuroginosa* to *C. cinereus* extracts.

The antimicrobial activity of extracts from *C. cinereus* grown on sisal waste without manure supplements against *E. coli* and *C. albicans*, strongly suggest that the Tanzanian *C. cinereus* produces bioactive secondary metabolites naturally. This observation agrees with the study by Ndyetabura et al. (2010) again which showed that *C. cinereus* exhibited antimicrobial activity against all Gram-negative bacteria and all fungi species tested, including *E. coli* and *C. albicans*. The observation also agrees with previous studies elsewhere which showed that Coprinus representatives are a possible source of antibiotics (Zenkova et al., 2003). Zenkova et al. (2003) study showed that, the Coprinus representatives were able to inhibit growth of all, common Gram-negative bacteria, Gram-positive bacteria and fungal strains contrary to Ndyetabura et al. (2010). The environmental growth conditions of the Tanzanian *C. cinereus* is different from where the study by Zenkova et al, took place, leading to different types and rate of biochemical processes (Sher et al., 2010) and so different types and/or quantities secondary metabolites were produced. The present study therefore shows that, supplementation of chicken manure had positive effects on the antimicrobial activity of the Tanzanian *C. cinereus* regardless of whether Gram-pisitive or Gram-negative bacteria or *C. albicans* was used and generally, the inhibition zones increased with increase in percentage of

chicken manure supplementation. These observation could be due to the fact that, increasing amount of chicken manure supplementation raises amounts of nutrients in the substrates which enhance the metabolic reactions that are taking place in the mushroom and therefore increase the production of secondary metabolites. The present study shows that, addition of free-range chicken droppings manure to the sisal waste substrate led to the production of active metabolites against Gram-positive bacteria *S. aureus* and *P. aeuroginosa* which were previously not sensitive to the mushroom extracts at all. The availability of high nitrogen and phosphorus contents 4.5 to 7.5% on dry weight basis of chicken manure (Nahm, 2003; Thomsen, 2004; Chen et al., 2009), Increased biochemical processes in the growing mushrooms, leading to the production of new types of or higher quantities of the existing active secondary metabolites. In addition, free range chickens eat a whole range of complex combination of different floral, fauna and innorganic material from their environments and so the manure may contain a variety of new other molecules which eventually led to the activation of other different pathways, leading to the production of active secondary metabolites. This study provides a baseline for scientist to unraval more facts on the subject. Based on the 301,366 tons of dry matter chicken manure per annum produced by free-range chicken populations in rural Tanzania (Thomsen, 2004), it is obvious that this novel use lead to income generation and improved health standards in communities.

More studies are underway to shed light on the mechanisms of the observed activities and will be published in the next article. It may be difficult to explain the differences in the sizes of the microbial growth inhibition zones among the different extracts because concentrations of active components in the crude extracts are not known yet, and compounds present in some of the crude extracts are unknown yet. However, the observed difference in the activities could also be due to differences in diffusion of an antimicrobial through the solidified medium, which could be the result of molecular size or chemical nature of the active compounds. Small molecules diffuse more easily than large molecules (Hewitt, 1977). The reason to why at 10% chicken manure supplementation, a reduced activity for *E. coli* (Figure 2a) or no activity for all other test microbes (Figure 2b to d) is a subject of further investigation. Mshandete and Cuff (2008) reported for the first time, the suitability of sisal organic waste as substrate for cultivation of local edible *C. cinereus*, *P. flabellatus* and *V. volvacea* mushrooms. The present study shows for the first time, the increased antimicrobial effects of *C. cinereus* by chicken manure supplements on non-composted sisal leaves decortications residue and sisal dust. However, the reasons for the observed broader spectra and higher activity of the mushrooms grown on sisal fiber dust than those grown on the sisal leaves remain to be investigated.

In conclusion, *C. cinereus* grown on sisal wastes without chicken manure supplementation exhibited activity against *E. coli* only, whereas for *S. aureus* and *P. aeuroginosa* inhibiton zones were observed after supplementation with chicken manure. The activity increased with increase in percentage of manure supplementation. These findings show that the Tanzanian *C. cinereus* mushrooms contain antimicrobial compounds and that chicken manure could be used in the mushroom cultivation to increase the production of active secondary metabolites, which could be used as lead compounds for discovery of new and more effective drugs against microbial infections. By manipulation of the growth conditions, it is possible for the production of a secondary metabolite to be prevented or a fungus may be induced to over-produce some of these compounds.

ACKNOWLEDGEMENTS

The financial support was provided by SABINA (Southern African Biochemistry and Informatics for Natural Products). Gratitude is expressed to the Department of Molecular Biology and Biotechnology (MBB), College of Natural and Applied Science (CoNAS), University of Dar es Salaam (UDSM), Tanzania for generously providing the facilities to carry out the research. The technical support provided by Mr. Mutemi Muthangya, Mr. Ferdinand Patrick and Mr. Prosper Raymond of the MBE-UDSM is appreciated.

REFERENCES

Abraham WR (2001). Bioactive sesquiterpenes produced by fungi: Are they useful for humans as well? Curr. Med. Chem., 8:583-606.

Bahl N (1994). Handbook on mushrooms. Oxford and IBH Publishing CO. Pvt. Ltd. Delhi, India.

Chen LJ, Xing L, Han LJ (2009). Quantitative determination of nutrient content in poultry manure by near infrared spectroscopy based on artificial neural networks. Poult. Sci., 88(12): 2496-2503.

Gu YH, Leonard J (2006). *In vitro* effects on proliferation, apoptosis and colony inhibition in ER-dependent and ER-independent human breast cancer cells by selected mushroom species. Oncol. Rep., 5(2): 417-23.

Härkönen M, Niemelä T, Mwasumbi L (2003). Tanzanian Mushrooms. Edible, harmful and other fungi. Botanical Museum, Finish Museum of Natural History, Helsinki.

Hewitt W (1977). Microbiological assay: An introduction to quantitative principles. New York: Academic Press.

Isaac S (1997). Fungi naturally form many diverse biochemical products, some of which are now commercially important: How and why do they do this? Mycol. Answers, 11(4): 182-183.

Li S, An L, Zhang H (2001). Effects of polysaccharide from Coprinus comatus on activity of serum lysozyme in Kunming mouse, China Edible Fungi China, 20(4): 36-8.

Moshi MJ, Mbwambo ZH, Kapingu MC, Mhozya VH and Marwa C (2006). Antimicrobial and Brine Shrimp Lethality of extracts of *Terminalia mollis* Laws. Afr. J. Trad. CAM, 3(3): 1-10.

Mshandete AM, Cuff J (2008). Cultivation of Three Types of Indigenous Wild Edible Mushrooms: *Coprinus Cinereus, Pleurotus flabellatus* and *Volvariella volvocea* on Composted Sisal Decortications Residue in Tanzania. Afr. J. Biotechnol., 7(24): 4551–4562.

Nahm KH (2003). Evaluation of the nitrogen content in poultry manure. World Poult. Sci. J., 59: 77-88.

Ndyetabura T, Lyantagaye SL, Mshandete AM (2010). Antimicrobial activity of ethyl acetate extracts from edible tanzanian *Coprinus cinereus* (schaeff) s. Gray s.lat. Cultivated on grasses supplemented with cow dung manure. Arpn. J. Agric. Biol. Sci., 5(5). Available online from September 2010. www.arpnjournals.com.

Ohtsuka S, Ueno S, Yoshikumi C, Hirose F, Ohmura Y, Wada T, Fujii T, Takahashi E (1973). Polysaccharides having an ant carcinogenic effect and a method of producing them from species of Basidiomycetes. UK Patent 1331513, 26 September, 1973.

Philips R (1981). Mushrooms and other fungi of Great Britain and Europe (Pan Books Ltd. 1981/Book Club Associates 1981).

Rojas JJ, Ochoa VJ, Ocampo SA, Munoz JF (2006). Screening for antimicrobial activity of ten medicinal plants used in Colombian folkloric medicine: A possible alternative in the treatment of non-nosocomial infections. BMC Complement. Altern. Med., 6: 2.

Sher H, Al-Yemeni M, Bakhali AHA, Sher H (2010). Effect of environmental factors on the yield of selected mushroom species growing in two different agro ecological zones of Pakistan. Saudi J. Biol. Sci., 17: 321-326.

Thomsen IK (2004). Nitrogen efficiency N-labelled poultry manure. Soil Sci. Soc. Am. J., 68: 538-544.

Ying jz, Mao XL, Ma QH, Zong YC, Wen HA (1987). Icons of Medicinal Fungi From China. Beijing: Science Press.

Zenkova VA, Efremenkova OV, Ershova EY, Tolstych IV, Dudnik YV (2003). Antimicrobial activity of medicinal mushrooms from the genus Coprinus (Fr.) S. F. Gray (Agaricomycetideae). Int. J. Med. Mushrooms, Begell Digital Library 5(1).

7

Antifungal activity of extracts obtained from actinomycetes

Harpreet Sharma and Leena Parihar*

Department of Biotechnology, Lovely Professional University, Phagwara-144402 Punjab, India.

In the present investigation an attempt has been done on isolation of actinomycetes from the soil, extracting the antifungal compounds from these isolated actinomycetes and then testing the extract against the growth of *Alternaria* sps, *Aspergillus niger*, *Aspergillus flavus*, *Fusarium* sps, and *Rhizopus stolonifer*. During the investigation it was found that nearly all the extracts were effective against the test fungi and the mycelial growth of fungi is inversely proportional to the concentration of extract.

Key words: Antifungal activity, actinomycetes extracts, plant pathogenic fungi.

INTRODUCTION

Fungal phytopathogens pose serious problems worldwide and cause a number of plants and animal diseases such as ringworm, athlete's foot, and several more serious diseases. Plant diseases caused by fungi include rusts, smuts, rots, and may cause severe damage to crops. Fungi are some of the world's largest and possibly oldest individuals. Some species of fungi produce mycotoxins that are very toxic to humans. For example, the fungus *Claviceps purpurea* causes the ergot poisoning. An individual infected with the mycotoxin experiences hallucination, gangrene, and blood flow restrictions in his limbs. Humans usually get infected with the fungus after eating cereal grains contaminated with *C. purpurea* (Bauman, 2007).

Excessive use of chemical fungicides in agriculture has led to deteriorating human health, environmental pollution, and development of pathogen resistance to fungicide. Because of these problems in fungal disease control, a serious search is needed to identify alternative methods for plant protection, which are less dependent on chemicals and are more environmentally friendly. Microbial antagonists are widely used for the biocontrol of fungal diseases. Actinomycetes are the main source of antifungal, hence, highly used pharmacologically and commercially. These are the secondary metabolites of actinomycetes. The antagonistic activity of actinomycetes to fungal pathogens is usually related to the production of antifungal compounds against *Fusarium oxysporum*,

Sclerotinia minor, and *Sclerotinia rolfsii* (Lim et al., 2000). Biological control of plant diseases has received worldwide attention in recent years mainly as a response to public concern about the use of hazardous chemicals in the environment. Soil actinomycetes particularly *Streptomyces* sp enhances soil fertility and have antagonistic activity against wide range of soil-borne plant pathogens (Aghighi et al., 2004).

In the present investigation the ability of extracellular antifungal metabolites of Actinomycetes against *Rhizopus stolonifer*, *Aspergillus flavus*, *F. oxysporum* and *Alternaria* sp has been reported. This study investigated the antifungal activity of the cell-free culture filtrate of this antagonist to determine secondary antifungal compounds. The antifungal potential of extracellular metabolites produced by soil-borne Actinomycetes could be exploited for its future use as an antifungal compounds (Figure 1).

MATERIALS AND METHODS

Collection of soil sample

The soil samples were collected from various locations from Lovely Professional University, Phagwara, from farms in Hoshiarpur (Punjab) and from the farms in Patiala. Several diverse habitats in different areas were selected for the isolation of Actinomycetes. These habitats included the Rhizosphere of plants, agricultural soil. The samples were taken up to a depth of 20 cm after removing approximately 3 cm of the soil surface (Vijaya et al., 2007). The samples were placed in polyethylene bags, closed tightly and stored in a refrigerator. Heat treatment and CaCO$_3$ treatment of soil

*Corresponding author. E-mail: leena.parihar@gmail.com.

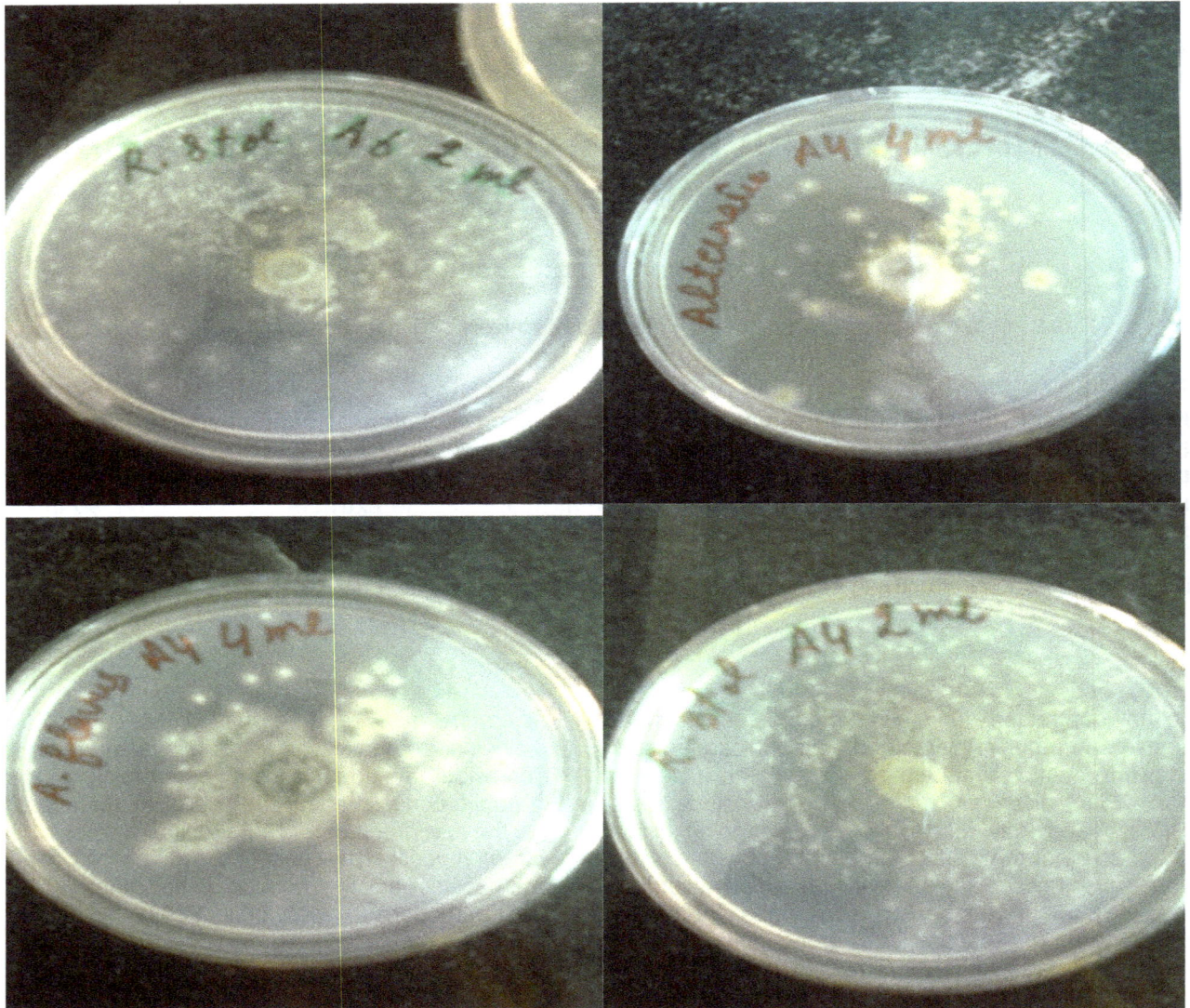

Figure 1. Antifungal activities shown by actinomycetes extract against fungi.

samples were done to isolate actinomycetes.

Media used for the isolation of actinomycetes

Different culture media used for isolation of actinomycetes were Starch casein agar medium, Actinomyces Hi Veg Agar medium, Actinomycete Isolation Agar medium, Streptomyces Agar medium

PROCEDURE FOR ISOLATION OF ACTINOMYCETES

Primary isolation

Serial dilution agar plating method is used for the isolation and enumeration of actinomycetes (Aneja, 2003). Prepared Starch casein agar medium is used for the isolation of Actinomycetes. Three starch casein agar media plates with the following dilutions: 1:10,000, 1:100,000 and 1:1,000,000 were used and in these plated 1 ml aliquots of various dilutions was added over cooled and

solidified agar medium. The plates were incubated at 28ºC for at least one week.

IDENTIFICATION OF ACTINOMYCETES

The identification of actinomycetes was done on the basis of morphology of spore chain, pigment production, color of aerial mycelium, color of substrate mycelium, consistency, gram's staining, growth on actinomyces media, growth on streptomyces media, etc. bio-chemical characterization of actinomycetes was done by esculin hydrolysis, starch hydrolysis, casein hydrolysis, glucose utilization and sucrose utilization.

The potent actinomycetes selected for further studies were characterized by morphological and biochemical methods. The microscopic characterization was done by gram's staining. The mycelium structure, color and arrangement of conidiospore and arthrospore on the mycelium were observed through the oil immersion (100X). The classical method described in the

Table 1. Screening of antifungal activity with extracted antibiotic.

Fungal species	Diameter of Fungal Growth			
	Extract of actinomycetes isolate (2 ml) (mm)	Extract of actinomycetes isolate (4 ml) (mm)	Extract of actinomycetes isolate (6 ml) (mm)	Extract of actinomycetes isolate (10 ml) (mm)
Aspergillus niger	23	18	15	14
Aspergillus flavus	20	No growth	16	15
Alternaria sps	17	15	14	13
Fusarium sps	19	No growth	14	No growth
Rhizopus stolonifer	21	17	15	15

identification key by Nonomura (1974) and Bergey's Manual of Determinative Bacteriology (1957) has been used for the identification of actinomycetes.

EXTRACTION OF ANTIBIOTIC FROM ISOLATED ACTINOMYCETES

Method for fermentation

Preparation of inoculums for fermentation process

Starch casein nitrate broth (100 ml) was prepared and then under aseptical conditions a loop full of purified growth added in the Starch casein nitrate broth. This broth was incubated at 28°C in shaking incubator at 150 rpm for 5 days. After 5 days the inoculums for fermentation process was ready for use.

Fermentation method 1: (In shaking incubator) 100 ml Starch casein broth was prepared and autoclaved. 10 ml of the prepared inoculum was added in the broth. Again incubated at 28°C for 5 days in shaking incubator at 150 rpm. The pH of the medium maintained at 5 - 7 after incubation extraction was done (Atta, 2009).

Fermentation method 2: (In fermenter).Prepared 1500 ml Starch casein nitrate broth and autoclaved. Prepared 150 ml inoculums for the fermentation process. Maintained the pH at 5 - 7. Temperature is maintained at 28°C. Followed the fermenter instructions. After five days take out the fermented broth and extract as explained here.

Method for extraction

After fermentation, the medium was harvested and centrifuged to remove cells and debris. Filtrate is collected in a sterilized screw cap bottle. Filter the fermented broth. The filtrate was mixed with ethylacetate in the ratio of 1:1 (v/v) and shaken vigorously for 1 h in a solvent extraction funnel.

The solvent phase that contains antibacterial compound was separated from the aqueous phase. Solvent phase was evaporated to dryness in water bath at 80 - 90°C and the residue is used to check antibacterial activity (Atta, 2009).

Antifungal activity of actinomycetes

Determination of antifungal activities of pure actinomycetes cultures were performed by using agar disc method. Potato dextrose agar plates were prepared and mixed with Actinomycetes culture extract of different concentrations such as 2, 4, 6 and 10%. Then the plates were inoculated with the agar disc of test fungi in the center of the Petri dish and incubated at 28°C for 4 days. The test fungi used are Alternaria, *A. niger*, *A. flavus*, *Fusarium*, *R. stolonifer*.

RESULTS AND DISCUSSION

The crude extract of antifungal compounds isolated from Actinomycetes was used to check the antifungal activity against test organism. Different concentrations such as 2, 4, 6, and 10% of extract were used to check antifungal activity and to check the minimum inhibitory concentration. On potato dextrose medium the different fungal strains show different zone of inhibition against crude extract of antifungal compound extracted from Actinomycetes species. The extract obtained from Actinomycetes had the ability to inhibit growth of pathogenic fungi at varying degree. From the Table 1 it has been observed that the growth of fungal mycelium decreases with the increase in the concentration of compound extracted from Actimomycetes species. It has been observed that the extract isolated have inhibited the growth of nearly all the test fungi. Similar results have been investigated by various authors including Khamna (2009) who reported that the crude extract of antifungal compounds was active against *R. stolonifer*, *A. flavus*, *F. oxysporum* and *Alternaria*. Lim et al. (2000) selected 32 Actinomycetes isolates, which showed the inhibitory activity against mycelial growth of plant pathogenic fungi like *Alternaria mali*, *Colletotrichum gloeosporides*, *F. oxysporum*, *cucumerinum*, *Magnaporthe grisea*, *Phytophthora capsici*, and *Rhizoctonia solani*.

In search for soil Actinomycetes having antifungal activity against plant fungal-pathogens, 110 isolates were screened by Aghighi et al. (2004), from which 14 isolates were found active against *A. solani*, *A. alternate*, *Fusarium solani*, *Phytophthora megasperma*, *V. dahlia* and *Sacchromyces cerevisiae*. Bonjar et al. (2005) assayed antifungal Actinomycetes strains for antagonistic activity against *V. dahlia*, *A. solani*, *F. solani* and *G. candidum* four worldwide phytopathogenic fungi. From 110 soil inhabitant strains that have been isolated from soil samples, 10 strains showed antifungal activity as determined through screening and bioassays by agar disk and well diffusion methods. Similar results have been found by El-mehalawy et al. (2005), Kathiresan et al. (2005), Gebreel et al. (2008), Anitha and Rebeeth (2009) and Kavitha et al. (2010). The Present work has resulted in selective isolation of novel soil Actinomycetes and their antifungal activity against some pathogenic fungi. But more precise work and further development in this field is required to produce more potent bioactive antifungal compounds from Actinomycetes which are easily available in the soil.

REFERENCES

Aghighi S, Bonjar GHS, Rawashdeh R, Batayneh S, Saadoun I (2004). First report of antifungal spectra of activity of Iranian Actinomycetes strains against *Alternaria solani, Alternaria alternate, Fusarium solani, Phytophthora megasperma, Verticillium dahlia* and *Saccharomyces cerevisiae*. Asian J. Plant Sci., 3(4): 463-471.

Aneja KR (2003). Experiments in Microbiology Plant Pathology and Biotechnology. Forth edition. New Age International Limited, Publishers, New Delhi.

Anitha A, Rebeeth M (2009). *In vitro* antifungal activity of *Streptomyces griseus* against phytopathogenic fungi of tomato field. Acad. J. Plant Sci., 2(2): 119-123.

Bauman R (2007). Microbiology: With diseases by taxonomy. San Francisco, Calif.: Pearson/Benjamin Cummings.

Bonjar GHS, Farrokhi PR, Aghighi S, Bonjar LS, Aghelizadeh A (2005). Antifungal characterization of Actinomycetes isolated from Kerman, Iran and their future prospects in biological control strategies in greenhouse and field conditions. Plant Pathol. J., 4(1): 78-84.

El-Mehalawy AA, Abd-Allah, NA, Mohamed RM, Abu-Shay MR (2005). Actinomycetes antagonizing plant and human pathogenic fungi. II. Factors affecting antifungal production and chemical characterization of the active components. Int. J. Agric. Biol., 7(2): 188-196.

Gebreel HM, El-Mehalawy AA, El-Kholy IM, Rifaat HM, Humid AA (2008). Antimicrobial activities of certain bacteria isolated from Egyptian soil against pathogenic fungi. Res. J. Agric. Biol. Sci., 4(4): 331-339.

Kavitha A, Vijayalakshmi M, Sudhakar P, Narasimha G (2010). Screening of Actinomycete strains for the production of antifungal metabolites. Afr. J. Microbiol. Res., 4(1): 027-032.

Khamna S, Yokota A, Peberdy JF, Lumyong S (2009). Antifungal activity of *Streptomyces* spp. isolated from rhizosphere of Thai medicinal plants. Int. J. Integr. Biol., 6(3): 143-147.

Lim SW, Kim JD, Kim BS, Hwang B. (2000). Isolation and numerical identification of Streptomyces humidus strain S5-55 antagonistic to plant pathogenic Fungi. Plant Pathol. J., 16(4): 189-199.

8

Determination of effective dose of garlic for controlling seedborne fungal disease of tomato

M. M. Rashid[1], A. B. M. Ruhul Amin[2] and F. Rahman[3]

[1]Bangladesh Rice Research Institute, Regional Station, Sonagazi, Feni, Bangladesh.
[2]Rupali Bank Ltd., Bangladesh.
[3]Student, Plant Pathology Division, Bangladesh Agricultural University, Mymensingh, Bangladesh.

Tomato seeds collected from farmer's were treated with garlic tablet at concentration 1:3 w/v, 1:4 w/v, 1:5 w/v and 1:6 w/v for controlling seedborne fungal diseases. Seed health status of treated seeds was evaluated following blotter incubation method. Three different fungal pathogens viz., *Aspergillus* spp., *Fusarium* spp. and *Penicillium* spp. were identified from seed samples by blotter incubation method. Garlic tablet at 1:3 w/v dose showed better performance in increasing seed germination and reducing prevalance of fungal pathogens over control treatment. The highest germination recorded was 71.25% at 1:3 w/v dose of garlic tablet which represents an increase of 11.25% over control. Germination percentage was increased 18.75% over control when treated seeds were sown in tray soil. Substantial importance in seed quality was noticed in reducing hard seed, damping off, blighted seedlings and tip over. In pot experiment, 1:3 w/v dose also performed best to yield the lowest percentage of hard seed, damping off, blighted seedlings, tip over and seedlings with highest seed germination.

Key words: Garlic, tomato, seed treatment, germination, seedborne pathogens, *Aspergillus*, *Fusarium*, *Penicillium*.

INTRODUCTION

Tomato *(Lycopersicon esculentum,* Mill) a member of the family Solanaceae, is the most important and popular vegetable in the world because of its taste and high nutritive value and also for its diversified use (Bose et al., 1986). It is widely grown in almost all countries of the world due to its adaptability to a wide range of soils and climate (Ahmad, 1976). The yield of the crop is very low in Bangladesh compared to those of other countries. Recent statistics show that in Bangladesh tomato was grown in 15,789 ha of land and the production was approximately 102,000 metric tons in 2002 to 2003, the average yield was 4.4 tons/ha compared to 6.67 ton/ha in India, 16.657 tons/ha in China, 20.00 ton/ha in Egypt, 60.00 ton/ha in Japan, and 14.54 ton/ha in USA (FAO, 2003). Now-a-days, the average yield of tomato is 2.8 tons per acre (Monthly Statistical Bulletin, 2006).

Diseases of tomato are among the main factors limiting its production. Over 200 diseases have been reported to affect the tomato plants in the world (Watterson, 1986). Among them the seedborne pathogens play a vital role in disease development (Fakir and Khan, 1992). Six seedborne fungal diseases of tomato viz. early blight (*Alternaria solani*), germination reduction (*Aspergillus flavus, Penicillium* spp.), seed discoloration (*A. fumigatus*), Fusarium wilt (*Fusarium oxisporum*), and late blight (*Phytophthora infestans*) have been detected in Bangladesh (Fakir, 2001).

Seedborne diseases create a great loss to the production of crops in Bangladesh. In order to reduce the loss, farmers use to treat seeds with chemicals. Chemicals are quite effective in reducing seedborne infection. Indiscriminate use of chemicals for controlling diseases of crop plants resulted environmental pollution, health hazards etc., all over the world. Moreover, the costly chemicals are being imported from overseas and farmers have to purchase with high price. As an alternate means of avoiding these problems, use of organic/plant

*Corresponding author. E-mail: mirza_26658@yahoo.com.

extract as control agents is now being considered in many developed countries for combating the disease with the aim of increasing food production. Effective and efficient use of garlic tablet can control seed-borne disease and subsequently crop yield can be increased using healthy seeds and seedlings. Incorporating garlic tablets in the existing IPM system of Bangladesh can save huge foreign currency and reduce pollution of the environment. Moreover, garlic tablets are easily available and low cost compare to chemical fungicides. Garlic tablet have been formulated in the IPM laboratory of Department of Plant Pathology and found effective against various diseases and pathogens. The present research was undertaken to achieve the following objectives:

1. To identify the fungal pathogens associated with tomato seeds.
2. To determine the effective dose of garlic tablet against seed germination and seedborne fungal pathogens of tomato.

MATERIALS AND METHODS

The experiments (in Laboratory and pot) were conducted between July, 2005 to September, 2006 in the Integrated Pest Management (IPM) laboratory and in the net house of the Department of Plant Pathology, Bangladesh Agricultural University (BAU), Mymensingh. Seed samples of tomato were collected from local market of Mymensingh. Four hundred seeds of tomato were randomly taken from seed sample for each treatment so that maximum incidence of seed-borne infection may be available. To prepare 1:3 solution, one part tablet (ground and powdered) was added into three parts distilled water (weight/volume). Prepared suspensions were used instantly. In the same way, 1:4 w/v, 1:5 w/v and 1:6 w/v suspensions were prepared. Four hundred seeds were treated in each dilution of garlic tablets for fifteen minutes. In case of control treatment seeds were soaked in distilled water and no tablets were used to know the actual status of the pathogen.

Plating the treated seeds

After washing the plastic petridish, surface sterilization was done by 70% alcohol and allowed to open for sometimes for volatilization. Two filter papers soaked in sterile water were set at the bottom of the petridish. Then, twenty five seeds were plated in each petridish. For one treatment sixteen petridishes were required. Each petridish with the seeds was considered as a replicate. The petridishes were incubated at room temperature (25±2°C) for one week. After one week of incubation period, percentage of germination, abnormal germination, hard seeds, rotten seeds and infected seeds were recorded from each petridish.

Net house experiment

The doses of garlic tablet which showed the best performance were further investigated through pot experiment in the net house. For both 1:3 w/v dose and control treatment, eight hundred seeds were sown in surface sterilized tray filled up with sterilized soil. One hundred seeds per tray were sown for treated and untreated conditions. Observations were made for seed germination and other parameters.

Collected data were analyzed using ANOVA following completely randomized design (CRD) using MSTAT-C computer program. Differences in means of different treatments were evaluated for significance following Duncan's Multiple Range Test (Gomez and Gomez, 1984).

RESULTS

Fresh garlic tablets were used in the present study to find out the effect on different parameters of seed quality, both in the blotter and in the pot experiment. The seed quality investigated were seed germination, abnormal seed germination, rotten seeds, hard seeds, seed infection and seedling diseases, such as damping off and tip over were recorded.

Effect of garlic tablet treatment on seed germination and other parameters seed germination

Germination percentage at different doses varied significantly. All doses of garlic tablet increased germination over control. The highest (80.0%) germination percentage was recorded at 1:3 w/v doses which were 18.0% higher than control. Statistically similar germination percentage was recorded at 1:4 (75%) and 1:5 (73%) w/v dose of garlic tablets. The lowest (68.0%) germination percentage was recorded in the control (Table 1).

Abnormal seed germination

Statistically significant abnormal seed germination was recorded in different treatments. Abnormal seed germination recorded the lowest (6.0%) in 1:3 w/v doses and highest abnormal seed germination (15.0%) was recorded in 1:6 w/v dose and control. All doses of garlic tablet decrease abnormal seed germination. At 1:3 w/v dose 60.0% lower abnormal seed germination was recorded (Table 1).

Dead seed

Significantly different hard seeds were recorded in different doses of garlic tablets. Highest (33.0%) percentage of hard seeds was found in the 1:6 w/v doses and in the control (32.0%). The lowest (20.0%) hard seeds were found at 1:3 w/v dose of garlic tablet. At 1:3 w/v dose 38.0% lower hard seeds were found compared to the control (Table 1).

Rotten seed

Percentage of rotten seeds was recorded highest (32.0%) in the control and lowest (2.0%) at 1:3 w/v

Table 1. Effect of different doses of garlic tablet on the quality of cucumber seeds.

Treatment	%Germinated seed	%Abnormal Germinated seed	%Dead seed	%Rotten seed	%Infected seed
1:3	80.0 a	6.0 d	20.0 d	2.0 e	5.0 e
1:4	75.0 b	8.0 c	25.0 c	7.0 d	10.0 d
1:5	73.0 b	14.0 b	27.0 b	11.0 c	29.0 c
1:6	67.0 c	15.0 a	33.0 a	22.0 b	47.0 b
Control	68.0 c	15.0 a	32.0 a	32.0 a	73.0 a
LSD (P≥0.05)	1.63	0.4	0.53	1.09	2.56
CV (%)	5.71	11.44	8.62	8.01	15.65

Figures in a column with common letter (s) do not differ significantly at P ≥ 0.05. Data were analyzed after transformation.

Table 2. Effect of different doses of garlic in controlling the seed-borne fungi.

Treatment	% Seed-borne infection					
	Aspergillus spp.	Decreased over control	*Fusarium* spp.	Decreased over control	*Penicillium* sp.	Decreased over control
1:3	3.0 c	90.7	2.0 c	89.7	1.25 c	84.5
1:4	4.0 c	87.7	2.5 c	87.2	1.25 c	87.5
1:5	17.5 b	46.2	10.25 b	47.4	7.5 b	25.0
1:6	30.5 a	6.2	18.75 a	3.8	8.75 ab	12.5
control	32.5 a	-	19.5 a	-	10 a	-
LSD (P≥0.05)	2.71		2.81		2.48	
CV	17.18		13.28		17.85	

Values within the same common with a common letter (s) do not differ significantly (P = 0.05). Data were analyzed after transformation.

doses. Lower rotten seeds (7.0%) were also observed at 1:4 w/v doses. Compared to control treatment 94.0% lower rotten seeds were recorded at 1:3 w/v dose (Table 1).

Seed infection

Highest percentage of seed infection was recorded (62.0%) among control seeds. Seed infection was obtained lowest (5.0 %) in 1:3 w/v doses. The percentage of seed infection increased with the decrease of the garlic tablet concentration. At 1:3 w/v dose 93.0% lower seed infection were observed over control (Table 1).

Effect of garlic tablet on the prevalence of seed-borne fungi

Aspergillus spp.

Prevalence of *Aspergillus* spp. was significantly different in different treatments. The lowest seed infection (3.0%) was recorded at 1:3 w/v doses statistically similar seed infection was found at 1:4 w/v doses (4.0%). The highest

seed infection (32.5%) was recorded in the control. Percentage of seed infection observed at 1:6 w/v doses was statistically similar to that of the control. Prevalence of *Aspergillus* increased with the increase of dilution. At 1:3 w/v doses, 90.7% reduction of *Aspergillus* spp. was observed over control. 87.7% reduction of *Aspergillus* spp. was also observed at 1:4 w/v dose (Table 2).

Fusarium spp.

In the case of *Fusarium* spp., lowest seed infection (2.0%) was recorded at 1:3 w/v dose followed by 1:4 w/v dose (2.5%). The highest prevalence (19.5%) was recorded in the control, while similar prevalence (18.75%) was also recorded at 1:6 w/v dose (Table 2).

Penicillium sp.

Prevalence of *Penicillium* sp. was significantly different among the treatments. The highest percentage of prevalence was observed in the control (10.0%). The lowest (1.25%) prevalence was observed at 1:3 w/v and 1:4 w/v doses. The prevalence was recorded 87.5% at

Table 3. Comparative performance of garlic (1:3 w/v doses) and control in tray soil.

Treatment	Seed germination (%)	Abnormal seed germination (%)	Hard seed (%)	Rotten seed %	Damping off (%)	Seedling blight (%)	Tip over (%)
1:3	71.25	3	28.75	13.25	1.25	1.0	0.75
Control	60.0	9.0	40.0	9.0	17.0	8.0	5.25
LSD (P≥0.05)	2.26	1.16	2.26	0.88	2.98	1.41	0.96
CV (%)	3.56	13.61	6.80	9.53	7.42	15.95	16.15

Values within the same common with a common letter (s) do not differ significantly (P=0.05). Data were analyzed after transformation.

1:3 w/v dose compared to control treatment (Table 2). Among three different fungi associated with tomato seeds, prevalence of *Aspergillus* spp. was higher followed by *Fusarium* sp. and *Penicillium* sp.

Net house experiment

In the net house, an experiment was conducted with the treatment that showed the best seed quality results in blotter incubation test. In general, seeds treated with 1:3 w/v dose garlic tablets gave luxuriant growth of seedlings in tray soil whereas in the control poor growth of seedling was observed.

In tray soil seed germination was recorded 71.25% at 1:3 w/v doses which was 18.75% over control and 60.0% at control treatment (Table 3). At 1:3 w/v dose, percentage of hard seed, damping off, seedling blight and tip over were significantly different from the control and they were 28.75, 1.25, 1.0 and 0.75 respectively (Table 3). On the other hand percentage of hard seed, damping off, seedling blight and tip over were 40.0, 17.0, 8.0 and 5.25 respectively in control treatment (Table 3).

Hard seed percentage was significantly lower in 1:3 w/v doses. 28.13% less hard seeds were observed in the tray experiment. Seed treatment with 1:3 w/v doses of garlic tablet showed significant influence on damping off disease in tray experiment. Percentage of damping off diseases were recorded 1.25% in 1:3 w/v dose of garlic tablet which was 92.64% lower than control treatment.

Percentage of seedling blight was recorded 1.0% in 1:3 dose of garlic tablet. It was estimated that 87.5% lower seedling blight was observed in treated seed over control treatment.

Tip over affected plants were recorded 0.75% in treated seeds which was significantly different from untreated control. 85.71% lower tip over affected seedlings was observed in treated seeds over control (Table 3).

DISCUSSION

The experiment was conducted to determine the effective dose of garlic tablet against seedborne fungi associated with tomato seeds. To achieve this objective, garlic tablet of different concentrations were applied to tomato seeds

collected from local market. In the experiment with tomato seeds it was found that percentage of seed germination was highest in seeds treated with garlic tablet at 1:3 w/v doses both in the blotter and in the pot experiment. Garlic tablet at 1:3 w/v dose reduced damping off, Seedling blight and tip over very significantly. Garlic tablet at 1:3 concentrations also enhanced seed germination over control treatment. The findings are in agreement with Awal (2005) where 30 days old garlic tablet showed similar effect at 1:3 w/v doses on seed germination and vigor index. Khan and Kumar (1992) also found that garlic extract inhibited spore germination of *Bipolaris sorokiniana* at concentration 1:3. Kuprashvile (1996) reported that extracts of garlic disinfected the seeds of Capsicum, Aubergine, Tomato, Cabbage, Carrot and Onion by *Peronospora destructor*, *Phomopsis vexans*, *Fusarium oxysporum*. The result showed that the plant extracts disinfected seeds. Alice and Rao (1987) also observed that seeds treated with garlic extract produced seedlings of higher shoots and roots than those of the untreated Checks. All these findings are limitedly in agreement with only concentration of extracts of garlic. Seed infection also decreased significantly at different concentration of solution especially 1:3 w/v doses. The findings are in agreement with Hawlader (2003) who reported that garlic bulb extract (1:1 w/v dose) effectively increased germination of egg plant seeds and significantly reduced damping off, seedling blight and tip over. In the net house experiment it was found that garlic tablet at 1:3 w/v dose showed significant improvement in seed germination. Seed infection, damping off, tip over and seedling blight decreased significantly in comparison with control seeds. Arun et al. (1995) reported that the extract of garlic bulbs was effective in suppressing redial growth of *Fusarium* and *Sclerotium* and was more effective when added after sterilization.

REFERENCES

Ahmad KU (1976). Phul Phal, O. Shak, Sabji. 3rd Edn. Alhaz Kamisuddin Ahmad. Banglow No. 2 Farm Gate, Dhaka-15, Bangladesh. p. 470

Alice D, Rao AV (1987). Antifungal effects of plant extracts on *Drechslera oryzae* in rice. Int. Rice Res. Newsl., 12(2): 28.

Arun A, Tekha C, Chitra A (1995). Effect of allicin and of garlic extract and begonia on two fungi. Indian J. Mycol. Plant Pathol. 25(3): 316-318.

Awal KJM (2005). Determination of effective dose of garlic tablet and its durability in controlling seedling diseases of eggplant. M. Sc. Thesis. Department of Plant Pathology, BAU, Mymensingh. 1-88 p.

Bose TK, Kabir J, Maity TK (1986). Vegetables of India. Nayaprakash, Calcutta, India. 312-334 pp.

Fakir GA (2001). An annotated list of seed borne disease in Bangladesh. Seed Pathology Laboratory. Department of Plant Pathology, BAU, Mymensingh. 41 p.

Fakir GA, Khan AA (1992). Control of some selected seed borne fungal pathogens of jute by seed treatment with garlic extract. Proc. BAU Res. Prog., 6: 176-180.

Food and Agricultural Organization (FAO) (2003). FAO prodiction Year Book. Basic Data unit, Statistics Division, FAO, Rome, Italy, 57: 147-148.

Gomez KA, Gomez AA (1984). "Statistical Procedures for Agricultural Research" (second Edn). John Wiley and Sons, New York, 680 pp.

Hawlader AN (2003). Effect of seed selection and seed treatment on the development of *Phomopsis* blight or fruit rot of egg plant. MSc Thesis, Dept. of Plant Pathol., BAU, Mymensingh. 40-68 p.

Khan MI, Kumar R (1992). Antifungal activity of leaf extract of neem on seed mycoflora of wheat. Indian J. Seed Abs. 15(7): 299.

Kuprashvile TD (1996). The use of phytoncydes for seed treatments. Zashchita-i-Karantin-Restenil, 5:31.

Monthly Statistical Bulletin (2006). Bangladesh J., p. 55.

Watterson JC (1986). Diseases, the tomato crops. Edited by Atherton and Rudich. Champan and Hall Ltd. Ny. p. 461-462.

9

Anti-mycobacterium activity from culture filtrates obtained from the dematiaceous fungus C10

João V. B. de Souza*, Alita M. Lima, Eveleise S. de J. Martins, Julia I. Salem

Instituto Nacional de Pesquisas da Amazônia, Manaus, AM, Brasil.

The aim of this work was to increase the concentration of substances with anti-mycobacterium activity in culture filtrates obtained from the dematiaceous fungus C10. An experimental design was employed to study the effect of glucose, potato infusion and *Senna reticulata* infusion. The anti-mycobacterium activity was determined by evaluating the growth of bacteria in culture medium containing "culture filtrate" (the products of the submerged fermentation of the fungus). It was observed that the concentrations of glucose 30 g/L, potato infusion 50% v/v and *S. reticulata* infusion 0% v/v (a better result was obtained not using *S. reticulata* infusion) were the best conditions for metabolites production. The influence of each variable was determined and it was possible to produce a mathematical model and a surface response to demonstrate the influence of the studied variables. In conclusion, we note that the culture medium had a great importance in the production of culture filtrate with anti-mycobacterial activity and that the experimental design showed to be a functional statistical tool for studying the influence of culture medium composition.

Key words: Culture medium, *Senna reticulata*, endophytic fungi, experimental design.

INTRODUCTION

Bioprospecting is a term currently used to refer to the search for novel products or organisms of economic importance from the world's biota. The Amazon rain forest is one of the most species-rich containing the greatest microbial diversity in the word (Stinson et al., 2003; Souza et al., 2004). It has the potential for supplying materials and metabolites for an infinite number of researches. The Rain Forest is well known internationally for it´s vegetal diversity however inside trees, the microbiological diversity is even more impressive (Lodge et al., 1992; Azevedo et al., 2000; Araújo et al., 2002).

Endophytes are microorganisms that live in the intercellular spaces of stems, petioles, roots and leaves of plants. They cause no discernible manifestation of their presence and have typically gone unnoticed (Lu et al., 2000). These microorganisms constitute a valuable source of secondary bioactive metabolites that are produced to protect the plant in ways such as: plant growth regulation, antibacterial, antifungal, antiviral and insecticidal. These substances have been systematically for their potential therapeutic use (Lingham et al., 1993; Dreyfussn and Chapela, 1994; Bills et al., 1994; Pelaez et al., 1998).

Recently, endophytic fungi that were isolated from species of tropical Amazon rainforest were identified for producing substances with biological activity (Carvalho, 2005; Lima, 2007). More specifically, a strain known as C10, still not identified, produced substances that inhibit the growth of *Mycobacterium tuberculosis* (Lima, 2007).

These anti-mycobacterium substances should be characterized and identified. However, it is necessary to increase the concentration of them, to do a more adequate analytic study. An alternative, to increase the concentration of these substances, is to optimize culture composition. The aim of this work was to increase the concentration of substances with anti-mycobacterium

*Corresponding author. E-mail: joaovicentebragasouza@yahoo.com.br.

Table 1. Level and factors used in the experimental design.

Level	-1	0	+1
Glucose g/L	0	15	30
Potato infusion (% v/v)	0	25	50
S. reticulata infusion (% v/v)	0	25	50

activity in culture filtrates obtained from the dematiaceous fungus C10.

MATERIALS AND METHODS

Microorganism

The strain C10 was isolated from *Senna reticulata* leaves as described by Pereira and Azevedo (1993). The tree was growing wild in National Institute for Research in Amazonia (INPA) land reserve, Latitude-south 3°10′43″ and Longitude-east 59°99′99″). The plant leaves were washed with sterile water and decontaminated with 70% ethanol for 1 min, 3.0% sodium hypochlorite for 4 min, 70% ethanol for 1 min and then rinsed with sterile water for three times. Discs of 7 mm in diameter were cut from the leaves using a sterile hole-punch and transferred to Petri dishes containing the culture media Potato Dextrose Agar. The developed colonies were transferred to PDA medium plates.

All the 62 isolates, including C10, were deposited in the "Microorganisms Collection of Medical Interest" from the National Institute for Research in Amazonia (INPA) and were investigated for the production of culture filtrates with anti-mycobacterium activity (Carvalho, 2005; Lima, 2007). The dematiaceous fungus C10 was one of the best producers of bioactive compounds.

Submerged fermentation bioprocess for producing the "culture filtrate"

The fungal inoculum was obtained by growing the C10 culture on potato dextrose agar (PDA) at room temperature for 7 days. The bioprocess was carried out in 125 ml Erlenmeyer flasks containing 50 ml of culture medium and 1×10^5 fungal cells. The concentration of glucose, potato infusion and *S. reticulata* infusion was defined according to an experimental design. The bioprocess incubation was carried out in a rotary shaker under orbital agitation of 100 rpm at room temperature for 2 weeks. After this period, the culture medium was filtered and the filtrate (the culture filtrate) was submitted to an anti-mycobacterium activity assay.

Experimental design for the optimization of the bioprocess

In order to verify the influence that the content of glucose (g/L), potato infusion (% v/v) and *S. reticulata* infusion (% v/v) had on the production of culture filtrates that inhibit *M. tuberculosis* growth, a 2^3 experimental design, with three repetitions on the central point, was employed (Barros et al., 1995). The infusions were obtained by using potato pieces (100 g/L, 10 cm² surface per piece), or sterns of *S. reticulata* (100 g/L, pulverized). Both types were boiled in water for 10 min and then filtered with gauze. Table 1 shows the level and the factors used in the design. All eight experiments were performed in the same conditions as the three assays representing the central point (coded value 0). A statistical model to determine the inhibitive properties to *M. tuberculosis* growth was determined by the response regression procedure.

The statistical analysis was performed by using the STATGRAPHICS statistical software version 6.0 and the STATISTICA program version 5.0.

Anti-mycobacterium activity assay

This bioassay quantified the influence of culture filtrates in the growth of *M. tuberculosis* H37Rv by turbidimetry (660 nm), using a GeneQuant spectrophotometer (Amersham Pharmacia Biotech). The bioassay consisted in 0.5 ml of *M. tuberculosis* inoculum (3x105 cell/ml), 0.5 ml of culture medium (Middlebrook 7H9GC culture x 2) and 1 ml of the culture filtrates. This bioassay was incubated at 37°C per 6 days. The turbidimetry values obtained in bioassays were compared with the values of the control (an assay containing only the *M. tuberculosis* inoculum and the culture medium). It was quantified the absorption in 660 nm from the culture filtrates, culture medium and initial inoculum in order to allow an adequate quantification (without interferences) of the bacterial growth.

RESULTS

In order to determine the influence of glucose (g/L), potato infusion (% v/v) and *S. reticulata* infusion (% v/v) content in the production of culture filtrates that present inhibitive properties to *M. tuberculosis* growth, a 2^3 experimental design was used. Table 2 shows the results of the experimental design. The inhibition of *M. tuberculosis* growth, revealed by the 2^3 experimental design, ranged from 38.6 to 76.4%.

The main effects and their respective interactions calculated from the data of Table 2 are shown in Table 3. The standard errors and the estimated effects are shown in Table 3. Barros et al. (1995) only considers significant (for 95% confidence) the effects with values higher than $tv \times \mu$. The tv value is t test for v freedom degree. In this study the t test, for 2 freedom degree (95% confidence), was 4.3025.

The linear effects of glucose (A), potato infusion (B) and *S. reticulata* infusion (C) were significant and a linear model was adjusted (Equation 1).

Inhibition of *M. tuberculosis* growth (%) = 45.4318 + 0.838333* (Glucose] + 0.16* (Potato infusion) − 0.177* (*S. reticulata* infusion) (Equation 1). Table 4 shows the variance analysis, ANOVA, for the linear effects shown in Equation 1.

The ANOVA test was used to evaluate the regression and the lack-of-fit of the model (Barros et al., 1995). The P-value for all the considered factors was near or inferior to 0.05, showing that these effects have significant

Table 2. Results of the 2^3 experimental design with three repetitions on the central point.

Test	Glucose (g/L)	Potato infusion (% v/v)	S. reticulata infusion (% v/v)	M. tuberculosis growth inhibition (%)
1	0	50	50	43.6
2	0	0	0	41.6
3	30	0	50	57.6
4	30	50	50	70.1
5	0	50	0	53.5
6	0	0	50	38,6
7	30	50	0	76.4
8	15	25	25	60.6
9	15	25	25	59.4
10	30	0	0	73.8
11	15	25	25	58.2

Table 3. Variables affecting the production of culture filtrates that present inhibitive properties to M. tuberculosis growth as revealed by the 2^3 experimental design.

Variables	Estimated effects ± Standard error
Average	57.5818 ± 0.361814
A: Glucose*	25.15 ± 0.848528
B: Potato infusion*	8.0 ± 0.848528
C: S. reticulata infusion*	-8.85 ± 0.848528
AB	-0.45 ± 0.848528
AC	-2.4 ± 0.848528
BC	0.75 ± 0.848528

Standard error estimated from pure error with 2 f.d. *Significant effects at the 5% level (t = 4.3025).
[A] = dextrose content; [B] = sulfate ammonium content.

Table 4. Analysis of variance for evaluation of the model (Equation 1).

Source of variation	Sum of squares	Df	Mean square (MQ)	F-ratio	P-value
A: Glucose	1265.04	1	1265.04	878.50	0.0011
B: Potato infusion	128.0	1	128.0	88.89	0.0110
C: S. reticulata infusion	156.645	1	156.645	108.78	0.0091
Lack-of-fit	61.9664	5	12.3933	8.61	0.1074
Pure error	2.88	2	1.44		
Total (corr.)	1614.54	10			

R-squared = 95.9836%, R-squared (adjusted for d.f.) = 94.2623%.

regression. The P-value of lack-of-fit was 0.1074 which shows that the lack-of-fit was not significant. The absence of lack-of-fit, the significant regression and the high variance percentage explained, showed that the model presented in Equation 1 could be used to explain the studied region. The surface response created using the model (Equation 1) is shown in Figure 1.

DISCUSSION

Bacterial and viral drug resistance and the spread of fungal and parasitic diseases necessitate the search for additional antibiotic compounds that present activity at low concentrations and with reasonably low toxicity to humans. It has been demonstrated that the search for these compounds among endophytic bacteria and lichens has been effective since they represent biological associations that provide protection against competing organisms (Francolini et al., 2004; Hoffman et al., 2007).

Specifically, endophytic fungi have been described as good producers of substances with antimicrobial activity (Hoffman et al., 2007; Rukachaisirikul et al., 2008). During the screening for antibiotic producers, the strain

Function = Z = 45.4318 + 0.838333*X + 0.16*Y − 0.177*25

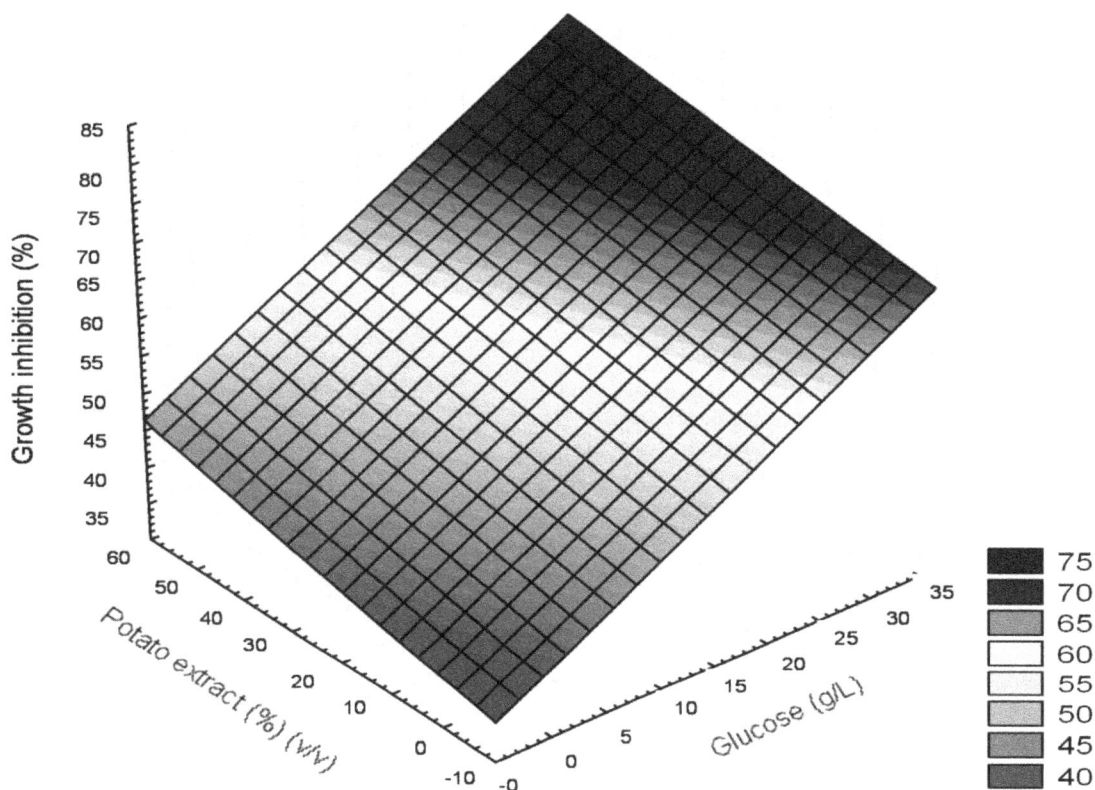

Figure 1. Surface response demonstrating the influence of glucose (g/L) and potato infusion (% v/v) in the production of culture filtrates that have inhibitive properties to *M. tuberculosis* growth.

C10 (that was isolated from *S. reticulata*) produced culture filtrate that was able to inhibit the growth of *M. tuberculosis*. However, it has been necessary to stimulate the production of these interesting substances to make their purification, identification and characterization easier.

The inhibition assays presented results that ranged from 38.6 to 76.4%, showing that the compositions of the medium used in the bioprocess have a significant influence in the production of anti-mycobaterium substances. This result was expected, because the literature has shown many works demonstrating the influence of the culture medium in the production of metabolic substances (Berdy, 1998; Brizuela et al., 1998; Lu et al., 2000).

In the presented experiments, the glucose and potato infusion had positive influence on the production of inhibitive substances. Conversely, the *S. reticulata* infusion had negative influence. The highest production of inhibitive substances was reached using 30 g/L glucose and 50% v/v potato infusion. That composition is similar to the conventional Potato Dextrose Agar medium.

The influence of *S. reticulata* infusion was investigated in order to find out if the plant components were necessary for the production (or the stimulation of the production) of culture filtrates that causes inhibition in *M. tuberculosis*. The data (Table 3) showed that the plant infusion caused a decrease in the production of the inhibitive substances. This result could be due to the presence of substances in the *S. reticulata* infusion that interfere in one, or more, of the phases involved in the fungal bioproduction of the antimicrobial substances. Another explanation could be the existence of anti-fungal metabolites in the *S. reticulata* infusion.

On the other hand, it is important to remember that literature shows (Borba and Rodrigues, 2000; Ryan et al., 2003; Girão et al., 2004) that some strains lost the ability to produce antimicrobial substances after being taken away from their natural botanic sources. So, it is necessary to indentify substances that are essential for the stimulation of the metabolic paths related to the production of the antimicrobials.

No statistically significant result was observed involving the interaction of glucose, potato infusion or *S. reticulata* infusion during the experiments. The three investigated components provide different nutrient sources to the microorganism: a) the glucose provides a fast carbon source for immediate utilization, b) the potato infusion provides starch that is used as a secondary carbon source and also provides protein and mineral sources that could be used as nitrogen sources, c) and the *S. reticulata* infusion provides soluble components present in the plant that could activate the production of antimicrobials.

The factorial design experiments and the surface response methodology demonstrated to be useful statistical tools that allowed for the calculating of the importance of each factor and their interaction. It also produced a mathematical model that, after validation, could be used to produce statistically significant data. Other factors such as: aeration, pH, and inoculums size should be included in a new experimental design, in order to identify increases in the production of substances that have bioactivity.

REFERENCES

Araújo WL, Lima AOS, Azevedo JL, Marcon J, Sobral JK, Lacava (2002). PT Manual: Isolamento de microrganismos endofíticos. Calq., pp. 9-18.

Azevedo JL, Maccheroni JW, Pereira JO, Araújo WL. (2000). Endophytic microorganisms: A rewiew on insect control and recent advances on tropical plants. Eletron. J. Biotechnol., 3(1):1-4.

Barros NB, Scarminio IS, Bruns RE (1995). Planejamento e otimização de experimento. Campinas, Editora da Unicamp.

Berdy J (1998). The discovery of new bioactive microbial metabolites. Progr. Ind. Microbiol., 27: 3-25.

Bills GF, Pelaez F, Polishook J, Diez-Matas M, Harris G, Clapp W (1994). Distribution of zaragozic acids (squalestatins) among filamentous Ascomycetes. Micol. Res., 23(5): 201-204.

Borba CM, Rodrigues KF (2000). Viability and sporulating capability of Coelomycetes preserved under a range of different storage regimes. Rev. Iberoam Mycol., 17: 142-145.

Brizuela MA, Garcia L, Pérez L, Mansur M (1998). Basidiomicetos: Nueva fuente de metabolitos secundários. Rev. Iberonam Micol., 15: 69-74.

Carvalho CM (2005) Recursos Naturais Amazônicos com Perspectivas de Uso Biotecnológico sobre o *Mycobacterium tuberculosis*. Dissertation (Masters in Inter Biotechnology). São Paulo: Universidade de São Paulo, Butantan Institute / Institute for Technological Research.

Dreyfussn MM, Chapela IH (1994). Potential of fungi in the discovery of novel, low-molecular weight pharmaceuticals. In: V.P. Gullo (ed.). The Discovery of Natural Products with Therapeutic Potential. Butterworth-Heinemann, Boston.

Francolini NP, Piozzi A, Donelli G, Stoodley P (2004). Usnic Acid, a Natural Antimicrobial Agent Able To Inhibit Bacterial Biofilm Formation on Polymer Surfaces. Antimicrob. Agents Chemother., 48(11): 4360-4365.

Girão MD, Prado MR, Brilhante RSN, Cordeiro RA, Monteiro AJ, Sidrim JJC (2004). Viabilidade de cepas de *Malassezia pachydermatis* mantidas em diferentes métodos de conservação. Revista da Sociedade Brasileira de Medicina Tropical; 37(3): 229-233.

Hoffman AM, Mayer SG, Strobel GA, Hess WM, Sovocool, GW Grange, AH (2007). Purification, identification and activity of phomodione, a furandione from an endophytic *Phoma* species. Phytochemistry, 69(4): 1049-1056.

Lima AM (2007). Estudo da Atividade de Fungos Endofíticos e Extratos de *Piper aduncum* L. (Piperaceae) sobre o *Mycobacterium tuberculosis*. Dissertation (MSc in Tropical Pathology). Manaus: Universidade Federal do Amazonas.

Lingham RB, Silverman KC, Bills GF, Cascales C, Sanchez M. Jankins RG, Gartner SE (1993). *Chaetomella acutiseta* produces chaetomellic acids A and B which are reversible inhibitors of farnesyl-protien transferase. Appl. Microbiol. Biotechnol., 40(2-3): 370-374.

Lodge DJ, Hawksworth DL, Ritchie BJ (1992). Microbial diversity and tropical forest functioning. Ecol. Stud., 122: 70-100.

Lu H, Zou WX, Meng JC, Hu J, Tan, RX (2000). New bioactive metabolites produced by *Colletotrichum* sp., an endophytic fungus in *Artemisia annua*. Plant Sci., 151: 67–73.

Pelaez F, Collado J, Arenal F, Basilio A, Cabello A, Matas MTD (1998). Endophytic fungi from plants living on gypsum soils as a source of secondary metabolites with antimicrobial activity. Mycol. Res. 102(6): 755-761.

Pereira JO, Azevedo JL (1993). Endophytic fungi of Stylosanthes: A first report. Mycologia, 85: 362-364.

Rukachaisirikul V, Sommart U, Phongpaichit S, Sakayaroj J, Kirtikara K (2008). Metabolites from the endophytic fungus *Phomopsis* sp. PSU-D15. Phytochemistry, 69: 783–787.

Ryan MJ, Smith D, Bridge PD, Jeffries P. (2003). The relationship between fungal preservation method and secondary metabolite production in *Metarhizium anisopliae* and *Fusarium oxysporum*. World J. Microbiol. Biotechnol., 19: 839–844.

Souza, AQL, Souza ADL, Astolfi SF, Pinheiro ML, Sarquis MIM, Pereira JO (2004). Atividade antimicrobiana de fungos endofíticos isolados de plantas tóxicas da amazônia: *Palicourea longiflora* (aubl.) rich e *Strychnos cogens* bentham. Acta amazônica, 34(2): 185-195.

Stinson M, Ezra D, Hess WM, Sears J, Strobel G (2003). An endophytic Gliocladium sp. of *Eucryphia cordifolia* producing selective volatile antimicrobial compounds. Plant Sci., 165: 913-922.

10

Preparation and application of a novel environmentally friendly cucumber seed coating agent

Zeng Defang*, Tu Renjie and Wu Shan

School of Resource and Environmental Engineering, Wuhan University of Technology, Wuhan, People's Republic of China.
Hubei Key Laboratory of Mineral Resources Processing and Environment, Wuhan 430070, People's Republic of China.

A novel cucumber seed coating agent was made from natural polysaccharide, fertilizer and microelement, etc., which achieved a good effect on increasing cucumber yield and antifeeding against pests. The foremost difference between this seed coating agent and conventional ones was that it controlled pests through the approach of repelling pests and enhancing the immunity of seeds, but not by killing pests. Results indicated that this seed coating agent had excellent control of pests and diseases and increased yield by 8.4 to 10.8%, while the material cost was decreased by 16.7% compared with the conventional toxic seed coating agent. In addition, results showed it was safe for man and livestock without causing any pollution and harm. Therefore it had three characteristics of high yield, less cost and friendly environment.

Key words: Environmentally friendly type, cucumber seed coating agent, polysaccharide, antifeedant, germination percentage.

INTRODUCTION

Cucumber is one of the main thermophilic vegetables, its planting area accounts for more than 15% of the total area planted to vegetables in China. Now the biggest problem is chilling injury and pests damaged in seedling stage, yet seed coating is the main method and key technology to control of pests and diseases (Xiong, et al 2004). Seed coating technology has developed rapidly during the past two decades and has provided an economical approach to seed enhancement, especially for larger seeded agronomic and horticultural crops (TeKrony, 2006). Studies have shown that a seed coating is effective in preventing and controlling mould-induced diseases and pests causing them, promoting seedling growth, and increasing yields (Gesch and Archer, 2005). But the conventional toxic seed coating agent unavoidably brought harm to the environment and man during usage and disposal. In addition, it leaves a persistent pollution that has been difficult to eliminate from the environment for many years, and it has become a great hidden danger in China's ecological agriculture

(Raveton et al., 2007). So we prepared a novel environmentally friendly cucumber seed coating agent which was non-toxic, non-polluting and achieved a good effect on increasing cucumber yield and had the excellent effect of antifeedant to pests. In addition, it was safe for man and livestock without causing any pollution and harm.

The novel seed coating agent (short for NSCA) was prepared through advanced biological chelating technology, with natural polysaccharide as the main material and high-activity plant growth regulator as the auxiliary material (Herbert, 2003; Qiu, et al, 2005). It could enhance plants disease-resistant ability and had strong antifungal bioactivity; in addition, it has special odor which can generate excellent antifeeding role to pests. With the application of the new coating agent, cucumber seeds had better quality and properties that resulted in an increased yield, showing a great market prospect in the major wheat producing areas.

MATERIALS AND METHODS

Main apparatus and reagents

The reagents used include natural polysaccharide (prepared in

---

*Corresponding author. E-mail: df5152@163.com.

laboratory), sodium hydroxide, acetic acid, borax, fertilizer, ethylene glycol, microelement, film former, penetrating agent (analytically pure, Hubei University Chemical Plant), violet pigment (Guangdong Shantou food additive of Mingde Co., Ltd.), conventional seed coating agent (2.5% Shileshi suspension concentrates seed coating agent, Switzerland), rats (Animal Testing Center of Tongji Medical College of Huazhong University of Science and Technology), *Sphaerotheca fuliginea* and *Fusarium oxysporium* (College of Plant Science and Technology, Huazhong Agricultural University) and new cucumber seeds (Jingyou number two, Wuhan seed Co., Ltd of Tianhong in China). The apparatus used includes constant temperature and humidity incubator (Model No.WS-01, Hubei Huang Shi Hengfeng Medical Inaszdxftrument Co. Ltd.), warm up hygrometer (Model No.STH950, SUMMIT, USA), and high pressure steam sterilizer (Model No.YXQ-SG46-48SA, Shanghai Boxun Industry Co. Ltd.). The major instruments and glassware used in present study were electronic balance (Model No.FA2004, Shanghai Yuefeng Instrument Appearance Ltd.), electron constant speed mixer (Model No.GS28B, Shanghai Anting Electronic Instruments Plant), and Petri dishes (90 cm diameter, Shanghai Yuejin Medical Treatment Instrument Plant).

Preparation of NSCA

The natural polysaccharide was prepared in 1wt% acetic acid to a final concentration of 1wt% and stirred at room temperature for approximately 3 h. The components have a molecular weight of from 30 to 1400 KD, deacetylation degree of from 80 to 90%.The optimal formulation of the novel seed coating agent (NSCA) was determined through orthogonal test. NSCA was prepared with the following components (wt%): Natural polysaccharide 55, fertilizer 15 , microelement 15, sodium hydroxide 4,ethylene glycol 2, borax 2, film former 1, penetrating agent 0.5, violet pigment 0.5, and water 5. Aqueous solutions of fertilizer, microelement, sodium hydroxide, ethylene glycol, borax, film former, penetrating agent and violet pigment were prepared respectively at a certain concentration. After mixing the entire component completely at room temperature, the working liquid of NSCA was a purple suspension.

Laboratory method for antifeeding test

According to the guideline for laboratory bioassay of pesticides, the antifeeding effect of NSCA was studied with the artificial mixed feeding method. The treatments were divided into four groups: Every 100 g of artificial feed containing 13.3, 10 or 8 mg of NSCA, and uncoated cucumber seeds (CK) to serve as control. Then the feeds as previously mentioned were put into each hole with diameter of 1.5 cm and depth of 1.5 cm. The consistent growth insects were chosen and placed side the hole and each treatment were replicated 3 times. A layer of blow moulding plate was placed between the pest-feeding plate and the cover to retain moisture and to prevent the tested pests from escaping. The antifeeding effect was determined at 48 h after cultivation in the incubator at a temperature of 25±1℃, and relative humidity of 75 to 85%, and then weights the artificial mixed feeds. The selective antifeedant rate was calculated by the following formula (1):

$$Antifeedant\ rate(\%) = \frac{A-B}{A} \times 100\% \quad (1)$$

where A is the weight of the control group, and B is the weight of the treatment group.

Laboratory method for year-on-year test

The year-on-year test in laboratory was done to determine the

enhancement effect of NSCA through comparing the germinability and germination percentage of NSCA, CA and CK. The cucumber seeds were coated in the proportion of 1:40 (w/w) and then dried by airing at room temperature for 20 to 30 min to prepare for use. Uncoated cucumber seeds (CK) were prepared as blank control group.

Laboratory method for the germination test

According to the rules for seed testing of International Seed Testing Association (Wu et al., 2003),100 seeds taken from each group were arranged on two layers of wet filter paper in each Petri dish filled with wet sand. Each Petri dish contained 25 seeds and each treatment replicated 3 times. All Petri dishes were incubated in the constant temperature and humidity incubator at 28±1℃ and air relative humidity of 85%. The germinability (GE) and germination percentage (GP) of cucumber seeds were investigated on the third day and seventh day, respectively. The calculation formulas were as follows:

$$GE = \frac{C}{E} \times 100\% \quad (2)$$

$$GP = \frac{D}{E} \times 100\% \quad (3)$$

where C is the number of germinated seeds on the third day, D is the number of germinated seeds on the seventh day, and E is the number of total seeds investigated.

Laboratory method for fungistatic text

According to microbiological test requirements for inhibitory rate experiment, by measuring the growth rate we evaluated the two seed coating agents of antifungal activity on mycelial growth in two phytopathogens including *Sphaerotheca fuliginea*, and *Fusarium oxysporium* on PDA; cultivated the phytopathogen for 2 days and plugged with cork borer to form a cake of 5 mm in diameter, then, inoculated the cake in another PDA medium containing seed coating agent and incubated at 25℃ for 7 days. The fungistatic effect was observed under the biomicroscope and the diameter of colony was measured twice. The average value was recorded. Each treatment was replicated three times. The inhibitory rate was calculated by the following formula (Sun et al., 2004):

$$Inhibitory\ rate = \frac{F-G}{F} \times 100\% \quad (4)$$

F is the colony diameter of the control group and G is the colony diameter of the treatment group.

Through the Formula (4) and the experimental results, we calculated the germinability, germination percentage, inhibitory rate, etc., and finally determined the optimal formula of NSCA through the result of field trial.

Laboratory method for toxicity test

According to the toxicological test methods of pesticides for registration (The Chinese State Standard GB15670-1995), the toxic effect of NCSA and CA were studied with the LD_{50} method. There were two major exposure routes by which toxicity materials may enter the body: Ingestion (gastrointestinal tract) and dermal contact.

Table 1. The comparison results of antifeeding test.

Treatment	Dosage of NSCA (mg)	Repetition	Weight of AF (g)	Weight of AF in 48 h (g)	Feeding volume (g)	Antifeedant rate (%)
NSCA	13.3	1	5.25a	4.77c	0.48a	78.76a
		2	6.92b	6.55a	0.37d	83.63b
		3	5.49a	5.08b	0.41a	81.86a
		Average				81.42a
	10	1	5.45c	4.73a	0.72b	68.14b
		2	5.84a	5.06b	0.78a	65.49c
		3	6.31c	5.44b	0.87a	61.50a
		Average				65.04a
	8	1	6.35b	5.30a	1.05d	53.54c
		2	6.03d	4.81b	1.22a	46.02a
		3	5.62a	4.53c	1.09a	51.77b
		Average				50.44a
CK	-	1	5.36a	3.15a	2.21a	-
		2	6.23b	4.00b	2.23b	-
		3	6.55c	4.21d	2.34b	-
		Average	6.05a	3.79b	2.26b	-

Values within a column followed by different letters are significantly different (p =0.05) according to the Duncan's multiple range test. AF means artificial feed; NSCA means novel seed coating agent; CA means conventional seed coating agent; CK means blank control.

The rats were fasted overnight before treatment, 14 days later, the toxic symptoms of rats were observed and the median lethal dose (LD_{50}) was determined after infection with seed coating agent at various dosages. According to the results of the toxicity test, we determined the differences in the aspects of toxicity and safety, etc. between NSCA and CA.

Field trial

The field trial was conducted in the plots of the experimental field in Hubei Provincial Seed Group Co., Ltd, China, in 2008 and 2009. In this test, the method of seed treatment was the same as that in the laboratory experiment. The cucumber seeds were coated with each of the seed coating agents in the proportion of 1:40 (w/w) and then dried by airing at room temperature. After spreading and airing for about 30 min, the seeds were sown into the field. The experiments were designed as a randomized block design with each treatment consisting of several plots separated from each other by a row 20 cm wide.

Each plot (4.0 m wide by 5.0 m long) contained 500 cucumber seeds. Each plot was sown either with coated seeds or with uncoated seeds to serve as control and field management was the same for all experimental plots. The emergence situation 7 days after sowing was examined, the growth situation in growth period investigated and whether the maturation delay before harvest was observed. Total yield for each plot was determined by weighing the 10 cucumber selected random and then investigating the yield of per plant. The best formulation was selected on the basis of yield performance and cost effectiveness.

RESULTS

Antifeedant effect of NSCA

Results of antifeeding test showed that the antifeedant effect of NSCA was increased with increasing amount of dosage in every 100 g of artificial feed. The antifeedant rate of containing 13.3 mg was best in antifeedant effect (Table 1 and Figure 1). Figure 1 showed excellent antifeedant effect by an analysis of weight and selective antifeedant rate of NSCA; the artificial feed without adding NSCA was nearly eaten up while the artificial feed containing NSCA was rarely bitten. It is found that the antifeedant rate of every 100 g of artificial feed containing NSCA 13.3, 10 and 8 mg on cutworm was 81.42, 65.04 and 50.44%. Compared with CK, the antifeedant effects of NSCA were significantly better than those of CA.

Effect of NSCA on seed germination

Results of year-on-year test showed different seed coating agents had different influence on cucumber seedling growth (Table 2). The average level of germinability of seeds coated with NSCA was 86.05%, with a range from 84.8 to 87.6%, about 8.75% higher

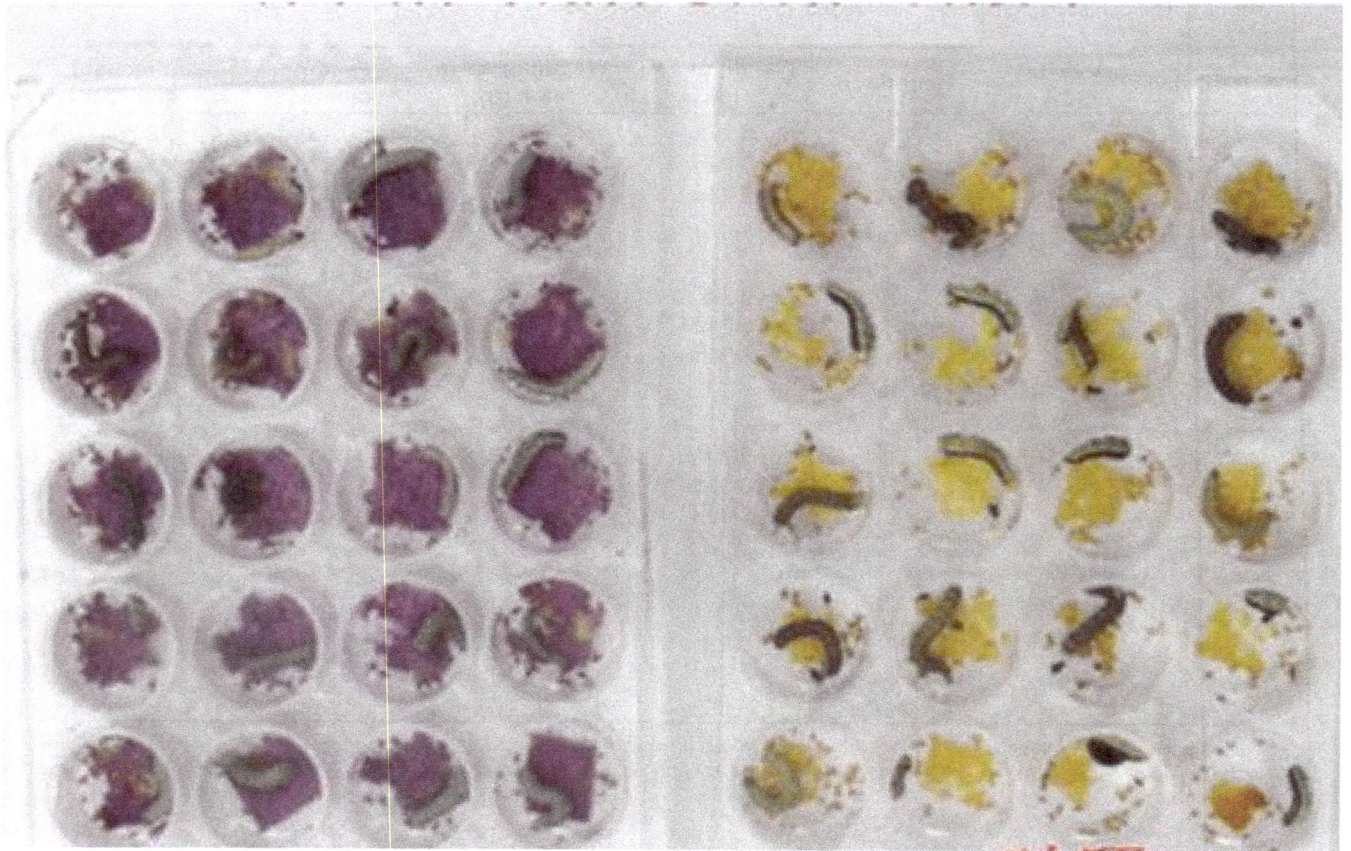

Figure 1. Antifeeding effect of NSCA and CK on cutworm in 48.

Table 2. The comparison results of year-on-year test in the laboratory.

Compare index	Treatment					
	NSCA1	NSCA2	NSCA3	NSCA4	CA	CK
GE/%	84.8a	86.3c	87.6a	85.5b	77.3b	70.5a
GP/%	91.6b	88.6a	92.5b	93.2a	84.9a	79.7d
inhibitory rate/%	89.4a	88.4b	93.5a	95.4a	85.5c	-

Value within a column followed by different letters are significant different (p=0.05) according to Duncan multiple range test. GE means germination energy; GP means germination percentage; NSCA means novel seed coating agent; CA means conventional.

than that of seeds coated with CA and almost 15.55% higher than that of CK. The average level of seeds coated with NSCA showed 6.58 and 11.78% increase in germination percentage compared with CA-treated seeds and CK, respectively. Similar results were recorded with respect to germinability and germination percentage. The average level of inhibitory rate of seeds coated with NSCA was 91.68%, about 6.18% higher than that of seeds coated with CA .The results showed that four types of NSCA were superior over CA and CK in terms of germinability, germination percentage and inhibitory rate and grew well.

Results of acute toxicity test

Results of acute toxicity test showed there were significant differences in the LD_{50} between these two seed coating agents (Table 3). In the acute oral toxicity test, rats infected with a high dosage of CA would exhibit toxic symptoms after 3 to 6 min such as systemic muscle spasm, salivation, and even convulsions, and sticky nasal and ocular secretions in the 12 h before death. The rats

Table 3. The comparison results of acute toxicity test.

Treatment	Rats' gender	Acute oral toxicity LD$_{50}$ (mgkg^{-1})	Acute skin toxicity LD$_{50}$(mgkg^{-1})	Toxicity classification
NSCA1	Male	843	2873	Low-toxic
	Female	754	2775	Low-toxic
NSCA 3	Male	863	2646	Low-toxic
	Female	764	2745	Low-toxic
NSCA 4	Male	872	2887	Low-toxic
	Female	787	2798	Low-toxic
CA	Male	216	763	Moderate-toxic
	Female	192	645	Moderate-toxic

Acute oral toxicity grading scale: Low-toxic: LD$_{50}$ > 500 mg·kg^{-1}; moderate-toxic: 50 < LD$_{50}$ < 500mg·kg^{-1}. Acute skin toxicity grading scale: Low-toxic: LD50 > 2000 mg·kg^{-1}; moderate-toxic: 200 < LD50 < 2000 mg·kg^{-1} (Chinese acute toxicity classification standard).

Table 4. The comparison results of field trial.

Treatment	Germination percentage (%)		Average output per plant (kg)		Yield (kg ha^{-1})		Cost (US $ kg^{-1})	Toxicity and pollution
	2008	2009	2008	2009	2008	2009		
NSCA1	90.2a	91.4b	3.4a	3.2c	222100a	209100b	1.5	None
NSCA 3	90.4c	89.3a	3.2b	2.9a	208700b	193950a	1.5	None
NSCA 4	91.6a	91.7b	3.7c	3.4a	242650b	223500a	1.5	None
CA	84.3a	85.6b	3.1d	2.9c	202550b	192750b	1.8	Serious
CK	78.5b	77.8a	2.9b	2.7c	189700b	177900b	-	None

Values within a column followed by different letters are significantly different (p =0.05) according to the Duncan's multiple range test. NSCA means novel seed coating agent; CA means conventional seed coating agent; CK means blank control.

infected with low dosages still showed slight muscle spasm. However, the rats infected with NSCA did not exhibit the toxic symptoms aforementioned. In the acute skin toxicity test, the rats were observed consecutively for two weeks after they were infected with the toxicant according to the conventional method. There is no significant difference between the median lethal dose (LD$_{50}$) to the male and that to the female rats for NSCA, and the acute toxicity of NSCA in this study is much lower than that of the traditional agent. Therefore, the novel seed-coating agent NSCA is safe for the public and meets the environmental requirements during usage and disposal.

Results of field trial

Results of field trial showed that NSCA noticeably improved the main performance indices such as the germination percentage, the average yield per plant and the yield per ha (Table 4), while the comparative results between NSCA and CA were observed in the plots of the experimental field in Hubei Provincial Seed Group Co.,

Ltd. It was concluded that NSCA-coated group was the best compared with the other groups for corn output, which showed the increase of 18.3% over CK and 10.8% over the CA-coated group in 2008. In 2009, the output of NSCA-coated group was 17.4% higher than that of CK and 8.4% higher than that of CA-coated group. Furthermore, the cost of NSCA was 16.7% less than that of CA. These results confirmed that seed coating agents can not only significantly improve seed germination but can also enhance cucumber yield. The statistical analysis showed that there was significant difference among the three treatments in cucumber yield. So the ratio of performance to price for NSCA is much higher than that of CA. Correlative analysis proved the main performance indexes in the field experiment had significant positive correlation with the laboratory experiment.

DISCUSSION

The mechanism of antifeedant effect

Because NSCA stimulates plants to produce systematic antibodies, once a certain part of the plant produces resistance it can induce the whole plant to resist pests and diseases, which can produce repellant effects to deter insect pests. As a plant disease resistance exciton, NSCA is also believed to be able to adjust enzymatic activity related to the disease resistance in plants. The perfect combination of polymer polysaccharide and receptor proteins on the plant cell membrane can bring resistant material such as plant antitoxin, polyphenol oxidase, peroxidase, catalase, etc. and antibacterial materials such as salicylic acid and jasmonic in the process of activating disease immunity, which produced obvious antifeedant effect. In addition, the NSCA with natural antifungal property can produce induce systemic resistance and produce significant repellent action to insect pests of rice seed (Shibuya and Minami, 2001; Ma and He, 2001). The restraint of feeding behavior might because information input sensor was interrupted. The odour from NSCA played a barrier function to feeding behavior or directly act on the nervous system of animals which caused the unusual discharge of the nervous system and prevented animals from getting correct information of taste, thus, it can make appropriate feeding reaction.

The mechanism of the yield increase effect

The natural polysaccharide has excellent film-forming property, making it easy to form a semi-permeable film on the seed surface which can maintain the seed moisture and absorb the soil moisture, and thus it can promote seed germination. The natural polysaccharide film is also considered to have a good selective permeability, which can prevent oxygen from entering the film, restrict loss of CO_2 and maintain a high concentration of CO_2 in the film, so as to restrain the seed's respiration, and thus, make the internal nutrient consumption of seeds fall to the lowest possible level (Furbank et al., 2004). Furthermore, a semi-permeable film on the seed surface can delay the release of fertilizer elements and reduce nutrient losses. This kind of semi-permeable film is believed to be able to maintain the seed moisture and absorb the soil moisture, and thus it can promote seed germination (Sigler and Turco, 2002).

Another reason for the cucumber corn yield increase may be the effect of the trace elements and fertilizer contained in NSCA, which can provide adequate nutrition and protection for plant growth. The plant growth regulator and trace fertilizer in NSCA can promote seedling growth, improve the germination rate of seed and quality of seedlings; it plays an increasing grain yield role finally. Also, natural polysaccharide as a novel plant disease inhibitor can induce and improve the disease resistance of plant, thus has a repellent effect to the pests in the soil (Chen and Xu, 2005). The natural polysaccharide possesses the natural antifungal role, which increased the permeability of the outer membrane and inner membrane and ultimately disrupted bacterial cell membranes, with the release of cellular contents (Liu et al., 2004; Nielsen et al., 1994). In addition, the film-forming property itself also hinder the direct contact between pathogens and host tissues or cells, thus effectively reduce the effect of plant diseases and insect pests.

ACKNOWLEDGEMENTS

The research was supported by the Scientific and Technological Project of Zhejiang Province, China (ZK20061103). We thanked Zhejiang Science and Technology Agency, as well as Wuhan University of Technology for their contribution.

REFERENCES

Chen, HP, Xu LL (2005).Progress of study on chitosan in regulating plants' growth and eliciting plants' defense responses. Acta Botanica Yunnanica, 27: 613-619.

Furbank RT, White R, Palta, JA (2004).Internal recycling of respiratory CO_2 in pods of chickpea: The role of pod wall, seed coat, and embryo. J. Exp. Bot., 55: 1687-1692.

Herbert RM (2003). Application of latex emulision plomears in seed coating technology. Pest. Formulations Appl. Syst., 23: 55-67.

Gesch RW, Archer DW (2005). Influence of sowing date on emergence characteristics of maize seed coated with a temperature-activated polymer. Agron. J., 97: 1543-1550.

Liu H, Du Y, Wang X, Sun L, (2004).Chitosan kills bacteria through cel membrane damage. Int. J. Food Microbiol., 95: 147-155.

Ma PP, He LQ (2001). Progress of chitosan in suppression of plant diseases. Nat. Prod. Res. Dev., 13: 82-86.

Nielsen K, Jorgense P, Mikkelsen JD, (1994). Antifungal activity of sugar beet chitinase against *Cercospora beticola*: An autoradiographic study on cell wall degradation. Plant Pathol., 43: 979-986.

Raveton M, Aajoud A, Willison J, Cherifi M, Tissut M, Ravanel P (2007). Soil distribution of fipronil and its metabolites originating from a seed-coated formulation. Chemosphere, 69: 1124-1129.

Russ WG, David WA (2005). Influence of sowing date on emergence characteristics of maize seed coated with a temperature-activated polymer. Agron. J., 97: 1543-1550.

Shibuya N, Minami E (2001). Oligosaccharide signalling for defence responses in plant. Physiol. Mol. Plant Pathol., 59: 223-233.

Sigler WV, Turco RF (2002). The impact of chlorothalonil application on soil bacterial and fungal populations as assessed by denaturing gradient gel electrophoresis. Appl. Soil Ecol., 21: 107-118.

Sun XH, Wang HF, Liu YF, Chen B, Ji, PJ, Yang JW (2004). Synthesis of a new fungicide pyrimethanil Salts. Chin. J. Org. Chem., 24: 506-510.

TeKrony DM (2006). Seeds: The delivery system for crop Science. Crop Sci., 46: 2263-2269.

Qiu J, Wang RM, Yan JZ, Hu J, (2005). Seed film coating with uniconazole improves rape seeding growth in relation to physiological changes under waterlogging stress. Plant Growth Regulation, 47: 75-81.

Wu XH, Zhang WH, Liu PF (2003). Research and development trend of the seed coating agent in China. China Plant Protect., 10: 36-38.

Xiong, YF, Wen ZY, Jiang JA, Xiong HR, Zhou YB, (2004).Advance of studies on seed coating agents for crops. J. Hunan Agric. Univ. Nat. Sci., 30: 187-192.

11

Effects of storage conditions and tillage operation on some fungal diseases and yield of maize

G. O. Ihejirika*, M. I. Nwufo and S. O. Anagboso

Department of Crop Science and Technology, Federal University of Technology, P. M. B. 1526, Owerri, Nigeria.

Maize (*Zea mays* L) belongs to the family *Poaceae* and it is mostly grown as food for man and feed for animal. A two-season experiment was conducted in 2006 and 2007, respectively, to determine the effect of storage condition and tillage operation on some fungal diseases and yield of maize. Analysis of variance indicated that storage condition significantly affected plant height (cm) 5.69; 6.26 at 9 weeks after planting. Tillage operation was statistically significant on leaf spot and blight infection at 9 weeks after planting at 5% probability level. Interaction of storage condition and tillage operation was also significant on blight infection at 9 weeks after planting 0.50; 0.58. Seeds from dehusked maize recorded highest plant height, leaf formation and grain yield, while shelled was lowest on both parameters in 2006 and 2007, respectively. However, the three storage conditions investigated were statistically similar on blight infection. Dehusked and undehusked plots recorded statistically similar result on leaf spot severity while shelled had the highest. Spot tilled plots recorded lowest leaf spot and blight severity but highest grain yields when zero-tilled plots (control) had the highest in all the field diseases investigated, but lowest grain yield in 2006 and 2007 respectively. Micro-organisms identified on infected leaves were *Helminthosporium* spp, *Spermospora* spp, while *Fusarium* spp, *Penicilium* spp, *Blastomyces* and *Aspergillus* species were identified with infected grains with *Fusarium* and *Penicilium* species occurring highest in 2006 and 2007, respectively.

Key words: Storage condition, tillage operation, fungal diseases, yield, maize.

NTRODUCTION

Maize (*Zea mays* L) is a native of America, although, origin has yet to be shown and Teosinte is a wild relative of maize. It is introduced by Portuguese early in the 10th century along the West Coast of Africa. It grows well in under the most varied conditions, unlike barley, wheat and rice which are limited by Climate Fischer and Palmer 1984. In Southern Nigeria, maize has been the principle cultivated cereal until the introduction and expansion of production of rice and has been used primarily as human food FAO (1990). It can be eaten whole and can also be processed into different products consumable by man, as animal feed and for industrial uses. In Nigeria, maize consumption (both man and animals) ranges from 26 - 28 kg per week. The seed is the nucleus of farmers' production activities, its activities quality, should be guaranteed at all times. In 1977, about 90,000 tons of

maize grain was imported into the country at a cost of over ₦25 million while 293,000 and 345,000 tons were imported in 1981 and 1982, respectively (FAO,1980).

In any crop production system, good quality seed inspires the confidence of farmers, because all other input will merely assist the seed to produce optimally. Germination percentages and purity in seed certification, seed maturity at harvest and method of drying are among the major deciding seed quality (Robinson, 1977). Maize quality are often reduced by drying injury, although the cause of impairment of lip body alignment along storage method has been associated with decrease in germination, quality and vigorous of maize seedling. The temperature of the seeds at storage greatly affects the germination and quality of a maize seed. Fungi accounts for about 75% (Anon, 1973; Obi et al., 1980) and it has been found to contribute to maximum damage in maize, such as abortion, rot, necrosis, discoloration and reduced germination and vigous. Grain moisture of 20% in cereals often causes corn rot and toxin production before harvest Essien (2000). *Fusarium* species is the most important field fungi

*Corresponding author. E-mail: ihegab@yahoo.com.

Table 1. Severity of leaf spot and blight diseases.

Severity estimation (%)	Scale	Interpretation
0	0	No infection
1 - 20	1	Slight infection
21 - 41	2	Moderate infection
41 - 60	3	Severe infection
61 - 80	4	Very severe infection
81 - 100	5	Complete infection

This was recorded on monthly basis in 2006 and 2007, respectively.

disease of maize worldwide and is known to produce over 100 secondary metabolites that can adversely affect human health (Visconti, 2000).

Fusasium moniliform has been found to be associated with pre-harvest and stored maize in Nigeria, (Hointink and Boehm, 1999), while Essien (2000) had reported that *Aspergillus flavus*, *Aspergillus nominus* and *Aspergillus parasitica* were isolated from cultures of most seed chips and cereal crops in Nigeria. Also, Bankole and Adebanjo (2003) observed heavy dietary health problems in inhabitants of sub-Sahara Africa as a result of mycotoxin particularly *Fumanisius* produced by *F. moniforum* and Aflatoxins from injection by *Aspergillus* specie. Tillage aims at creating soil environment favourable to plant growth. Lal (1973) observed that tillage is physical, chemical or biological soil manipulation to optimize condition for germination, seedling establishment and crop growth. Tillage system include no- tillage or slot planting, mulch tillage, strip or zonal tillage, ridge till include no-till include on ridges and reduced tillage. (Parr et al., 1990; Gajiri et al., 1994; Opara-Nnadi, 1990). Hence, the objective of this project is to determine the effects of storage conditions and tillage operation on some fungal diseases and yield of maize as well as to identify micro-organisms associated with these diseases.

MATERIALS AND METHODS

The experiment was conducted at Federal University of Technology, Owerri Research Farm. It is situated at Latitude 3˚N and Longitude 7˚E in Utisols of South Eastern Nigeria. The annual rainfall range was 2400-2500 mm and temperature range of 26-27˚C in 2006 and 2007 seasons, respectively. The experiment comprised of two treatments. These include storage conditions which appear as main plot while tillage operation occurs as sub plot. Storage conditions include shelled, unshelled and threshed maize seeds and tillage operations consist of three combinations which were zero (control), spot tilled and strip tilled plots, respectively. The treatment management was 3 x 3 factorial fitted into a randomized complete block design (RCBD) giving 3 x 3 = 9 treatments at 4 replications giving 9 x 4 = 36 experimental plots in 2006 and 2007, respectively.

Planting was carried out at the rate of 2 seeds per hole at three tillage operations, with spacing of 25 x 75 cm, giving 53,333 plants per hectare. Data were collected on leaf formation, plant height, leaf formation, grain yield, leaf spot and blight diseases.

Plant height (cm)

This was collected by measuring with a tape from the base to the apex of newly formed leaf at 3, 6 and 9 weeks after planting.

Leaf formation

This was obtained by counting the leaves formed by the sampled plants one after the other at 3, 6 and 9 weeks after planting.

Grain yield

This was obtained by threshing the grains from sampled plots and weighing in a precision balance and expressed in kilogram per hectare in 2006 and 2007, respectively.

Diseases severity

Severity of leaf spot and blight diseases were obtained monthly by visual observation and scoring according to the following format as proposed by Ford and Hewitt (1980) (Table 1).

Cultural, isolation and identification of some fungi species

The diseased leaves and grains were subjected to microbial isolation and identification. 280 g of fresh Irish potatoes were weighed, peeled and chopped into small pieces. Then, it was boiled in beaker using distilled water. After 30 min, the whole content was transferred to a muslin cloth in a beaker and the extract squeezed into a beaker. It was then made up to one liter with distilled water. 20 g of glucose and 20 g of agar powder were added. It was stirred on the hot plate and transferred into a conical flask. It was then covered with a cotton wool and foil to avoid contamination. The contents were poured into Petri-dishes at 20 ml each and allowed to cool to solidification, in line with Bankole and Adebayo (2003) as well as Barnett and Hunter (1998).

Samples were collected from each of the sampled diseased leaves and grain and they were inoculated into the nutrient potato dextrose agar medium. Growth was observed after 48 h; it was stained and observed using binocular microscope. The pathogens were identified using the laboratory manual by Barnett and Hunter (1998).

Data analysis

Data were analyzed using methods of Steels and Torrie (1981) and means were separated using the Fishers protected Least Significant Difference (LSD) according to Statistical Analysis System (SAS, 1999).

RESULTS

The result of soil analysis prior to planting showed that nitrogen is 0.09, potassium 0.238, calcium 0.38, aluminum 1.38, hydrogen 0.41% (Centimol/gram), according to AOAC (1990). The result also showed that storage condition was not significant at all stages in the

Table 2. Effect of storage condition and tillage operation on plant height, leaf formation and leaf spot disease in 2006 and 2007.

	Plant height		Leaf formation		Leaf blight		Leaf spot disease	
	2006	2007	2006	2007	2006	2007	2006	2007
Shelled	44.73	42.1	10.05	9.51	1.17	1.35	2.75	3.20
Undehusked	47.31	46.54	9.01	8.25	1.17	1.25	2.92	2.80
Dehusked	64.00	56.20	13.9	14.35	1.17	1.26	2.92	2.40
LSD $_{0.05}$	0.92	0.85	1.95	1.85	0.07	0.14	0.66	0.72
Zero tillage	64.87	68.9	11.08	13.00	1.58	1.88	3.42	3.65
Strip tillage	32.24	40.20	8.55	8.62	1.08	1.16	2.83	2.55
Spot tillage	56.03	60.54	9.17	9.38	0.83	0.78	2.3	2.00
LSD$_{0.05}$	0.265	0.31	2.84	2.02	0.56	0.52	0.58	0.66

Table 3. Effects of storage condition and tillage operation on cob weight and grain yield (kg/ha) in 2006 and 2007.

Sources	Cob weight (kg/ha)		Ai yield (kg/ha)	
	2006	2007	2006	2007
Shelled	980.10	110.22	2627.02	2636.1
Undehusked	8478.43	7570.4	3519.84	3620.6
Dehusked	11571.93	16805.5	7019.84	6875.41
LSD$_{0.05}$	6.95	8.55	2.72	3.15
Zero tillage	14738.84	13728.2	9731.27	9840.1
Strip tillage	6867.42	7080.54	2629.85	2752.5
Spot tillage	10968.6	11042.1	5847.96	6020.5
LSD$_{0.05}$	130.47	155.35	34.52	35.40

severity of leaf blight and leaf spot diseases when tillage operation significantly influenced leaf spot (3.53; 4.44) and blight (1.75; 1.88) disease infections at 9 weeks after planting in 2006 and 2007, respectively (Table 2). Subsequently, the interaction of storage condition and tillage operation was significant on cob weight and grain yield.

Seeds from dehusked maize recorded the highest plant height (64.00; 56.20) with shelled (44.70; 42.10) recording the lowest. Dehusked seeds also had the highest leaf formation (13.90; 14.35) while undehusked recorded the lowest (9.01; 8.25). Storage condition hayed statistically similar result on leaf blight and leaf spot infection in 2006 and 2007, respectively (Table 2). Considering the tillage operation, zero tilled plots recorded the highest plant height (64.87; 68.90) with strip tilled plots recording the lowest (32.24; 40.06), while similar results were obtained on leaf formation. Zero tilled plots recorded highest leaf spot disease severity (3.42; 3.654) as well as highest leaf blight (1.58; 1.88) with spot tilled plots recording lowest in all the diseases investigated in 2006 and 2007, respectively (Table 2). Dehusked recorded the highest cob weight (11571.93; 10805.50) and grain yield (7019.82; 6875.65) when shelled had the lowest cob weight (980.92; 110.22) as

well as grain yield (2627.02; 2636.10) (Table 3).

The interaction of storage conditions showed that zero tilled plots recorded the highest severity of blight and leaf spot diseases, while shelled and strip tillage interaction recorded the lowest leaf spot and blight diseases. The interaction of dehusked and zero tillage had the highest severity of all the diseases investigated (Table 4). Dehusked interaction with strip tillage recorded the highest cob weight as well as grain yield, while undehusked and spot tillage recorded the highest cob weight and grain yield in 2006 and 2007, respectively (Table 5). Micro-organisms identified with infected leaves were *Helminthosporum maydis*, *Spermospora*, while that with grains were *Aspergillus spp*, *Penincilium spp*, *Blastomyces* and *Fusarium spp* with *Fusarium* and *Penincdillum* species occurring most in all the seasons investigated.

DISCUSSION

The low level of nitrogen, phosphorus and potassium may be attributed to the typical fragile nature of the tropical soils, marked with intense leaching and volatilization due to the high rainfall and temperature leading to low fertility in soil status. The acidic nature of the soil may be due to low potassium and basic nutrient, hence low level of phosphorus impairs fixation. This is in agreement with AOAC (1990). The significant difference recorded by storage condition on plant weight with shelled recording lowest may be attributed to the complete separation of the seeds from husks. They are then subjected to high stress created by moisture stress, reduction in seed constituents. This may lead to high dormancy and reduced metabolic and biochemical processes resulting to low plant height.

The significant difference recorded by tillage operation on leaf spot and blight diseases especially at 9 weeks after planting, with zero tilled plots recording highest severity of all the diseases may be attributed to the fact that accumulation of organic matter in the soil which enhanced the activities of the micro-organisms in line

Table 4. Mean values of storage condition and tillage operation on leaf spot and blight severity in 2006 and 2007.

		Leaf spot		Leaf blight	
		2006	2007	2006	2007
Shelled	Z	3.25	2.92	2.0	2.4
	St	3	3.25	1.0	1.1
	Sp	2.0	1.8	0.5	0.58
Undehusked	Z	3.25	3.7	1.25	1.50
	St	3.0	3.25	1.25	1.44
	Sp	2.5	2.6	1.0	0.96
Dehusked	Z	3.75	3.86	1.5	1.67
	St	2.5	2.62	1.0	1.11
	Sp	2.5	2.0	1.0	1.18
	$LSD_{0.05}$ storage condition	1.18	1.2	0.07	0.012
	$LSD_{0.05}$ tillage operation	0.58	0.48	0.56	0.6

Table 5. Means of main effects on storage condition and tillage operation on cob weight and grain yield in 2006 and 2007.

		Cob weight (kg/ha)		Grain yield (kg/ha)		
		2006	2007	2006	2007	
SH	Z	6666.625	6725.6	333.3	3416.7	
	St	2319.99	2410.38	386.67	390.11	
	Sp	3533.31	3644.5	1879.98	1957.8	
UND	Z	3433.32	3455.42	2613.32	2570.04	
	St	2693.35	2711.98	1199.99	1216.2	
	Sp	5333.3	5475.1	3199.86	2894	
DEH	Z	4519.97	4428.7	1773.32	1560.36	
	St	1839.99	1754.66	1026.66	1088.25	
	Sp	2039.96	2128.5	746.64	785.9	
$LSD_{0.05}$ storage condition		6.95	5.26	7.70	5.88	2.7
$LSD_{0.05}$ tillage operation		130.47	20.18	12.55	26.74	34.52

with Lal (1976), Ahn and Hintz (1990) as well as Hoitink and Boehm (1999). Storage condition recorded significant cob weight as well as grain yield with shelled recording the lowest in line with same reasons assumed above for plant performance. Significant difference recorded by tillage operation with zero tilled plots recording highest cob weight and grain yield may be related to the sandy nature of the tropical soil, which is prone to degradation due to adverse effects of weather, thus, continued tillage decreases the level of organic matter in the soil by improving condition for its oxidation, hence no tillage operation encourages organic matter reduction, soil matter losses, soil erosion, less destruction of soil structure, decrease in labour and energy

consumption (Ahmed and Young, 1982). Moreover, the interaction of zero tillage and virtually all the storage conditions recorded highest disease severity with spot tillage and its interaction with any of the storage condition low, may be as a result of non-exposure of the sub-soil to the surface to the action of the ultraviolet rays. This helps to eliminate most of those pathogens and their spores as no-tilled plots encourage the colonization of maize root fungi which is in line with Ahmed and Young (1982). The presence of *H. maydis*, *Spermospora* spp on maize leaves, as well as *Aspergillus* spp, *Penincilium* spp, *Blastomyces* and *Fusarium* species on the grain with *Fusarium* and *Penincdillum* species occurring most, may be due to the fact that these micro-organisms are responsible for the disease development and spoilage of tropical crops, in line with Blancard (1994); Essien (2000); Bankole and Adebanjo (2003).

In conclusion, spot tilled plots recorded the lowest severity of leaf blight and leaf spot diseases, while zero tilled plots recorded the highest on both seasons investigated. Seeds from shelled maize recorded the lowest plant performance, while zero tilled plots had the highest. Same trend was observed in its interaction with any of the storage condition in all the seasons investigated.

REFERENCES

Ahmed EM, Young CT (1982). Composition, Quality and Flavour of Peanuts, In HE. Pattee and CT. Young. (ed.), Peanut Science and Technology, American Peanut Research and Education Society. Inc., Yoakum, Tex, pp. 655-688.

Ahn PM, Hinzte B (1990). No tillage, minimum tillage and their influence on soil properties. In: Organic management and Tillage in humid and sub-humid Africa. pp. 341-349, IBSRAM Proceeding of 9th ISTRO Conference Osijeth, Yugslavia, pp. 72-78.

Anon (1973). A Compendium of Corn Disease. The Cooperative Extension Services. United States Department of Soc. Inc. 3340 Pilot Road, St. Paul. Minnesota, SB121, USA, p. 54.

AOAC (1990). Association of Official Analytical Chemist. Official methods of Analysis, 15th ed. Washington DC.

Bankole SA, Adebanjo A (2003). Aflatoxin contamination of dried yam chips marketed in Nigeria. Trop. Sci. 43(3/4): 20. In press.

Barnett HL, Hunter BB (1998). Descriptions and Illustrations of genera. Illustrated Genera of Imperfect fungi. The American Phytopathological Society press. 3rd ed. St. Paul Minnesota, p. 218.

Blancard D (1994). A Colour Atlas of Tomato Diseases. Observation, Identification and Control. Manson Publishing Ltd. INRA. UK, pp. 157-164.

Essien SP (2000). Mycotoxigenic Moulds in Maize in Nigeria. Mud thumbus Trop. Sci. 40: 154-158.

FAO (1980). Food and Agricultural Organization of the United Nations. Monthly Bulletin of Statistics, 7(12): 76.

FAO (1990). Food and Agricultural Organization of the United Nations. Botary Maize Cereal crop. Corn. 10(2): 5.

Ford JE, Hewitt D (1980). In Vicia Foba. Bond, DA (ed. Feeding Values, Processing and Viruses. The Hage. Martinus Njhoff, pp. 125-139.

Gajiri PR, Arora VK, Chaudhary MR (1994). Maize growth responses to deep tillage straw mulching and farm yard manure in coarse textured soils of N.W. India. Soil Use Manage. 10: 15-20.

Hointink HAJ, Boehm MJ (1999). Biocontrol within the context of soil microbial communities; A substrate-dependent Phenomenon. Ann. Rev. Phytopath. 37: 437-446.

Lal R (1973). Effects of Seedbed preparation and Time of planting on Maize (Zea mays). In: Western Nigeria, Exper. Agric. 9: 321-333.

Lal R (1976). No-tillage effects on Soil Properties under Different Crops in Western Nigeria. Soil Sci. Soc. Am., J. 40: 752-768.

Obi IU, Hooker AL, Lim SM (1980). Phytoalexin production in Zea mays L. to Helminthosporium fungi. Ph.D thesis, University of Illinois, Urban-champaign, III. USA. IX, p. 119.

Opara-Nnadi OA (1990). Tillage Practices and their effect on Soil Productivity in Nigeria. In Organic matter Managemen6t and Tillage in Humid and Sub-humid Africa. pp. 87-111. IBSRA Proceedings No. 10. Bangkok: IBSRAM.

Parr JD, Rapendick RL, Hornick SB, Meyer RE (1990). The use of Cover crops, Mulches and Tillage for soil water conservation and weed control. In: Organic matter Management and Tillage in Humid and Sub-humid Africa, pp.261-366. IBSRAM Proceedings. No. 10, Bangkok, IBSRAM.

Robinson RW (1977). Tomato. Encyclopedia of Food Agriculture and nutrition. 4th ed. McGraw Hill Inc. New York, pp. 650-652.

SAS (1999). Statistical Analysis System. User's guide. Statistics SAS Institute Inc. Cary N.C., USA.

Steels RG, Torrie JH (1981). Analysis of Experiment. 2nd ed. London: Macmillan, p. 420-452.

Visconti A (2000). Problems associated with Fusarium Mycotoxin in Cereals. Bulletin of the Institute for Comprehensive Agricultural Sciences. Kinky UniversityX,109: 39-55.

12

Antimicrobial activity and physicochemical properties of oils from tropical macrofungi

David, O. M.[1]*, Fagbohun, E. D.[1], Oluyege, A. O.[1] and Adegbuyi, A.[2]

[1]Dapartment of Microbiology, Ekiti State University, Ado-Ekiti, Nigeria.
[2]Department of Pharmacy, University Teaching Hospital, Ado-Ekiti, Nigeria.

Nutriceutics potential and physicochemical properties of the oil extract from five species of macrofungi which include *Ganoderma lucidium*, *Pleurotus tuberregium*, *Termytomyces robustus*, *Schizophyllum commune* and *Trametes versicolor* were investigated using standard chemical and microbiological methods. The oil was extracted using Soxhlet method of extraction. Disc diffusion and agar dilution methods were used to test for the antibacterial and antifungal properties of the samples respectively. The extracted oils were tested against five clinical bacterial isolates: *Pseudomonas aeruginosa*, *Staphylococcus aureus*, *Bacillus cereus*, *Serratia marcescens*, *Enterococcus faecalis* and *Escherichia coli*. The susceptibility of three fungi genera to the oil samples was also tested. The potency of the extracts oils was determined at different concentrations. All the oils extracted from the mushroom were liquid at room temperature. The acid values of the oils ranged between 0.9 and 6.7 mg KOH/g in *T. robustus* and *T. versicolor* respectively. The high iodine values ranged between 39.8 in *T. versicolor* and 127.0 mg I$_2$/100 g. The saponification value was above 100 mg KOH/g except *T. robustus*. The aromabiogram of the oils from the mushroom had a pronounced effect on the Gram negative bacteria. Oil from *S. commune* has the least inhibitory effect on the bacteria. The antifungal assay of *P. tuberregium* was most effective against *Aspergillus parasiticus*, followed by that of *G. lucidium*. The least effective was oil from *T. versicolor*. The performance of oil from *G. lucidium* was the best out of all the samples. The inhibitory activities of the oils were concentration dependent. The oils tested were good sources of antimicrobials.

Key words: Mushroom, oil, nutriceutics, antimicrobial, pathogens, macrofungi.

INTRODUCTION

Macrofungi (mushrooms) are multicellular structures formed from the differentiation of vegetative mycelial cells (Chang and Miles, 1992). A mushroom is divided into two different tissues: the stipe (stem) and the pileus (cap). Mushrooms are the earliest known fungal organisms used as food for their taste and aroma. They are enjoyed in soups or other food preparations for all season (Hestbjerg et al., 2003; Sasek, 2003). They are very good sources of protein, vitamins, lipids and mineral elements. In searching for new therapeutic alternatives, scientists have studied many kinds of mushrooms and have found variable therapeutic activity such as anticarcinogenic, anti-inflammatory, immuno-suppressor, anti-plasmodium and antibiotic, among others (Asfors and Ley, 1993; Longvah and Deosthale, 1998; Tabata et al., 1981; Kapoor, 2010).

Oils are valuable natural products used as raw materials in many fields, including perfumes, cosmetics, aromatherapy, phytotherapy, spices and nutrition (Buchbauer, 2000). Aromatherapy is the therapeutic use of fragrances or at least mere volatiles to cure, mitigate or prevent diseases, infections and indispositions by means of inhalation (Buchbauer et al., 1993). This has recently attracted the attention of many scientists and encouraged them to screen plants to study the biological activities of their oils from chemical and pharmacological investigations to therapeutic aspects. Hopefully, this will lead

*Corresponding author. E-mail: davidgenerationng@yahoo.com.

Table 1. Physicochemical properties of five Nigeria mushroom.

Parameter	Macrofungi (mushrooms)				
	G. lucidium	*P. tuberregium*	*T. robustus*	*S. commune*	*T. versicolor*
Peroxide value (meq/kg)	7.6	6.4	5.0	4.4	4.0
Iodine value (Mg I$_2$/100 g)	127.0	159.4	39.8	127.5	51.3
Acid value (mg KOH/g)	5.61	1.1	6.7	2.4	0.9
Saponification value (mg KOH/g)	150.1	108.0	196.4	260.9	67.3
Free fatty acid (% Oleic acid)	2.8	0.6	3.4	1.2	1.0

to new information on plant applications and new perspective on the potential use of these natural products.

Antibacterial-resistance infections are on the high side in the recent time, and oils of plant origin have frequently been reported to be antimicrobial (Tantaoui-Elaraki et al., 1992, 1993; Lattaoui and Tantaoui-Elaraki, 1994). The search for new alternatives, of fungal origin, to antibiotic drugs to prevent the proliferation of pathogenic microbes and infections they caused is essential. There is paucity of information on the antimicrobial activity of oils from the Nigeria macrofungi which informed the objective of this study.

MATERIALS AND METHODS

Collection of samples and extraction of oils

Five macrofungi (mushroom) samples which include (*Pleurotus toberrigium, Termytomyces robustus, Schizophyllum commune, Ganoderma lucidium* and *Trametes versicolor* were collected from Ado-Ekiti and Iworoko-Ekiti in Ekiti State, Nigeria between September and November, 2010 and identification. Fresh mushroom were randomly divided into three samples of 400 g and air dried. The air dried samples were blended separately to a powdered using an electric blender. The powdered samples were store in different containers and labelled appropriately. Oils were extracted from the macrofungi according to the method of Dusk and Gokel (1987).

Determination of physiochemical properties of extracted oil

The physicochemical properties of the oil samples which include saponification, free fatty acid and acid values were determined using the AOCS official method of analysis (AOCS, 1997).

Source of test organisms

Bacterial isolates which include *Esherichia coli, Pseudomonas aeruginosa, Salmonella typhi, Shigella dysenteriae* and *Enterobacter* sp. were collected from Clinical Microbiology Laboratory, Lagos State University Teaching Hospital (LUTH), Idi-Araba, Lagos, Nigeria while the fungal isolates were collected from the Department of Microbiology, University of Ado-Ekiti, Nigeria.

Determination of antibacterial activities of extracted oil sample

The bacterial isolates used for this test were standardized

according to Bauer et al. (1966) while the method of Oloke (2000) was employed to screen the antibacterial and antifungal property of the oil samples.

RESULTS AND DISCUSSION

All the oils extracted from the mushroom were liquid at room temperature with colour varied from light golden yellow to dark brown. The oil yield ranged between 25.02 to 36.03%. This is lower compared with the oil yield of palm fruit (48.65%), groundnut (49.0%), and pumpkin leave (47.4%) (Fagbemi and Oshodi, 1991). The acid values of the oils ranged between 0.9 and 6.7 mg KOH/g in *T. robustus* and *T. versicolor* respectively. The high iodine values ranged between 39.8 in *T. versicolor* and 127.0 mg I$_2$/100 g. This indicates that the oils could be used for the manufacture of cosmetics (Dawodu, 2009). The oil also has saponification value above 100 mg KOH/g except *T. robustus* with 67.3 mg KOH/g. Except for *T. versicolor* (196.4 mg KOH/g) the saponification values of the oils from the mushroom were not within the values for the oils of most plant origin value reported to fall within 188 to 196 (Akanni et al., 2005). The lower the saponification value the larger the molecular weight (Agatemor, 2006). Peroxide value of the oils samples was highest in *G. lucidium* (7.6) followed by *P. tuberregium* (6.4) the least value was recorded in *T. robustus* 4.0. The values were less than 10 documented for majority of the vegetable oils (Ibrahim and Fagbohun, 2011). This implies that the samples will have a better keeping quality than oils from plants (Kolapo et al., 2007).

Search for natural products for a new antimicrobial agent that would be able to curb the menace of increasing resistance to antibiotics is on the increase. In this study, mushroom oil exhibited varying antibacterial activities against test clinical bacterial isolates. The aromabiogram of the oils from the mushroom is shown in Table 1. Gram positive bacteria were relatively resistant to the oils from mushroom than the Oil from *G. Lucidium*, which was most effective on *Bacillus cereus* which is an opportunist pathogen that causes food poisoning. *P. aeruginosa* exhibited the least resistance to the extract. *Enterococcus faecalis* was entirely resistant to oil from *P. tuberregium* even at the highest tested concentration. Oil from *S. commune* has the least inhibitory effect on the

Table 2. Antibacterial screening of oils from common Nigeria macrofungi on some medically important pathogens.

Samples	Concentration (mg/ml)	Pathogens					
		Pseudomonas sp	*E. coli*	*Serratia* sp	*S. aureus*	*Bacillus* sp	*Enterococcus* sp
G. lucidium	100	8	12	19	20	20	15
	10	7	10	17	12	20	16
	1	7	10	18	11	19	12
	0.1	3	9	11	9	13	12
	0.01	1	1	6	3	12	7
P. tuberigium	100	12	7	19	20	10	0
	10	11	7	15	15	10	0
	1	10	3	19	12	4	0
	0.1	11	1	12	15	1	0
	0.01	3	1	9	12	10	0
Trametes sp	100	25	12	10	20	12	0
	10	20	10	5	10	10	0
	1	11	0	4	10	0	0
	0.1	10	0	2	7	0	0
	0.01	9	0	1	2	0	0
T. robustus	100	24	17	14	15	15	10
	10	20	10	9	15	15	7
	1	15	11	6	10	12	5
	0.1	8	6	6	10	11	3
	0.01	7	2	4	7	5	1
S. commune	100	12	11	6	0	0	0
	10	9	10	0	0	0	0
	1	0	11	0	0	0	0
	0.1	0	7	0	0	0	0
	0.01	0	3	0	0	0	0

pathogens. In that it has little effect on the Gram negative pathogen and no effect on Gram negative bacteria. *P. aeruginosa* was the most sensitive pathogen to Trameter followed by *Staphylococcus aureus*. *E. faecalis*, on the other hand was completely resistant. Results of the antibacterial activity of the oil however, does not agree with the work of Burt (2004) who reported Gram positive bacteria to be more sensitive to oil than Gram negative bacteria. The antifungal assay was presented on Tables 2 to 6. The relative decrease in the radial mycelia growth was assumed to be as a result of inhibitory effect of the oils. The effect of essential oil on the radial fungal growth was supported by Iwalokun et al. (2007) who observed that the growth of tested fungi isolate was decreased by increasing the oil concentrations.

One of the major models of mechanism of anti-fungal properties of oils is to diffuse into cell membranes and cause them to expand, thereby increasing their fluidity disordering membrane embedded enzymes (Mendoza et al., 1997). Oils that have high phenol contents have a pronounced effect on the membranes transport, nutrient uptake, nucleic acid synthesis and lipase activities (Baydar et al., 2004; Ipek et al., 2005). The comparative low radial mycelia growth in the tested fungi in the presence of varying concentrations of oils from fungi agreed with the report of Iwalokun et al. (2007). This suggests their potential as nutriceutics and pharmaceutics. *P. tuberregium* was most effective against *Aspergillus parasiticus*, followed by *G. Lucidium*, the least effective among the tested oils was oil extracted from *T. versicolor*. *Aspergillus* causes systematic mycoses (Denning, 1996) and its infection is generally symptomatic (Durry et al., 1997). *G. lucidium* inhibited the growth of *Aspergillus niger* best, followed by *P. tuberregium*, *T. robustus* had the least effect on the fungus. *Rhyzopus* was the most susceptible among the test fungi.

P. tuberregium was the most effective against the

Table 3. Antimicrobial effects of oils of macrofungi on the growth of *Aspergillus parasiticus*.

Hours	Concentration (mg/ml)	Mushroom from where the oils were extracted					Control
		G. lucidium	*P. tuberregium*	*T. robustus*	*S. commune*	*T. versicolor*	
24	100	10.0	5.0	7.0	ND	6.0	13
	10	11.0	5.0	12.0	ND	6.0	
	1.0	13.0	13.0	17.0	ND	6.0	
	0.1	15.0	13.0	17.0	ND	15.0	
48	100	15.0	8.0	21.0	ND	14.0	Out grown
	10	15.0	10.0	20.0	ND	19.0	
	1.0	16.0	10.0	25.0	ND	23.0	
	0.1	17.0	25.0	26.0	ND	34.0	
72	100	21.0	15.0	20.0	ND	17.0	Out grown
	10	22.0	13.0	30.0	ND	25.0	
	1.0	25.0	10.0	32.0	ND	37.0	
	0.1	30.0	34.0	Out grown	ND	Out grown	

ND=Not determined.

Table 4. Antifungal activity of macrofungal oils on *A. niger*.

Hours	Concentration (mg/ml)	Mushroom from where the oils were extracted					Control
		G. lucidium	*P. tuberregium*	*T. robustus*	*S. commune*	*T. versicolor*	
24	100	0	0.0	2.0	ND	2.0	13
	10	2.0	0.0	4.0	ND	1.0	
	1.0	2.0	2.0	3.0	ND	6.0	
	0.1	2.0	8.0	9.0	ND	10.0	
48	100	0.0	5.0	12.0	ND	3.0	Out grown
	10	4.0	5.0	14.0	ND	12.0	
	1.0	8.0	8.0	20.0	ND	15.0	
	0.1	10.0	20.0	22.0	ND	20.0	
72	100	4.0	10.0	20.0	ND	12	Out grown
	10	14	20	23.0	ND	20.0	
	1.0	20.0	26.0	31.0	ND	20.0	
	0.1	30.0	30.0	out grown	ND	28.0	

ND=Not determined.

fungus while *G. lucidium* exhibited the least inhibitory effect. Effect of *G. lucidium* was highest on *Botryodiplodia* sp. Followed by *P. tuberregium*. *T. versicolor* had the least effect on the fungus. The performance of *G. lucidium* was the best out of all the samples. This justifies its use in folk medicine (Wasser and Weis, 1999). Its effectiveness may be due to its cellular components and secondary metabolites that have been used to treat a variety of disease states (Gan et al., 1998; Chen et al., 1995). Moreover, it is cultivated and consumed for its pharmaceutical value rather than as food (Jong and Birmingham, 1992).

In this study, we demonstrated the antimicrobial activities of oils isolated from different macrofungi. We believe that the activities of these oils are similar to the activities of other antimicrobials from plants. The inhibitory activities of the oils were concentration dependent. We show in this paper that oils from mushroom could lead to novel therapeutic agents with both antibacterial and antifungal attributes.

Table 5. Antifungal activity of macrofungal oils on *Rhyzopus* sp.

Hours	Concentration (mg/ml)	Mushroom from where the oils were extracted					Control
		G. lucidium	P. tuberregium	T. robustus	S. commune	T. versicolor	
24	100.0	2.0	0.0	2.0	ND	2.0	13
	10.0	6.0	0.0	5.0	ND	3.0	
	1.0	5.0	3.0	8.0	ND	4.0	
	0.1	12.0	6.0	9.0	ND	5.0	
48	100.0	4.0	2.0	2.0	ND	5.0	Out grown
	10.0	8.0	10.0	6.0	ND	8.0	
	1.0	12.0	6.0	10.0	ND	11.0	
	0.1	25.0	8.0	10.0	ND	14.0	
72	100.0	8.0	6.0	3.0	ND	6.0	Out grown
	10.0	9.0	10.0	8.0	ND	10.0	
	1.0	14.0	8.0	14.0	ND	13.0	
	0.1	25.0	18.0	15.0	ND	14.0	

ND=Not determined.

Table 6. Antifungal activity of macrofungal oils on *Botryodiplodia* sp

Hours	Concentration (mg/ml)	Mushroom from where the oils were extracted					Control
		G. lucidium	P. tuberregium	T. robustus	S. commune	T. versicolor	
24	100.0	4.0	10	6	ND	6	10
	10.0	5.0	10	8	ND	10	
	1.0	5.0	10	8	ND	12	
	0.1	7.0	10	12	ND	12	
48	100.0	4.0	15	10	ND	15	Out grown
	10.0	5.0	14	13	ND	14	
	1.0	6.0	15	12	ND	15	
	0.1	5.0	15	14	ND	20	
72	100.0	5.0	20	Out grown	ND	Out grown	Out grown
	10.0	5.0	22	Out grown	ND	Out grown	
	1.0	8.0	23	Out grown	ND	Out grown	
	0.1	15.0	20	Out grown	ND	Out grown	

ND=Not determined.

REFERENCES

Agatemor C (2006). Studies of selected physicochemical properties of fluted pumpkin (*Telfairia occidentalis* Hook F.) seed oil and tropical almond (*Terminalia catappia* L.) seed oil. Pak. J. Nutr., 5(4): 306-307.

Akanni AS, Adekunle SA, Oluyemi EA (2005). Physico-chemical properties of some non-conventional oil seed. J. Food Technol., 3: 177-181.

AOCS (1997). Sampling and Analysis of Commercial Fats and oils American Oil Chemists Society Official Methods. AOCS Press, Champaign, Illinois, USA.

Asfors KE, Ley K (1993). Sulfated polysaccharides in inflammation. J. Lab. Clin. Med., 121: 201-202.

Bauer A, Kirby WMM, Sharris JC, Truck M (1966). Antibiotic susceptibility testing by standardized single disk method. Am. J. Clin. Path., 45: 493-496.

Baydar H, Sagdic O, Ozkan G, Karadogan T (2004). Antibacterial activity and composition of essential oils from *Origanum*, *Thymbra* and *Satureja* species with commercial importance in Turkey. Food Control, 15: 169-172.

Buchbauer G (2000). The detailed analysis of essential oils leads to the understanding of their properties. Perfumer and Flavorist, 25: 64-67.

Buchbauer G, Jirovetz L, Jager W (1993). Fragrance compounds and essential oils with sedative effects upon inhalation. J. Pharm. Sci., 82: 660-664.

Burt S (2004). Essential oils: Their antibacterial properties and potential applications in foods - A review. Int. J. Food Microbiol., 94: 223-253.

Chang ST, Miles PG (1992). Mushroom biology-a new discipline. The Mycologist, 6: 64-65.

Chen WC, Hau DM, Lee SS (1995). Effects of *Ganoderma lucidum* and Krestin on cellular immuno-competence in gamma-ray irradiated mice. Am. J. Chin. Med., 23: 71-80.

Denning DW (1996). Therapeutic outcome in invasive aspergillosis. Clin Infec. Dis., 23: 608-615

Durry E, Pappagianis D, Werner SB (1997). Re-emergence of an endemic disease. J. Med. Vet. Mycol., 35: 321-326.

Dusk HD, Gokel GW (1987). Experimental Organic Chemistry. pp. 113-117. New York: McGraw Hill Book Company.

Fagbemi TN, Oshodi AA (1991). Chemical composition and functional properties of full fat fluted pumpkin seed flour. Nig. Food J., 9: 26-32

Gan KH, Fann YF, Hsu SH, Kuo KW, Lin CN (1998). Mediation of the cytotoxicity of lanostanoids and steroids of Ganoderma tsugae through apoptosis and cell cycle. J. Nat. Prod. (Lloydia), 61: 485-487.

Hestbjerg H, Willumsen P, Christensen M, Andersen O, Jacobsen C (2003). Bioaugmentation of tar-contaminated soils under field conditions using Pleurotus ostreatus refuse from commercial mushroom production. Environ. Toxicol. Chem., 22(4): 692-698.

Ibrahim TA, Fagbohun ED (2011). Physicochemical properties and in vitro antibacterial activity of Corchorus olitorius Linn. seed oil. Life Sci. Leaflets, 15: 499-505.

Ipek E, Zeytinoglu H, Okay S, Tuylu BA, Kürkcüoglu M, Başer KHC (2005). Genotoxicity and antigenotoxicity of Origanum oil and carvacrol evaluated by Ames Salmonella/microsomal test. Food Chem., 93: 551-556.

Iwalokun BA, Usen UA, Otunba AA, Olukoya DK (2007). Comparative phytochemical evaluation, antimicrobial and antioxidant properties of Pleurotus ostreatus. Afr. J. Biotechnol., 6(15): 1732-1739.

Jong SC, Birmingham JM (1992). Medicinal benefits of the mushroom Ganoderma. Adv. Appl. Microbiol., 37: 101-134.

Kapoor K (2010). Illustrated Dictionary of Microbiology. Oxford Book Company Jaipur New Delhi.

Kolapo AL, Popoola TOS, Sann MO (2007). Evaluation of biochemical deterioration of locust bean daddawa and soybean daddawa-two Nigerian condiments. Am. J. Food Technol., 2: 440-445.

Lattaoui N, Tantaoui-Elaraki A (1994). Individual and combined antibacterial activity of the main components of three thyme essential oils. Rivista Italiana, 13: 13-19.

Longvah T, Deosthale YG (1998). Compositional and nutritional studies on edible wild mushroom from northeast India. Food Chem., 63: 331-334.

Oloke JK (2000). Activity pattern of natural and synthetic antibacterial agents among hospital isolates. Microbios, 102: 175-181.

Sasek V (2003). Why mycoremediations have not yet come to practice. In The utilization of bioremediation to reduce soil contamination: Problems and solutions. Sasek V. et al. (Eds.). Kluwer Academis Publishers, pp. 247-276.

Tabata K, Ito W, Kojima T, Kawabata S, Misaki A (1981). Ultrasonic degradation of schizophyllan, an antitumor polysaccharide produced by Schizophyllum commune Fries. Carbohydr. Res., 89: 121-135.

Tantaoui-Elaraki A, Errifi A, Benjilali B, Lattaoui N (1992). Antimicrobial activity of four chemically different essential oils. Rivista Italiana, 6: 13-22.

Tantaoui-Elaraki A, Lattaoui N, Errifi A, Benjilali B (1993). Composition and antimicrobial activity of the essential oils of Thymus broussonettii, T. zygis and T. satureioides. J. Ess. Oil Res., 5: 45-53.

Wasser SP, Weis AL (1999). Therapeutic effects of substances occurring in higher basidiomycetes mushrooms: a modern perspective. Critical Rev. Immunol., 19: 65-69.

13

Fungicide toxicity against the growth of lineages of the fungus *Metarhizium anisopliae* var. *anisopliae* (Mestch.) Sorokin

Sideney B. Onofre*, Cristiane R. Kasburg, Danusa de Freitas, Silvana Damin, Andréia Vilani, Jéssica A. Queiroz and Francini Y. Kagimura

Departamento de Ciências Biológicas, Laboratório de Microbiologia, Universidade Paranaense, UNIPAR, Unidade-Campus Francisco Beltrão, Av. Júlio Assis Cavalheiro, 2000 – Bairro Centro - 85601-010,Francisco Beltrão – Paraná – Brasil.

In this work the fungicidal action of three agrochemicals (Sphere®, Nativo® and Bendazol®) used in soybean for control of fungal diseases on the lineages CG-28 and CG-30 of *Metarhizium anisopliae* var. *anisopliae* was evaluated. It was found that the fungicides inhibited the vegetative growth of the lineages at the concentrations indicated for the field, thereby showing its antifungal effect.

Key words: Biocontrol, crops, pesticides.

INTRODUCTION

Fungi are microorganisms most frequently found attacking insects, responsible for 80% of the epizootic surges occurring in agro-ecosystems. Among the fungi used is *Metarhizium anisopliae*, the first agent used in the microbial control described in the literature and occurring naturally in more than 300 species of insects (Alves and Lopes 2008; Fernandes et al., 2010).

In Brazil, this fungus has been found infecting several species of insects, especially *Mahanarva posticata, Diatraea saccharalis, Nezara viridula, Piezodorus guildini, Deois* sp., *Zulia* sp., *Bonagota salubricola* and *Anticarsia gemmatalis*, with a large potential for use in biological control. The major example of the successful use of *M. anisopliae* is in the biological control of leafhoppers, mainly in sugar cane and pastures. Its pathogenicity has been demonstrated in ticks of several genera and species (Schrank et al., 2001, Onofre et al., 2002).

An issue that is becoming more serious in the last years is the use of fungicides in the control of fungal diseases in agro-ecosystems, which is causing concern because of their interference with agents of biological control such as the fungi. The impact of the application of phytosanitary chemicals on entomopathogens can vary according to the species and lineage of the pathogen, the chemical nature of the products and the concentrations employed. These products can act by inhibiting vegetative growth, conidial genesis, sporulation and by causing genetic mutations that can lead to reduced virulence against the plagues. Therefore, it is necessary to use the selective products that do not affect the balance between the plagues and their predators, parasites and pathogens (Alves and Lopes, 2008).

Among the commercial products indicated for the control of fungal diseases of soybean we highlight Nativo®, Sphere® and Bendazol®, the first of which having trifloxystrobin and tebuconazole as active agents. Sphere® has trifloxystrobin and cyproconazole as active agents. Both fungicides are from the chemical groups of strobirulins and triazoles. The strobirulins inhibit respiratory chain by inhibiting complex II and III, interrupting the oxidative phosphorylation and interfering with the action of the ATP-synthase. The triazoles inhibit the biosynthesis of ergosterol (a fungal lipid), an important substance for the maintenance of the cell membrane integrity of the fungal cells (Bayer, 2008). Bendazol® has carbendazim as active agent and belongs to the benzimidazoles, affecting specifically the cell

---
*Corresponding author. E-mail: sideney@unipar.br

Table 1. Observed growth of two lineages of *M. anisopliae* var. *anisopliae* in Potato + Dextrose + Agar (PDA) medium, under controlled conditions.

Fungus	96*	144*	192*	240*	288*	336*
CG-30	$14.50\pm0.16^{a\#}$	21.00 ± 0.08^a	28.00 ± 0.07^a	33.70 ± 0.06^a	39.00 ± 0.12^a	43.80 ± 1.43^a
CG-28	$13.50\pm0.09^{a\#}$	20.50 ± 0.17^a	28.00 ± 0.10^a	32.00 ± 0.10^a	36.30 ± 0.22^a	39.30 ± 0.22^a

[1]Halos of growth in mm. *Time in hours. #Data followed by the same lower case letter in the column do not differ by Tukey's test at the level of 5%.

division by inhibiting the biosynthesis of tubulins (Milenia, 2010).

Moino and Alves (1998) put forward the hypothesis that the microorganism, through a mechanism of physiological resistance, can metabolize the toxic principles of the active chemicals, using the molecules resulting from this process and released in the culture medium as secondary nutrients, promoting its vegetative growth and conidial genesis. Still another possibility is that the fungus, in an activity comparable to that occurring with any living organism, uses its reproductive effort in the presence of a toxic compound that changes its environment and impairs its development, resulting in greater vegetative growth and conidial genesis.

MATERIALS AND METHODS

The fungal lineages assessed were: (a) *M. anisopliae* var. *anisopliae* lieage CG-28(AL), supplied by ESALQ (AL), isolated from *Mahanarva posticata* (Homoptera: Cercopidae) and (b) *M. anisopliae* var. *anisopliae* lineage CG-30(E-6) supplied by ESALQ (E6), isolated from *Deois flavopicta* (Homoptera: Cercopidae). The fungal lineages were supplied by ESALQ/USP (Superior School of Agriculture Luiz de Queiroz – University of São Paulo – Piracicaba – São Paulo – Brazil), lyophilized and conserved at low temperature. The lineages were invigorated using the bovine tick *Boophilus microplus*, later stored for evaluation.

The essays were conducted *in vitro*, in potato-dextrose-agar (PDA) culture medium, adding the phytosanitary products to the fused culture medium, not solidified at 45°C, at the concentration pre-established for the field. Next, the mixture was poured into Petri dishes. The control treatment was prepared inoculating each fungal lineage to the PDA medium in the absence of fungicides. After the solidification of the culture medium containing the fungicide, the inoculation of the fungal lineages to the center of the Petri dishes was proceeded and they were incubated for 336 h at 28°C (Alves and Lopes, 2008).

Three commercial products recommended for the control of fungal diseases of soybean culture in Brazil were tested: Nativo®, whose active agents are trifloxystrobin + tebuconazole, at the concentrations of 2.5 ml/L; 1,250, 650, 350, 150, 70 and 40 µl/L; Sphere®, whose active principles are trifloxystrobin + cyproconazole, at the concentrations of 2.0 ml/L; 1,000, 500, 250, 125, 62.5 and 32 µl/L; and Bendazol®, whose active compound is carbendazim, at the concentrations of 1.43 ml/L; 750; 350; 170, 85, 43 and 22 µl/L.

The determination of the toxic effect was made by evaluating the parameters of vegetative growth, using the model of classification of phytosanitary products for the toxicity against entomopathogenic fungi proposed by Alves and Lopes (2008). The vegetative growth

was determined measuring the diameter of the colonies in two orthogonal lines at the surface of the culture medium. The mean diameter of the colonies was considered. All assays were made in triplicate, and the data were submitted to ANOVA. The comparison of the means was made through Tukey's test at the level of significance of 5%.

RESULTS AND DISCUSSION

The data obtained after 336 h of growth of both lineages of *M. anisopliae* var. *anisopliae* in PDA medium, at 28°C, used as controls, are presented in Table 1 and Figure 1, while the data concerning the behavior of the lineages in the presence of the three fungicides are summarized in Table 2. The analysis of the data of Table 1 reveals that the halos obtained with both lineages are similar, because at all times they did not differ at the level of 5%. From these results, it is verified that both lineages had the same behavior in what concerns their growth under the conditions used.

When the data contained in Table 1 and Figure 1 were analyzed, it was noticed that the halos of growth showed similar kinetics, because the following linear equations did not differ: $y = 16.354Ln(x) + 12.067$ and $y = 14.642Ln(x) + 12.195$, for CG-28(AL) and CG-30(E6), respectively. These results suggest that these two lineages of *M. anisopliae* var. *anisopliae* were very similar. The data contained in Table 2 show that both fungal lineages under study had diverse behaviors in the presence of the three fungicides tested, because after 360 h of growth the fungicide Sphere® inhibited the vegetative growth of the fungus *M. anisopliae* var. *anisopliae* lineage CG-28(AL) at the field concentration of 2.0 ml/L and at the dilutions of 1,000 and 500 µl/L. At the dilutions of 250, 125, 62.5 and 32 µl/L there was mycelial growth of the fungus of 5.30±0.06, 9.20±0.10, 9.80±0.07 and 11.00±0.04 mm, respectively.

With the fungicide Nativo® it was observed that the growth of the lineage CG-28 was inhibited at the field concentration of 2.5 ml/L and at the dilutions of 1,250 and 650 µl/L, while at 350, 150, 70 and 40 µl/L there was development of the fungal mycelia, with halos measuring 1.80±0.02, 3.50±0.04, 4.70±0.06 and 7.50±0.04 mm, respectively. It was also noticed that the fungicide Bendazol® did not allow the development of the fungus *M. anisopliae* var. *anisopliae* lineage CG-28(AL) at the

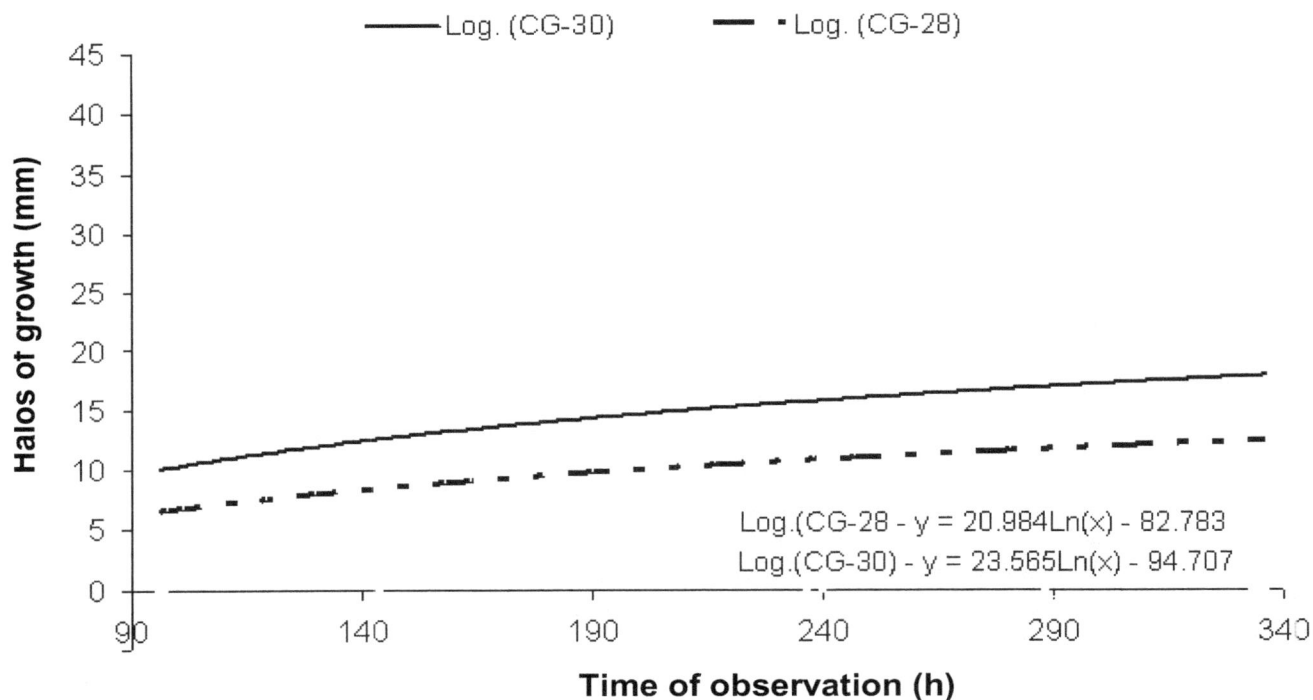

Figure 1. Kinetics of microbial growth of *M. anisopliae* var. *anisopliae*, lineages CG-28(AL) and CG-30(E6), represented by the linear equations y=16.354Ln(x) + 12.067 and y=14.642Ln(x) + 12.195, respectively.

field concentration of 1.4 ml/L, neither at the dilution of 750 µl/L. On the other hand, at the concentrations of 350, 170, 85, 43 and 22 µl/L, there was fungal growth, forming halos of 4.50±0.02, 5.00±0.04, 5.50±0.04, 5.00±0.10 and 5.50±0.06 mm, respectively.

It was observed that the minimum inhibitory concentration (MIC) obtained for the fungicide Sphere® was 500 µl/L; for Nativo® it was 650 µl/L and 750 µl/L for Bendazol®. However, the fungal growth obtained in the presence of the fungicides was lower than that obtained with the control group, which was 41.00±0.22 mm at the same time interval. Therefore, it can be ascertained that the fungicides assayed inhibited the vegetative growth of *M. anisopliae* var. *anisopliae* lineage CG-28(AL). In the presence of the fungicide Sphere® at the concentration of 2.0 ml/L, recommended for field use, the lineage CG-30(E6) of the fungus *M. anisopliae* var. *anisopliae* did not grow. At lower concentrations the fungus did grow, showing halos of 4.80±0.06, 5.30±0.02, 6.50±0.04, 10.10±0.04, 11.00±0.06 and 17.00±0.04 mm for the concentrations of 1,000, 500, 250, 125, 62.5 and 32 µl/L, respectively. Under the action of the fungicide Nativo®, there was no development of the lineage CG-30(E6) at the concentration of 2.5 ml/L, prescribed for field use. At the dilutions of 1,250, 650, 350, 150, 70 and 40 µl/L it was verified that there was fungal growth, with growth

halos of 2.00±0.04, 3.00±0.04, 6.60±0.06, 8.50±0.06, 14.50±0.04 and 17.70±0.02 mm, respectively.

As for the fungicide Bendazol®, it was observed that the lineage CG-30(E6) was also inhibited, because it did not develop at the field concentration of 1.4 ml/L and at 750, 350 and 170 µl/L. Only at the concentrations of 85, 43 and 22 µl/L the fungus developed, showing halos of vegetative growth of 2.00±0.04, 3.20±0.08 and 4.50±0.06, respectively. After the evaluation of the data it can be verified that the MIC of the concentrations assayed was 2.0 ml/L for the fungicide Sphere®, 2.5 ml/L for the fungicide Nativo® and 170 µl/L for the fungicide Bendazol®. It must be stressed that, despite the fungal growth in the presence of the fungicides Sphere®, Nativo® and Bendazol®, the values of the halos of growth are lower than that obtained with the control group, which was 46.20±0.14 mm at the same time interval. Therefore, it can be stated that there is a strong inhibition of the fungal growth of the lineage CG-30(E6) of *M. anisopliae* var. *anisopliae* by these fungicides.

The data obtained in this work are compatible with those of Yañez and France (2010), because in the evaluation of the fungicides azoxystrobin, benomyl, captan, chlorothalonil, fenhexamid, fludioxonil, iprodione and metalaxyl on five lineages of the fungus *M. anisopliae* var. *anisopliae*, it was verified that the

Table 2. Growth of *M. anisopliae* var. *anisopliae*, lineages CG 28(AL) and CG 30(E6) in control culture medium and under the action of three commercial fungicides, after 336 h of growth.

Fungicide sphere – CG-30(E6)						
2.0 ml/L	1,000 µl/L	500 µl/L	250 µl/L	125 µl/L	62.5 µl/L	32 µl/L
NG*	4.80±0.06	5.30±0.02	6.50±0.04	10.10±0.04	11.00±0.06	17.00±0.04
Fungicide bendazol – CG-30(E6)						
1.43 ml/L	750 µl/L	350 µl/L	170 µl/L	85 µl/L	43 µl/L	22 µl/L
NG	NG	NG	NG	2.00±0.04	3.20±0.08	4.50±0.06
Fungicide Nativo – CG-30(E6)						
2.5 ml/L	1,250 µl/L	650 µl/L	350 µl/L	150 µl/L	70 µl/L	40 µl/L
NG	2.00±0.04	3.00±0.04	6.60±0.06	8.50±0.06	14.50±0.04	17.70±0.02
Fungicide sphere – CG-28(AL)						
2.0 ml/L	1,000 µl/L	500 µl/L	250 µl/L	125 µl/L	62.5 µl/L	32 µl/L
NG	NG	NG	5.30±0.06	9.20±0.10	9.80±0.07	11.00±0.04
Fungicide bendazol – CG-28(AL)						
1.43 ml/L	750 µl/L	350 µl/L	170 µl/L	85 µl/L	43 µl/L	22 µl/L
NG	NG	4.50±0.02	5.00±0.04	5.50±0.04	5.00±0.10	5.50±0.06
Fungicide nativo – CG-28(AL)						
2.5 ml/L	1,250 µl/L	650 µl/L	350 µl/L	150 µl/L	70 µl/L	40 µl/L
NG	NG	NG	1.80±0.02	3.50±0.04	4.70±0.06	7.50±0.04

NG – No growth.

fungicides benomyl and fenhexamide did not inhibit the fungal growth, while azoxystrobin and fludioxonil inhibited the growth of the five lineages evaluated. The colonies of *M. anisopliae* that grew in the presence of the fungicides Sphere®, Nativo® e Bendazol® showed marked morphologic modifications when compared with the control colonies, such as the presence of irregular edges, vertical growth, change in the color of the conidia and also circular regions with cotton-like mycelium. Although not the focus of this work, it was noticed that the growing colonies showed reduced production of conidia.

According to Oliveira et al., (2002), the vegetative growth of *M. anisopliae* in treatments containing the agrochemicals Vertimec® (abamectin-based), Savey® (hexythioazox-based) and Rufast® (acrinathrin-based) at the concentrations of 1.0 and 0.5 ml/L did not differ statistically compared with the control treatment. However, the formulations of the agrochemicals Parsec® (amitraz) and Sanmite® (pyridaben) at the field concentration of 1.0 and of 0.5 ml/L differed significantly from the control, causing a reduction of the vegetative growth greater than 62%. It must be pointed out that these results coincide with those obtained in this work, although the active agents are different. The active agents abamectin, hexythioazox and acrinathrin did not inhibit the vegetative growth of the fungus *M. anisopliae* var. *anisopliae*, and thus can be recommended for integrated agricultural programs of pest control.

With the data obtained, it is verified that, from the MIC determined, halos of growth are observed which are much smaller that those of the control group at the same time interval. Therefore, it can be inferred that the three fungicides evaluated were capable of inhibiting the fungal growth of the CG-28(AL) and CG-30(E6) lineages of *M. anisopliae* var. *anisopliae*.

When comparing the mechanism of action of each active agent evaluated, it is verified that the strobirulins has translaminar action, specific for the pathogen, and high risk of resistance when compared with the triazoles (cyproconazole and tebuconazole). Strobirulins interfere with mitochondrial respiration by blocking the electron transfer through the cytochrome bc1 complex, at the Qo site, interfering with ATP production and inhibiting the cell respiration of the fungus (Ghini and Kimati, 2002).

The triazoles act by inhibiting the biosynthesis of ergosterol, a fungal lipid important for the maintenance of the cell membrane of the fungi. With the membrane rupture, there is leakage of ionic solutes and the cell death ensues (Ghini and Kimati, 2002).

The benzimidazoles affect specifically the cell division

by inhibiting the biosynthesis of tubulin, which is a microtubular protein (Kendall et al., 1994). Therefore, the formation of microtubules is impaired, and nuclear division and separation do not occur (Ghini and Kimati, 2002).

The *in vitro* studies have the advantage of exposing the microorganism maximally to the action of the chemical, as it does not take place in the field, where several factors prevent this exposure. In this way, once the safety of a product is checked in the laboratory, it is expected to the selective in the field. On the other hand, the high *in vitro* toxicity of a product does not always point to a high toxicity in the field, but the possibility of occurrence of such damages (Moino and Alves, 1998). It has to be stressed as well that the attempt of standardization of the compatibility tests of entomopathogenic fungi with phytosanitary products is recent and amenable to improvement. Factors such as conidial viability and their pathogenicity in the presence of phytosanitary products should also be taken into account in the choice of the most selective products.

Fungicide toxicity is one of the most limiting factors in the use of the group of chemical control; therefore, in a strategy of joint introduction of these with entomopathogenic fungi, it is suggested that the use of sub-lethal concentrations, or sub-concentrations of fungicides is equal or lower than the MIC, thus reducing the amount of residues released into the environment and foods, the toxic effect of these phytosanitary chemicals against entomopathogens and farm workers during preparation and application.

Conclusion

It is concluded that the fungicides Nativo®, composed of trifloxystrobin + tebuconazol; Sphere®, composed of trifloxystrobin + cyproconazole, and Bendazol®, with carbendazim as active agent, inhibit the growth of the lineages CG-28(AL) and CG-30(E6) de *M. anisopliae* var. *anisopliae* at their field concentrations. For the lineage CG-28 of *M. anisopliae* var. *anisopliae* the MICs for the fungicides were the following: Sphere®: MIC of 500 µl/L; Nativo®: MIC of 650 µl/L and Bendazol®: MIC of 750 µl/L. For the CG-30 lineage of *M. anisopliae* var. *anisopliae* the MICs were as follows: Sphere®: MIC of 2.0 ml/L (concentration indicated for field use); Nativo®: MIC of 2.5 ml/L (concentration indicated for field use); and Bendazol®: MIC of 170 µl/L. Therefore, it is suggested that studies should be carried out in the field to corroborate the *in vitro* results, once the environmental conditions are variable when compared to laboratory tests. Tests for the determination of conidial genesis are also suggested.

ACKNOWLEDGEMENTS

The authors thank EMBRAPA/CENARGEN – Brazilian Enterprise of Agriculture and Livestock Research/ National Center of Genetic Resources, and ESALQ/USP – Superior School of Agriculture Luiz de Queiroz of the University of São Paulo, for the supply of the fungal lineages, and Bayer of Brazil S.A. and Milênia S.A. for the supply of the fungicides.

REFERENCES

Alves SB, Lopes RB (2008). Controle Microbiano de Pragas na América Latina: Avanços e Desafios. FAELQ: Piracicaba – SP, 414pp.
Bayer CropScience Brasil, (2008) Fungicida Sphere. Disponível em: www.bayercropscinece.com.br/site/nossosprodutos/protecaodecultiv osebiotecnologia/detalhedoproduto.fss?produto. Acessado em 26/02/2010.
Fernandes EKK, Keyser CA, Rangel DEN, Foster RN, Roberts DW (2010) CTC medium: A novel dodine-free selective medium for isolating entomopathogenic fungi, especially *Metarhizium acridum* from soil. Biol. Control, 54: 197-205.
Ghini R, Kimati H (2002) Resistência de fungos a fungicidas Jaguariúna: Embrapa Meio Ambiente, p. 78.
Kendall S, Hollomon DW, Ishii H, Heaney SP (1994) Characterization of benzimidazole-resistant strains of *Rhynchosporium secalis*. Pestic. Sci., 40: 175-181.
Milenia - AgroCiência SA (2010) Fungicida Bendazol. Disponível em: www.milenia.com.br. Acessado em: 17 mar 2010.
Moino Jr. A, Alves SB (1998) Efeito de imidacloprid e fipronil sobre *Beauveria bassiana* (Bals.) Vuill. e *Metarhizium anisopliae* (Metsch.) Sorok. e no comportamento de limpeza de *Heterotermes tennuis* (Hagen). An. Soc. Entomol. Brasil, 27: 611-620.
Oliveira RC, Neves PMOJ, Guzzo EC, Alves VS (2002). Compatibilidade de fungos entomopatogênicos com agroquímicos. *Semina: Ciências Agrárias*, 23(2): 211-216.
Onofre SB, Vargas LRB, Rossato M, Barros NM, Boldo JT, Nunes ARF, Azevedo JL (2002). Controle biológico de pragas na agropecuária por meio de fungos entomopatogênicos. In. SERAFINI, L.A., BARROS, N.M., AZEVEDO, J.L. Biotecnologia: avanços na agricultura e na agroindústria. Caxias do Sul: EDUCS, p. 433.
Schrank A, Franceschini M, Guimarães AP, Camassola M, Frazzon AP, Baratto CM, Kogler V, Silva MV, Dutra V, Nakazoto L, Castro L, Santi L, Vainstein MH (2001). Biotecnologia aplicada ao Controle Biológico. O entomopatógeno *Metarhizium anisopliae*. Biotecnologia Ciência & Desenvolvimento, 23: 32-35.
Yañez M, France A (2010). Effects of fungicides on the development of the entomopathogenic fungus *Metarhizium anisopliae* var. *anisopliae*. Chilean J. Agric. Res., 70(3): 390-398.

14

In vitro antimicrobial studies of *Nodulisporium* specie: An endophytic fungus

S. Rehman[1]*, Tariq Mir[2], A. Kour[1], P. H. Qazi[1], P. Sultan[1] and A. S. Shawl[1]

[1]Indian Institute of Integrative Medicine (CSIR), Sanatnagar, Srinagar, India.
[2]Maternity and Child Hospital, Dammam, India

Nodulisporium sp. an endophytic fungus identified by 28s ribosomal gene sequencing isolated from a medicinal plant, *Nothapodytes foetida* was studied for its *in vitro* antimicrobial activity. Dual culture studies were carried out for antifungal activity where maximum antagonistic activity was against *Alternaria alternata* and *Colletotrichum gleosporoides*. For antibacterial studies, Gram positive and Gram negative human pathogens strains were used. The minimum inhibitory concentration (MIC) of ethyl acetate and methanol fractions of *Nodulipsporium* showed appreciable growth inhibition mainly active against disease causing Gram positive bacteria.

Key words: *Nothapodytes foetida,* endophytes, antagonism, antimicrobial, *Nodulisporium,* sp., pathogen, bacteria, minimum inhibitory concentration.

INTRODUCTION

Plants commonly act as hosts to a multitude of microbes including parasites, symbionts, endophytes, epiphytes and mycorrhizal fungi (Fisher and Petrini, 1990). These microorganisms may also influence the production of secondary metabolites. Endophytes, the microorganisms that reside in the intercellular spaces of stems, petioles, roots and leave of plants causing no discernible manifestation by their presence have typically gone unnoticed (Strobel and Long 1998). Our search for studying endophyte is driven by the fact that the contribution of the endophytes to the plant may be to provide protection to it by virtue of anti-microbial compounds that it produces. Some of these endophytes may be of interest to agricultural sciences, since they possess anti-fungal, anti-bacterial, anti-malarial and a host of other biological activities.

Here we report an endophyte from the medicinal plant *Nothapodyte foetida*, which grows widely throughout India including north-western Himalayan region. Two naturally occurring alkaloids, Nothapodytines A and Nothapodytines B have been isolated from the stem of *N.*

foetida having the antimicrobial property as well (Wu et al., 1996). Moreover *N. foetida* is used as a source of anticancer compound, camptothecin. Several hundreds of compounds with antibiotic activity have been isolated from microorganisms over the years (Harrison et al 1991). The phenomenal success of penicillin led to the search for other antibiotic-producing microorganisms. In the present study, isolated endophytic fungus was also found to possess antagonistic activity against few phytopathogens which could be used as potential biocontrol agent in disease management. In addition the endophytic extracts were found active against important Gram positive bacterial pathogens.

MATERIALS AND METHODS

Isolation and Identification of an endophyte

The endophytic fungus was isolated from twigs of *N. foetida* obtained from the Jodia forest of Karnataka. The organism was isolated by using the method described by (Strobel et al., 1996).

Homology modeling

Total genomic fungal DNA was extracted by cetyl trimethyl

*Corresponding author. E-mail: suriyamir@yahoo.com.

Table 1. Five different test plant pathogens used in dual culture method.

Test pathogen	Disease caused
Penicillium citrinum	Leaf spot, fruit rot, cucumber disease, minor foliar disease.
Asperillus niger	Crown rot, black mold, bole rot, canker.
Drechslera tetramera	Small brown spots.
Alternaria alternate	Black rot, leaf spot, potato early blight.
Colletotrichum gleosporoides	Anthracnose, leaf spot, stem spot.

ammonium bromide (CTAB) method. Briefly the endophyte was grown in 100 ml Sabouraud dextrose broth at 28 °C with constant shaking for 5 days. Hundred milligrams of mycelial biomass was taken following washing (two times) with sterile Tris-EDTA (TE) buffer, 6 ml of CTAB extraction buffer and 60 µl of β-mercaptoethanol were added. The mixture was incubated at 65°C for 45 min, and cooled down to room temperature. This was followed by extraction with equal volume of chloroform and centrifuging at 10,000 x g for 10 min. Subsequently, equal volume of isopropanol was added to the supernatant and mixed gently. The obtained DNA pellet was washed with ice cold 70% (v/v) ethanol, vaccum dried and dissolved in 100 µl of TE (pH 8.0).

Small subunit gene sequencing and analysis

The endophytic fungus was identified by the ribosomal gene analysis. The small subunit ribosomal gene was amplified using the D2 LSU Microseq ki (ABI, USA). The amplified products were purified using Microcon columns (Millipore, USA), and sequenced using ABI Prism310 genetic analyzer (ABI, USA) as per the manufacturer's instructions. The DNA sequence 280 bases (GenBank Acc. No. EU284592) was analyzed for homology studies by BLASTN program (Altschul et al., 1997). The ribosomal gene database (http//ncbi.nim.nih.gov) was accessed and sequence alignment was used as an underlying basis to identify the fungus.

Antimicrobial studies

Dual culture method

This experiment was performed by dual culture method. In this method endophytic fungus was studied for antagonism against fungal strains. Five different test plant pathogens were used for the studies which were obtained from National Fungal Research Institute (NFRI) New Delhi Table 1.

Endophyte and all the test pathogens were grown on Sabouraud's agar plates and incubated at 28 °C for seven days. 2 mm dia plugs of newly grown endophyte and test pathogens were taken with the help of cork borer and were aseptically placed 80 mm away from each other on the opposite sides of 90 mm Petri plates containing about 30 ml of fresh Sabouraud's agar medium and incubated at 28°C (Munshi and Dar, 2004). Simultaneously the disc of an endophyte and each test pathogen were placed separately on Sabouraud agar plate which served as control. All the inoculated plates were allowed to grow. Three replicates were used for each pathogen. After few days the plates were observed and growth of an endophyte and test pathogen was measured. The antagonistic activity was analyzed biostatistically. This experiment was carried out thrice.

MIC determination

This method was performed to test endophytic fungus against

human pathogens. The chloroform: methanol extract of mycelia (6 g) was prepared. It was subjected to usual silica gel column chromatography and the elution was carried out with benzene, ethyl acetate, methanol to obtain fractions ZPF-1, ZPF-2, ZPF-3 respectively (Wall and Wani 1977). The extracts were tested for antimicrobial activity against human pathogens using microdilution method defined by Clinical and Laboratory Standards Institute (CLSI) formerly known as National Committee for Clinical Laboratory Standards (NCCLS, 1996). A panel of laboratory standard pathogenic strains were used. Staphylococcus aureus was obtained from the American Type Culture Collection (Manassas, Va.). Methelene resistant S. aureus strains were obtained as a gift sample from Ranbaxy Research laboratories (New Delhi, India). Escherchia coli, Pseudomonas aeruginosa from Indian Institute of Integrative Medicines, Jammu. Ciprofloxacin which was used as reference drug was obtained from Cadila Pharmaceuticals, Gujarat India. These pathogenic strains were maintained on Mueller Hinton agar medium at 37°C in stationary phase and subcultured fortnight.

Stock solution of 20 mgml⁻¹ was prepared in DMSO. The stock solutions were serially diluted to obtain working test solutions with suitable growth medium. The final concentrations ranged from 4000 to 7.8125 µg/ml for test material and from 0.03 to 64 µg/ml for ciprofloxacin.

Bacterial suspensions were prepared from overnight grown cultures in Mueller Hinton agar medium. The turbidity of the suspensions was adjusted to a McFarland no. 1 in sterile normal saline and was further diluted to 1:50 in Mueller Hinton broth.

100 µl of sterilized growth medium was added in the wells of sterile 96-well plates (U-bottom) from columns 2 to 11. In row A 200 µl of 2-fold concentrated reference drug solution was added to the wells in column 1, 100 µl was transferred from column 1 to 2 and serially diluted upto 10 column. Column 11 and 12 containing 100 and 200 µl of medium without drug served as growth and medium control respectively. In other rows, test material was processed by same procedure. 100 µl of bacterial suspension was added to the wells of microplates from column 1 to 11 to achieve a final volume of 200 µl per well. Plates were sealed with parafilm and incubated overnight at 37°C. Visually clear well with no growth at bottom of the well was taken as MIC of that particular drug or extract.

RESULTS

Isolation and Identification of endophyte

N. foetida (family: Olacaceae) was chosen as a source plant for isolating the endophyte, since this plant has been reported to be one of the important medicinal plant. Stem twig of plant that is N. foetida was used for isolation. Isolated Endophyte typically possesses 3 to 4 µm in diameter which spread as white mat on solid media within 5 to 7 days (Figure 1).

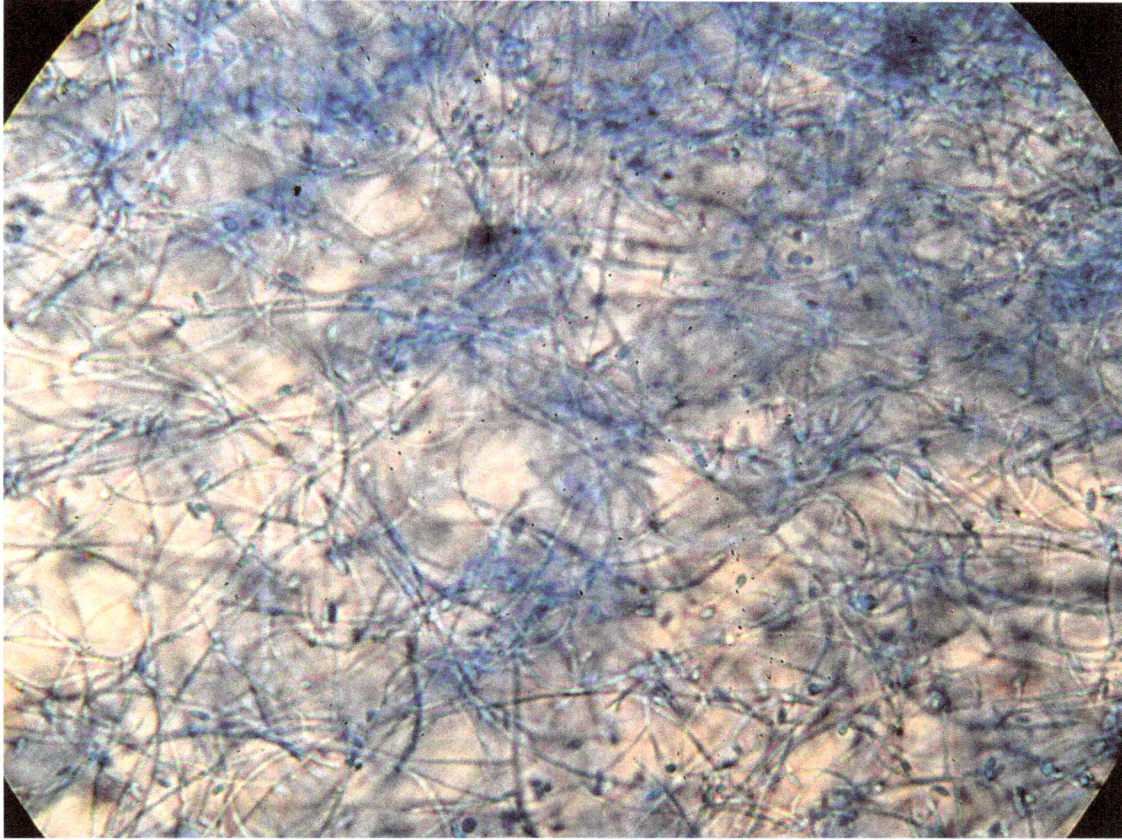

Figure 1. An endophytic fungus from *N. foetida*.

The fungus isolated from the inner bark of *N. foetida* was identified by 28s ribosomal gene sequencing. After 48 h of growth in Sabouraud Dextrose broth (with constant shaking) at 28°C, mycelial biomass could be collected in gram quantities. This collected biomass fraction was further used for DNA isolation, and the pellet obtained using Tris-EDTA buffer was vaccum dried and dissolved in CTAB buffer. Figure 2 shows the distance tree constructed on the basis of homology of 28s ribosomal gene sequence of endophytic strain with close members in GenBank. The isolate showed highest sequence similarity of 95% with *Nodulisporium* sp. CL108.

Antimicrobial studies

Dual culture

µg/ml against Gram positive strains *S. aureus*, MRSA whereas Benzene fraction did not show any activity. MIC of all fraction against Gram negative microorganism was observed as >4000 µg/ml and not very active. Ciprofloxacin was used as standard antibiotic in this study (Table 3).

In dual culture study, growth of endophyte (*Nodulisporium*) covered the entire medium surface and restricted the growth of all test pathogens (Figure 3). Endophyte inhibited all tested pathogens but maximum antagonistic activity was shown against *Alternaria alternata* and *Colletotrichum gleosporoides*, by covering the entire medium surface. The biostatistical analysis revealed that degree of antagonism against *A. alternata* and *C. gleosporoides*, *Asperigillus niger* and *Penicillium citrinum* is almost similar (Table 2).

MIC determination

Antibacterial activity was determined against human pathogens *S. aureus*, MRSA, *E. coli*, *P. aeruginosa*. The MIC of ethyl acetate and methanol fractions showed appreciable growth inhibition with MIC of 125 and 250

DISCUSSION

The endophytic fungus identified as *Nodulisporium* sp. identified by 28s DNA typing was obtained from inner

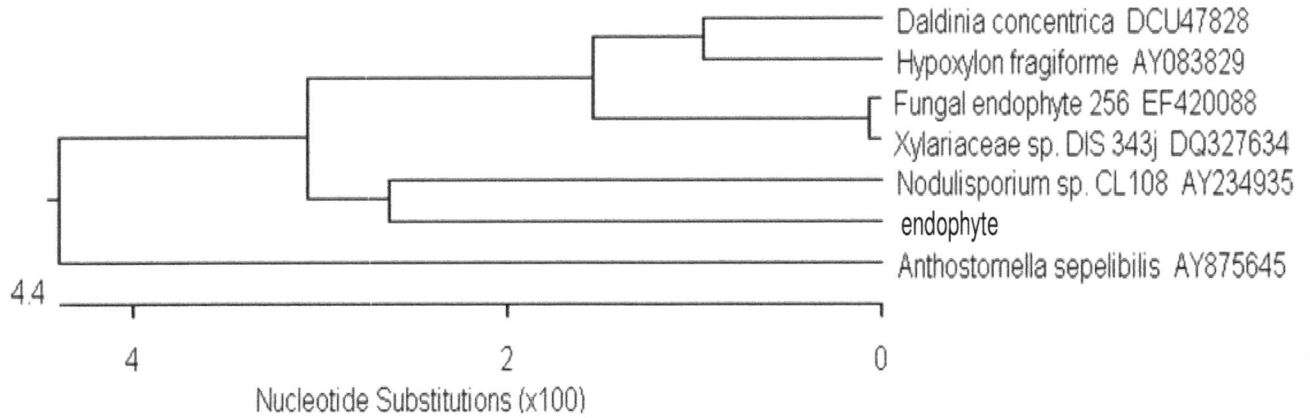

Figure 2. Phylogenetic position of an endophyte.

Table 2. Evaluation of endophyte against different plant pathogens by dual culture method.

Phytopathogens used	Mean growth value of endophyte (mm^2) against pathogens
Penicillium citrinum,	61.33* c
Alternaria alternata	81.66^{NS} a
Colletotrichum gleosporoides	77.66^{NS} a
Drechslera tetramera	68.0** b
Asperigillus niger	61.0* c

(Note- Means followed by similar letter(s) are identical.) NS= Not significant; * = 1%; **= 5%; SEM= 1.77; CD (P=0.05) = 5.57.

bark of *N. foetida* plant collected from Joida forest Karnataka. Until now the only recognized means of controlling plant diseases were to use chemical substances. These methods have attracted huge criticism from environmental groups and thus other means of control have to be investigated. Research on biocontrol through the application of endophytes has the goal of promoting internal fungi from resident to necrotrophic status by stimulating the fungi themselves. Endophytes themselves may also predispose their hosts to environmental damage by reducing the damage threshold. In the present study results of the dual culture test showed that endophytic "*Nodulisporium*" is antagonistic to all the test pathogens. *Nodulisporium* grew rapidly and covered

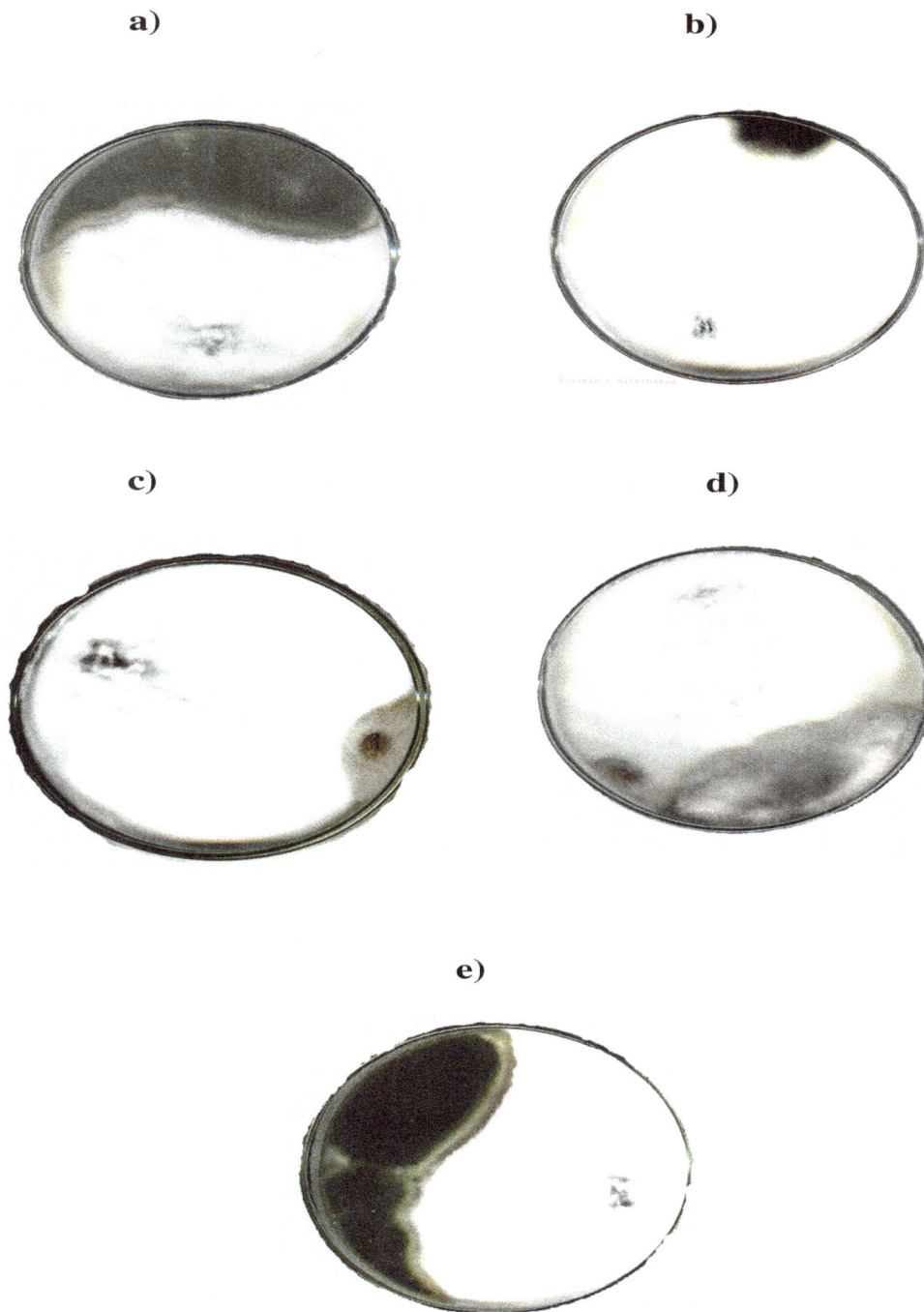

Figure 3. Antagonistic activity of endophyte against (a) *P. citrinum* (b) *A. alternata* (c) *C. gleosporoides* (d) *D. tetramera* (e) *A. niger*

the entire agar surface of Petri dishes after incubation for few days and restricted the growth of all test pathogens but maximum antagonism was shown against *A.alternate* (black rot, leaf spot, potato early blight) and *C. gleosporoides* (anthracnose, leaf spot, stem spot). Therefore, it is interesting to find antagonism in the endophytic fungus isolated from *N. foetida* plant. This may help to

emphasize the use of fungus as an important alternate source of biocontrols. The productions of anti-fungal compounds by endophytes have been previously reported by other workers (Liu et al. 2001).

MIC of endophytic fractions were determined against human pathogens. This sensitivity test was performed by microdilution method. The edge of the zone of inhibition

Table 3. MIC determination of organic extract/fractions of fungal mycelia against few bacterial strains.

Fractions from chloroform: methanol (4:1) extract of mycelia	MIC µg/ml			
	S. aureus	MRSA	E. coli	P. aeruginosa
ZPF-1	>4000	>4000	>4000	>4000
ZPF-2	125	125	>4000	>4000
ZPF-3	250	125	>4000	>4000
Cipro	0.025	8	0.03	0.06

ZPF-1: Benzene; ZPF-2: ethyl acetate; ZPF-3: methanol; Cipro: Ciprofloxacin (reference drug).

correlates with the MIC for that particular bacterium/ antimicrobial combination. The present investigation confirms that there is a moderate degree of antibacterial activity against human pathogens as well particularly against Gram positive bacteria *(S aureus and MRSA)* in the ethyl acetate and methanol fraction of *Nodulisporium* sp.

REFERENCES

Altschul SF, Madden TL, Schaffer AA, Zhang J, Zhang Z, Miller W, Lipman DJ (1997). Gapped BLAST and PSI-BLAST, A new generation of protein data base search programs, Nucl. Acids Res., 25: 3389-3402.

Fisher PJ, Petrini O (1990). A comparative study of fungal endophytes in xylem and bark of *Alnus species* in England and Switzerland. Mycol. Res., 94: 313-319.

Harrison L, Teplow D, Rinaldi M, Strobel GA (1991). Pseudomycins, a family of novel peptides from *Pseudomonas syringae*, possessing broad spectrum antifungal activity. J. Gen. Microbiol., 137: 2857-2865.

Liu CH, Zou WX, Lu H. Tan RX (2001). Antifungal activity of *Artemisia annua* endophyte cultures against phytopathogenic fungi, J. Biotechnol., 88: 277-82.

Munshi NA, Hassan DGH (2004). *Invitro* evaluation of some localy isolated microfungi for antagonism against mulberry Fusarial blight pathogen (*Fusarium pallidoroseum* (cook) SAAC.). J. Seriologia, 44(3): 341-346.

National Committee for Clinical Laboratory Standards. 1996. Methods for determining bactericidal activity of antimicrobial agents. Approved guideline M26-A. National Committee for Clinical Laboratory Standards, Wayne, Pa.USA.

Strobel GA, Yang X, Sears J, Kramer R, Sidhu RS, Hess WM (1996). Taxol from *Pestalotiopsis microspore*, an endophytic fungus cf *Taxus wallachiana*. J Microbiol., 142: 435-142.

Strobel GA, Long DM (1998). Endophytic microbes embocy pharmaceutical potential. ASM News. 64: 263-268.

Wu TS, Leu YL, Hsu HC, Ou L F (1996). Nothpodytine A and B from *Nothapodytes foetida*. Phytochemistry, 42 (3): 907-908.

Wall ME, Wani MC (1977). Antineoplastic agents from plants. Ann. Rev. Pharmocol. Toxicol., 17: 117-132.

15

Comparative analysis of bioethanol production by different strains of immobilized marine yeast

P. Senthilraja*, K. Kathiresan and K. Saravanakumar

Department of Zoology, Faculty of Science, Annamalai University, Annamalai nagar, India.

Yeasts are well known for bioethanol production. However, marine yeasts are less known for the activity. In the present context of increasing demand for energy and biofuel, the microbial synthesis of ethanol using cellulosic waste materials has gained recent importance. The present study deals with the identification of potential marine yeasts for ethanol production. Ten species of marine yeasts were cultured for 24, 48, 72, and 96 h for bioethanol production. Of the ten species, *Candida albicans* exhibited the maximum production of ethanol (47.3±3.1 g/L) within 96 h, when glucose was used as carbon source. The ethanol production by this species was found higher when the yeast cells were immobilized in sodium alginate compared to suspension culture. This experiment was also conducted with both immobilized yeast cells and non-immobilized cells. The experiment revealed that the marine yeast *C. albicans* is efficient in bioethanol production, when it is immobilized.

Key words: Bioethanol, marine yeast species, monoclonal antibodies, thermotolerant.

INTRODUCTION

Immobilization in biotechnology is the technique used for the physical or chemical fixation of cells, organelles, enzymes or other proteins (e.g. monoclonal antibodies) onto a solid support, into a solid matrix or retained by a membrane, in order to increase their stability and make possible their repeated or continued use. Therefore, it is expected that the microenvironment surrounding the immobilized cells is not necessarily the same experienced by their free-cell counterparts. Immobilization of microbial cells in biological processes can occur either as a natural phenomenon or through artificial process. While the attached cells in the natural habitat exhibit significant growth, the artificially immobilized cells are allowed restricted growth. Since the first report of successful application of immobilized cells in industrial applications, several research groups worldwide have attempted whole-cell immobilization as a viable alternative to conventional microbial fermentations. Using immobilized cells, different bioreactor configurations were reported with variable success rate. The study on the physiology of immobilized cells and development of noninvasive measuring techniques have remarkably improved our understanding on microbial metabolism under immobilized state. We have presented an overview of this field.

Saccharomyces cerevisiae was immobilized in Hollow-Fiber Membrane Bioreactors for ethanol production by following the method of Inloes et al. (2008). The ethanol production by free and Ca-alginate immobilized cultures of the thermotolerant yeast was compared. It was found that initial yields produced by the immobilized culture lagged behind those produced by cultures in free suspension. However, in subsequent batch-feed experiments it was demonstrated that the ethanol-producing ability of the immobilized preparation increased with successive feeds, while production by the free suspension reduced significantly (Inloes et al., 2008).

MATERIALS AND METHODS

Marine yeast species

Ten species, *Candida albicans, Candida tropicalis, Debaryomyces hansenii, Geotrichum* sp., *Pichia capsulata, Pichia fermentans, Pichia salicaria, Rhodotorula minuta, Cryptococcus dimennae* and *Yarrowia lipolylica* isolated from mangrove sediments were used in the present study. After identification, screening was done to

*Corresponding author. E-mail: lionbioinfo@gmail.com.

Table 1. Ethanol production in culture filtrates of immobilized marine yeast species.

S/N	Name of marine yeast	Ethanol production by non-immobilised marine yeasts (g\L)	Ethanol production by immobilised marine yeasts (g\L)
1.	Candida albicans	28.12±2.14	47.3±3.1
2.	Candida tropicalis	14.13±1.89	25.00±1.7
3.	Debaryomyces hansenii	18.76±2.65	24.00±0.7
4.	Geotrichum.sp	26.79±3.65	33.00±0.2
5.	Pichia capsulata	9.98±1.23	14.00±1.6
6.	Pichia fermentans	13.98±2.54	22.00±2.0
7.	Pichia salicaria	28.50±4.32	38.00±1.2
8.	Rhodotorula minuta	12.34±3.54	16.00±1.9
9.	Cryptococcus dimennaea	14.32±2.98	24.00±1.6
10.	Yarrowia lipolytica	9.80±1.32	13.00±2.2

identify the potential strain for ethanol production; *C. albicans*, and it sequenced their 18s rDNA and it conformed with *C. albicans* (AC No: Jf292449) with the sequence analysis, submitted to the NCBI.

Immobilization of yeast cells for ethanol production

The calcium alginate gel-entrapping method was used in the present study (Inloes et al., 2008). The spherical gel method was employed for the preparation of calcium alginate gels. This spherical gels can be readily obtained by adding sodium alginate solution to calcium chloride solution using a nozzle. No special granulation apparatus was used at the time of equipment assembling, but a gel-dropping nozzle was provided at the top of the fermenter. The fermenter was filled with a calcium chloride solution and substrate was pretreated as innoculum prior to fermentation, and sodium alginate solution was added drop wise to form granules. The culture medium was then supplied to the fermenter to initiate the fermentation. This procedure was carried out to simplify the gel preparation process.

Production of bioethanol

The production of bioethanol was done using immobilized marine yeast fermentation by following the method outlined by Inloes et al. (2008). 1 ml of the yeast was enriched in yeast malt broth (dextrose-5.0 g, peptone-5.0 g, yeast extract-3.0 g and malt extract-3.0 g in 1000 distilled water added with 50% seawater). The fermentation was carried out in 500 ml Erlenmeyer flasks using 100 ml of medium. It was kept for fermentation at 28°C for 120 h on a shaker at 120 rpm. The level of ethanol in all the flasks was estimated at every 24 h time interval of incubation and effect of pH on bioethanol production was also analyzed.

Estimation of bioethanol by using gas chromatography

Ethanol concentration in the samples was estimated using a Hewlett Packard 5890 Series II gas chromatography with nitrogen as a carrier gas. The temperature of the injection port, oven and detection port were 250,120 and 250°C, respectively. For the analysis, 2 ml of liquid samples was withdrawn from the fermentation broth by using gas tight syringes and then the sample was injected into gas chromatography. The ethanol concentration

was determined by using ethanol standard plot and was expressed in percentage of ethanol.

RESULTS

Determination of ethanol production by immobilized and non-immobilized marine yeasts

Ethanol production in culture filtrates of immobilized and non-immobilized marine yeast species was determined. Results are tabulated in Table 1 which shows the comparison between immobilized and non immobilized yeasts. Maximum ethanol production was obtained from *C. albicans* (47.3±3.1) which was subjected to calcium alginate entrapment immobilization.

Changes in pH during the screening of marine yeasts for alcohol production

The changes in pH of ten species of marine yeasts, which were screened for ethanol production during various time intervals, are depicted in Figure 1. The results revealed that *C. albicans* was showing the lowest pH and the highest ethanol production.

Effect of ethanol production in immobilized yeast and incubation period

Effect of incubation period on immobilization was determined. Maximum ethanol production was obtained in immobilized marine yeast *C. albicans*, maximum bioethanol production was recorded at 96 h of incubation, but there was no significant changes after 96 h incubation (Figure 2).

DISCUSSION

Alcohol is a source of energy used for heating, cooking, lighting and as a motor fuel. Many researches are at progress in finding an alternative fuel through biological

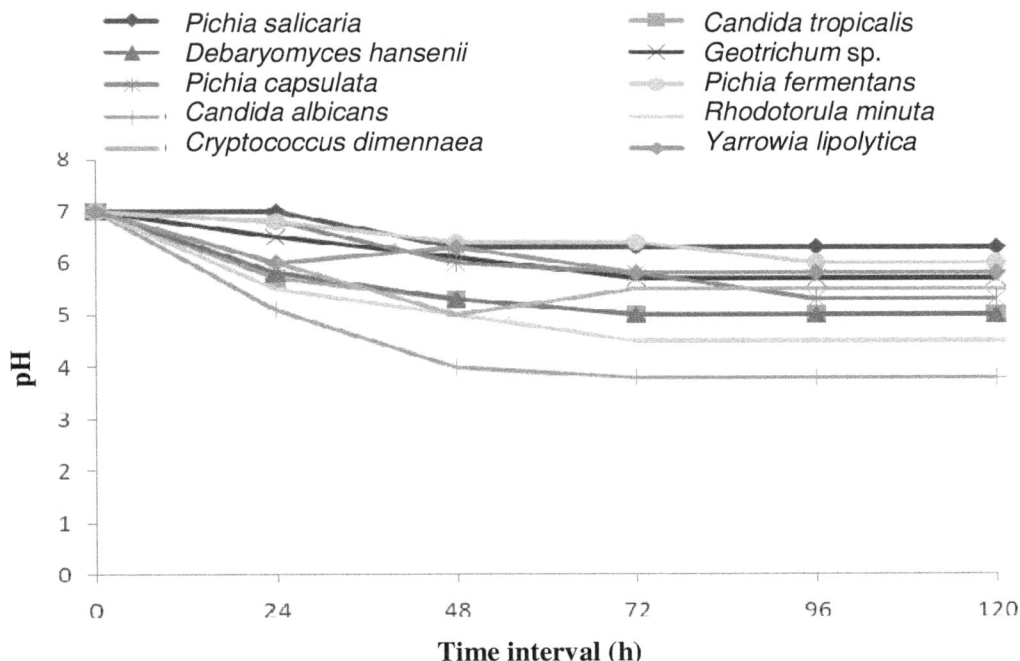

Figure 1. Change in pH of yeast strain cell filtrate.

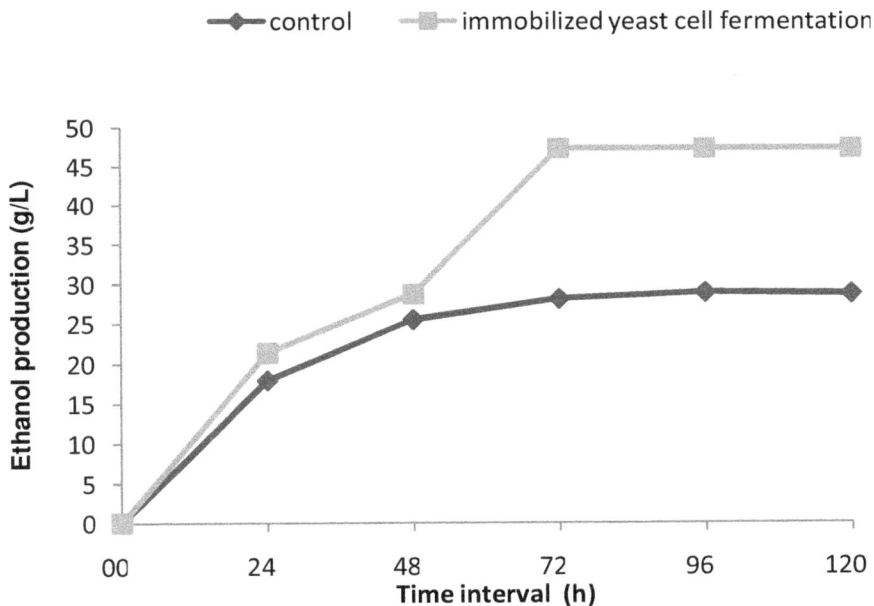

Figure 2. Effect of incubation period on ethanol production by immobilized yeast.

ways. The marine microorganism would be a potential source of alcohol, but they are largely ignored for biofuel studies. Therefore, this investigation was aimed to study the feasibility of utilizing the marine yeast for ethanol production. Among ten species of the marine yeast screened for alcohol production, *C. albicans* showed maximum production of alcohol. It was 28.12±2.14 and 47.3±3.1 g/L under non immobilized and immobilized conditions respectively. There are many studies on bioethanol production from yeast fermentation (Lark et al., 1997; Zhang and Lynd, 2007). Viruthagiri and Sasikumar (2007) produced ethanol using *Trichoderma viride* and thermotolerant yeast *Kluyveromyces marxianus*. *S. cereviseae* is capable of converting only hexose sugars to ethanol. The most promising yeasts that have the ability to use both C5 and C6 sugars are *Pichia stipitis*, *Candida shehatae* and *Pachysolan tannophilus*.

However, ethanol production from sugars derived from starch and sucrose has been commercially dominated by the yeast *S. cereviseae* (Lin and Tanaka, 2006). Thermotolerant yeast could be more suitable for ethanol production at industrial level. In high temperature processes energy savings can be achieved through a reduction in cooling costs. The results of the present study revealed that the bioethanol production varied significantly between the species (P<0.05). The changes in pH of ten species of marine yeasts that were screened for ethanol production during various time intervals revealed that *C. albicans* was showing the lowest pH, hence highest ethanol production. Optimization is an important aspect to be considered in the development of fermentation technology. Sree et al. (2000) reported ethanol production of 3.24% (v/v) at 30°C, whereas at 40°C, it produced 1.92% (v/v) by *Zymomonas mobilis*. Jyothi et al. (2005) reported the ethanol production from *Candida intermedia*. Benschoter and Ingram (1986) reported that *Z. mobilis* showed maximum ethanol production and sugar utilization at 30°C.

In the present study, incubation period of 96 to 120 h did not show alcohol production in any of the yeast strains. The calcium-alginate gel is an efficient matrix for the entrapment of yeast cells. These studies and others demonstrate that *C. albicans* cells maintained about 10.0% viability. Less than 1.0% of the entrapped cells were viable after 1 month. The enzyme systems were responsible for alcohol production from glucose function in the yeast cells for as long as 90 days in the fermentation. Our data strongly suggest that the efficient use of immobilized microorganisms for ethanol production has outstanding potential for making future alternative industrial and domestic fuels.

ACKNOWLEDGMENT

We are thankful to the authority of Annamalai University for providing necessary facilities to carry out this work.

REFERENCES

Inloes DS, Taylor DP, Cohen SN, Michaels AS, Robertson CR (2000). Ethanol production by *Saccharomyces cerevisiae* immobilized in hollow-fiber membrane bioreactors. Appl. Environ. Microbiol., 46: 264-278.

Lark N, Xia Y, Qin C, Gong CS, Tsao GT (1997). Production of ethanol from recycled paper sludge using cellulase and yeast, *Klyuveromyces marxianus*. Biomass and Bioenergy, 12: 135-143.

Zhang YHP, Lynd LR (2007). Cellulose utilization by Clostridium thermocellum:bioenergetics and hydrolysis product assimilation. PNAS, 102: 7321-7325.

Viruthagiri T, Sasikumar E (2007). Bioenergy Research. Optimization of Process conditions Using Response Surface Methodology (RSM) for Ethanol Production from Pretreated Sugarcane Bagasse: Kinet. and Model., 1: 3-4

Lin Y, Tanaka S (2006). Ethanol fermentation from biomass resources:Current state and prospects. Appl. Microbiol. Biotechnol, 69: 627-642.

Sree NK, Sridhar M, Suresh K, Banat IM and Rao LV (2000). Isolation of thermotolerant, smotolerant, flocculating *Saccharomyces cerevisiae* for ethanol production. Biores. Technol., 72: 43-46.

Jyothi AN, Sasikiran K, Nambisan B, Balagopalan C (2005). Optimization of glutamic acid production from starch factory residue using Brevibacterium divaricatum. Process Biochem., 40:3576-3579

Benschoter AS, Ingram LO (1986). Thermal Tolerance of *Zymomonas mobilis*: Temperature-Induced Changes in Membrane Composition. Appl. Environ. Microbiol., 51: 1278-1284.

16

Benzo[a]pyrene degradation by soil filamentous fungi

M. Cristina Romero[1]*, **M. Inés Urrutia**[2], **H. Enso Reinoso**[1] **and M. Moreno Kiernan**[3]

[1]Facultad de Ciencias Veterinarias, calle 60 y 119, s/nº, Universidad Nacional de La Plata, 1900 La Plata, Argentina.
[2]Facultad de Ciencias Agrariasy Forestales, calle 60 y 119, s/nº, Universidad Nacional de La Plata, 1900 La Plata, Argentina.
[3]Minist. Salud. Prov. B.A., calle 60 y 119, s/nº, Universidad Nacional de La Plata, 1900 La Plata, Argentina.

The fungal ability to biotransform xenobiotics had received attention due to their dominance, ubiquity and different pathways that detoxificate aromatic hydrocarbons. The filamentous fungi *Aspergillus flavus* and *Paecilomyces farinosus* showed a significant degradation activity on benzo[a]pyrene with and without C_{16} as cosubstrate. $^{14}CO_2$, ^{14}C-volatile organic, ^{14}C-extractable, ^{14}C-nonextractable, ^{14}C-biomass and ^{14}C-aqueous fractions were determined with [7, 10]^{14}C-BaP assays, with *A. flavus, Cladosporium cladosporioides, Gliocladium viride, P. farinosus* and *Talaromyces rotundus*. However, the activity of *A. flavus* and *P. farinosus* were higher. These non-ligninolytic fungi degraded BaP by cometabolism in C_{16} presence, were adapted to toxicants and dominant in polluted habitats, so they could play an important role in self- bioremediation processes.

Key words: Benzo[a]pyrene biodegradation, cometabolism, mycoremediation, PAHs mixture, polluted sites, soil filamentous fungi.

INTRODUCTION

Polycyclic aromatic hydrocarbons (PAHs) and their derivatives are widespread products of incomplete combustion of organic materials arising from natural combustion such as forest fires and volcanic eruptions (Da Silva et al., 2003; Pang et al., 2003). The major PAH pollutions are industrial activities, transportation, refuse burning, gasification and plastic waste incineration (Mrozik et al., 2003). Benzo[a]pyrene (BaP), 5-ring hydrocarbon, is formed during pyrolysis of organics being a petroleum, coal tar and fuels-oil component (Kanaly, 1997). Due to its chemical structure is highly recalcitrant and resistant to microbial degradation (Sutherland et al., 1995; Sack et al., 1997).

The fungal PAHs-degradation is an effective strategy to remove pollutants from the environment by bioremediation (Lowborn and Ekwonu, 2009). Diverse ligninolytic fungi had been confirmed as BaP degradersjavascript: popRef ('end-a1') (Bogan and Lamar, 1996; Kotterman et al., 1998; Pointing, 2001; Zheng and Obbard, 2002). These fungi did not compete for prolonged time and were not frequents species in heavy polluted habitats, so, their contribution to BaP detoxification was limited (Steffen, 2002; Tortella et al., 2005).

The knowledge about non-ligninolytic fungi is scarce, although the biomass and diversity of SFF were higher in contaminated sites (Sack and Gunther, 1993; Romero et al., 2001). Therefore, the aims of this study were to assess the ability of SFF isolated from industrial polluted sediments to metabolize BaP and to evaluate the incorporation into biomass, $^{14}CO_2$, extractable metabolites, nonextractable, volatile organic and aqueous phases.

*Corresponding author. E-mail: cmriar@yahoo.com.ar.

Abbreviations: BaP, Benzo[a]pyrene; **BM,** basal medium; **C₁₆,** *n*-hexadecane; **DMF,** N,N-dimethylformamide; **PAHs,** polycyclic aromatic hydrocarbons; **SFF,** soil filamentous fungi.

MATERIALS AND METHODS

Sampling area, isolation and identification of BaP degrading fungi

SFF were isolated from contaminated sediments of the industrial area, near an oil refinery, La Plata, Argentina; the sediments features were previously published (Romero et al., 1998). The isolate methodology, the basal medium (BM) and culture conditions were described by Massaccesi et al. (2002). The SFF were

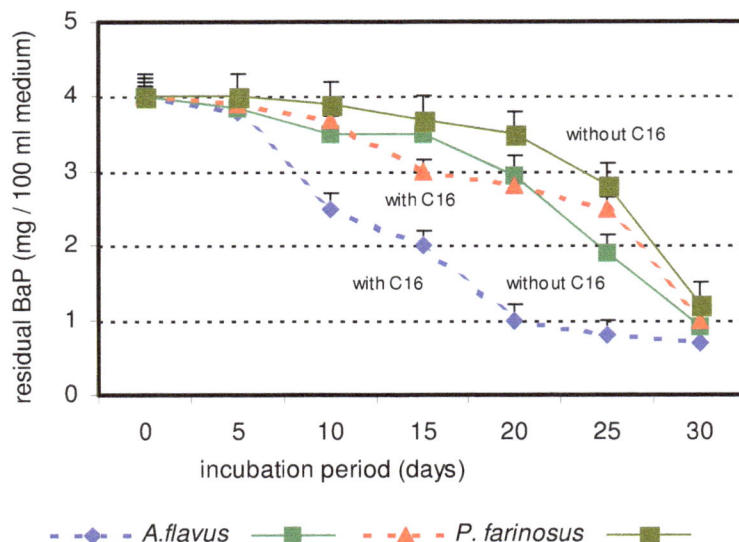

Figure 1. Residual BaP in *A. flavus* and *P. farinosus* cultures with and without C_{16} during the incubation period.

cultured in BM plus 100 µl of 40 mg BaP/2.5 ml N, N-dimethylformamide (DMF) solution, aseptically added on the agar-plate, pH 6.5.

The sporularing isolates were identified by different culture media according to Hanschke and Schauer (1996) and Romero et al. (2005). Non sporulating strains were rejected for further studies as in bio-remediation technologies are better to know the assayed species.

Culture conditions

Fungi were incubated in 100 ml flask with 20 ml BM plus 100 µl of 40 mg BaP/ l stock solution: (1) with 0.5% *n*-hexadecane (C_{16}) or (2) without C_{16} as cosubstrate, pH 6.5 and inoculated with 1 cm^2 plugs of well fungi developed on BaP-agar. The remaining BaP were deter-mined by extracting with 200 ml ethyl acetate at 5, 10, 15, 20 and 30 incubation days. The HPLC methodology was described by Romero et al. (2002). Benzo[*a*]pyrene was identified by its retention time (36.8 min), with a detection limit at 1 µg/ml solvent extract. The BaP-peak identity was confirmed by comparison with an authentic standard, by mass spectrometric identification of the molecular ion (m/z 252) and characteristic fragments (m/z 126) carried at Mass Spectrometry Center, UBA. BaP identification was also controlled with a calibration curve with different BaP levels in DMF (SD 1.5%).

Mass balance analysis

These assays were performed with 0.1 µCi of [7, 10^{14}C]-BaP (56 mCi./mmol specific radioactivity, radiochemical purity 98%) added to each fungal culture and the ^{14}C in the volatile organic substances, solvent extractable metabolites, nonextractable fractions, biomass and $^{14}CO_2$ fractions were obtained (Romero et al., 2002). The fungal growth was measured as mycelia dry weight, after vacuum filtration and dried at 90°C for 24 h.

Three controls were implemented; an inoculated and sterilized flask with BaP, a non-inoculated one and a third culture without BaP. The first and second controls showed the abiotic processes

(Pal et al., 1994) and the third one controlled the contamination of the cultures. All the assays were done by triplicate and expressed at mean values.

CHEMICALS

[7, 10-^{14}C]-BaP (50 mCi./mmol specific radioactivity, radiochemical purity 97.1%) was purchased from Marshal Buchler (Braunschweig, Germany); BaP was obtained from Aldrich (Steinheim, Germany); DMF and chemicals were reagent grade and of the highest purity available from Merck (Darmstadt, Germany) and Fluka Chemie AG (Neu Ulm, Germany). The scintillation cocktail Optic-Fluor and Carbo Sorb were purchased from Packard (Meriden, Conn., USA) and Quickszint 212 from Zinsser Analytik (Frankfurt/Main, Germany).

RESULTS

Soil filamentous species were isolated and selected from chronical polluted sediments of an oil industrial area; *Aspergillus flavus*, *Cladosporium cladosporioides*, *Gliocladium viride*, *Paecilomyces farinosus* and *Talaromyces rotundus* were isolated on the basis of their prevailing growth on subsequent plantings on BaP-plates. But *A. flavus* and *P. farinosus* were only selected and assayed as BaP-degraders, due to its higher rates and increased activity after the 10th and 15th incubation day (Figure 1). Without C_{16}, BaP levels declined at 15th - 23rd day, the cosubstrate increased the SFF abilities. *A. flavus* BaP-uptake rate increased early respect to the *P. farinosus* ones; however, these differences diminished at the end of the experiment. The BaP consumed was 92.5 - 77.5% and 75.0 - 70.0% of the initial concentration, 4 mg BaP/100 ml medium, with or without cosubstrate, respectively.

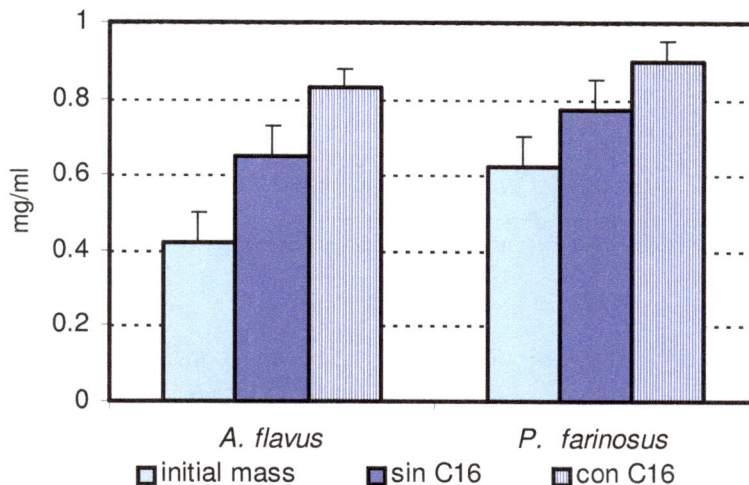

Figure 2. Biomass increase of *A. flavus* and *P. farinosus* with and without cosubstrate (C_{16}) in the BaP assays.

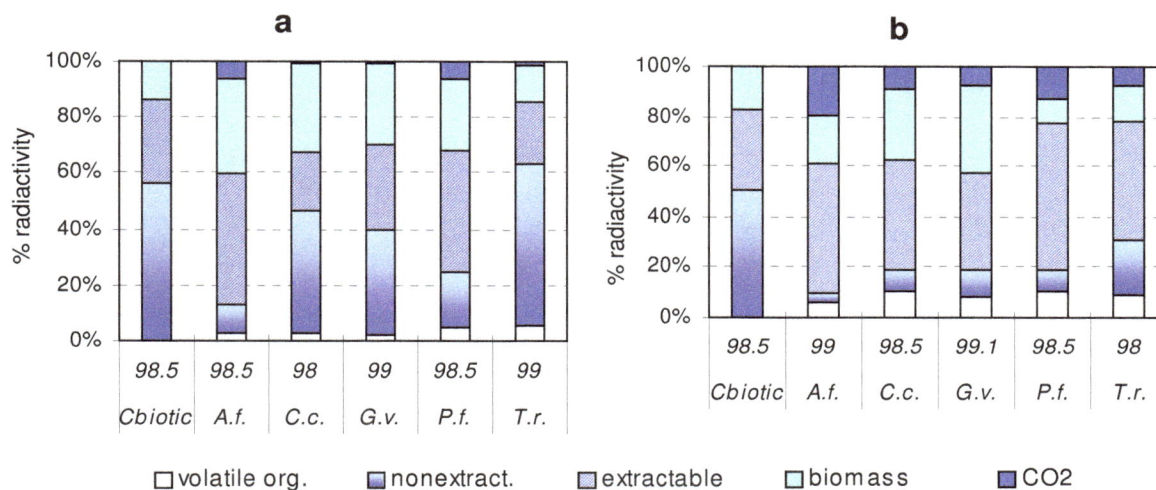

Figure 3. Balance activities among different fractions with ^{14}C (a) without C_{16}, (b) with C_{16} (C: biotic control; *A.f. A. flavus, C.c. C. cladosporioides, G.v. G. viride, P.f. P. farinosus* and *T.r. T. rotundus*).

A. flavus and *P. farinosus* initial biomass were 0.42 and 0.62 mg/ml, this parameter increased to 0.65 - 0.83 and to 0.77 - 0.90 mg/ml with and without C_{16}; therefore, the increment were 97 - 54% and 45 - 24% (Figure 2). The cosubstrate enhanced the mass production and BaP-uptake as C-source.

Although, *G. viride, T. rotundus* and *C. cladosporioides* grew on BaP-plates; they showed significantly less activity in relation to *A. flavus* and *P. farinosus* (Figure 3). Similar patterns were obtained between extractable and mineralization rates, being greater in C_{16} presence. If any BaP metabolites were produced during the incubation period they should be contained in this fraction, and volatile organics increased in all flasks.

^{14}C-biomass counts were a significant recovery fraction, although combusted mycelia could be due to unreacted BaP associated with fungal mass; so, this parameter was rather ambiguous to explain. Nonextract-able parts represented the ^{14}C-BaP solubilized in BM without biological reactions; so the BM had BaP enough to go on the hydrocarbon uptake.

The higher values of the C_{16}-fractions confirmed that these two species mineralized BaP by cometabolism (Figure 3b). Cultures of single species oxidized BaP but not mineralized data had been reported, so this ring cleavage intermediates were produced cometabolically in the presence of other PAHs. To date, BaP had not been reported to support microbial growth (Cerniglia, 1993),

Table 1. Distribution of ^{14}C-BaP in the abiotic controls[a].

Day	Extractable fraction	Irreversibly bound	Reversibly bound	% recovery
0	94.3	1.2	2.5	98.0
10	92.7	2.6	1.7	97.0
20	92.7	3.6	1.2	97.5
30	91.8	3.7	1.0	96.2

[a] Averages of the triplicate measurements expressed in percentages (%) of the total ^{14}C-BaP added (4 mgBaP).

therefore a time-dependent enrichment of BaP degraders is necessary to biotransform this PAH.

The c.a. 100% recovery of the originally ^{14}C-BaP added in both biotic and abiotic controls and triplicates indicated the general validity of the approach. Day-0 ^{14}C-binding in the biotic assays consisted of 7% tie but 6 - 5 - 5% were reversible binding, so only a 1 - 1.5% remained fixed in the biomass. In contrast, the abiotic controls showed 7.5% strong tie and 2.7% irreversible one. These values increased during the assays, probably due to the increment of dead mass particles (Table 1). The BaP amount did not change in non-inoculated controls, and a negligible BaP-portion (< 1%) adsorbed to the autoclaved media.

DISCUSSION

Based on our findings and previous reports, increasing biodegradation of xenobiotics by cometabolisms had been confirmed, even though these SFF lack ligninolytic activities and few of them were able to utilize the compounds as sole source of carbon and energy (Da Silva, 2003; Da Silva et al., 2004; Johnsen, 2005). BaP metabolism had been shown for several molds, deuteromycetes and zygomycetes. *Aspergillus ochraceae* (Datta and Samanta, 1988) and *Penicillium* spp. hydroxylated BaP via cytochrome P-450-dependent mono-oxygenases (Kapoor and Lin, 1984; Launen et al., 1995).

Other authors evaluated the PAHs-transformation by natural communities with chronical pollution and this systems degraded mixed PAHs (Oleszczuk and Baran, 2003; Owabor, 2007). In contaminated areas, the presence of different pollutants was the real situation, so, a natural simple organic could be the cosubstrate that contributed to detoxificate high condensed aromatic hydrocarbons by the autochthonous micro flora (Tortella, 2005; Najafi and Hajinezhad, 2008).

Therefore, our results showed that *A. flavus, C. cladosporioides, G. viride, P. farinosus* and *T. rotundus*, non-ligninolytic fungi, degraded benzo[a]pyrene by cometabolisms. *A. flavus* and *P. farinosus* were adapted to mixed hydrocarbons and could be used in bioremedia-tion techniques. Even more, these SFF are active sure enough in natural decontamination and could play an important role for self- bioremediation processes.

In conclusion, fungal degradation of xenobiotics is looked upon as an effective method to remove chronic pollutants by remediation strategies. This study confirmed that a wide variety of SFF without ligninolytic enzymes, are capable of degrading an equally wide range of PAHs via cometabolism reactions.

ACKNOWLEDGEMENTS

This work was supported by grants from the National Council of Scientific and Technological Research - CONICET and from National University of La Plata, Argentina.

REFERENCES

Bogan BW, Lamar RT (1996). Polycyclic aromatic hydrocarbons capabilities of *Phaenerochaete laevis* HHB-1625 and its extracellular ligninolytic enzymes. Appl. Environ. Microbiol. 62: 1597 -1603.
Cerniglia CE (1993). Biodegradation of polycyclic aromatic hydrocarbons. Biodegradation 3: 351-368.
Da Silva M, Cerniglia CE, Pothuluri JV, Canhos VP, Espoisto E (2003). Screening filamentous fungi isolated from estuarine sediments for ability to oxidize polycyclic aromatic hydrocarbons. World J Microbiol. Biotechnol. 19: 399-405.
Da Silva M, Esposito E, Moody JD, Canhos VP, Cerniglia CE (2004). Metabolism of aromatic hydrocarbons by the filamentous fungus *Cyclothyrium* sp. Chemosphere 57: 943-952.
Datta D, Samanta TB (1998). Effect of inducers on metabolism of benzo[a]pyrene *in-vivo* and *in-vitro*: analysis by high pressure liquid chromatography. Biochem. Biophys. Res. Chemosp. 28: 683-692.
Haemmerli SD, Leisola MSA, Sanglard D, Fiechter A (1986). Oxidation of benzo[a]pyrene by extracellular ligninases of *Phanerochaete chrysosporium*. J. Biol. Chem. 261: 16 948 - 16 952.
Hanschke R, Schauer F (1996). Improved ultrastructural preservation of yeast cells for scanning electron microscopy. J. Micros. 184: 81-87.
Johnsen AR, Wick LY, Harms H (2005). Principles of microbial degradation in soil. Environ. Pollut. 133: 71-84.
Kanaly R, Bartha R, Fogel S, Findlay M (1997). Biodegradation of [^{14}C]benzo[a]pyrene added in crude oil to uncontaminated soil. Appl. Environ. Microbiol. 63: 4511-4515.
Kapoor M, Lin WS (1984). Studied on the induction of aryl hydrocarbon (benzo[a]pyrene) hydrolase in *Neurospora crassa*, and its suppression by sodium selenite. Xenobiotica 14: 903-915.
Kotterman MJ, Vis EH, Field JA (1998). Successive mineralization and detoxification of benzo[a]pyrene by the white-rot fungus *Bjerkandera* spp strain BOS55 and indigenous microflora. Appl. Environ. Microbiol. 64: 2853-2858.
Launen L, Pinto LJ, Wiebe C, Kiehlmann E, Moore MM (1995) The oxidation of pyrene and benzo[a]pyrene by nonbasidiomycete soil fungi. Canad. J. Microbiol. 41: 477-488.
Massaccesi G, Romero MC, Cazau MC, Bucsinszky AM (2002). Cadmium removal capacities of filamentous soil fungi isolated from

ndustrially polluted sediments in La Plata (Argentina). World J. Microbiol. Biotech. 18: 817-820.

Mrozik A, Piotrowska-seget Z, Labuek S (2003). Bacterial degradation and bioremediation of polycyclic aromatic hydrocarbons. Polish J. Environ. Stud. 12: 15-25.

Najafi SH, Hajinezhad H (2008). Solving one-dimensional advection-dispersion with reaction using some finite-difference methods. Appl. Math. Sci. 2: 2611-2618.

Oleszczuk P, Baran S (2003). Degradation of individual polycyclic aromatic hydrocarbons (PAHs) in soil polluted with aircraft fuel. Polish J. Environ. Stud. 12: 431-437.

Owabor CN (2007). Substrate bioavailability and biodegradation of naph-thalene, anthracene and pyrene on contaminated aqueous soil matrix Ph. D Thesis, University of Lagos. Nigeria.

Pal S, Bollag JM, Huang PM (1994). Role of abiotic and biotic catalysis in the transformation of phenolic compounds through oxidative coupling reactions. Soil Biol. Biochem. 26: 813-820.

Pang L, Goltz M, Close M (2003). Application of the method of temporal moments to interpret solute transport with sorption and degradation. J. Contam. Hydrol. 60: 123-134.

Pointing SB (2001). Feasibility of bioremediation by white-rot fungi. Appl. Microbiol. Biotech. 57: 20-33.

Romero MC, Cazau MC, Giorgieri S, Arambarri AM (1998). Phenanthrene degradation by microorganisms isolated from a contaminated stream. Environ. Pollut. 101: 355-359.

Romero MC, Hammer E, Cazau MC, Arambarri AM (2001). Selection of autochthonous yeast strains able to degrade biphenyl. World J. Microbiol. Biotech. 17: 591-594.

Romero MC, Hammer E, Hanschke R, Arambarri AM, Schauer F (2005). Biotransformation of biphenyl by filamentous fungus Talaromyces helicus. World J. Microbiol. Biotech. 21: 101-106.

Romero MC, Salvioli ML, Cazau MC, Arambarri AM (2002). Pyrene degradation by filamentous soil fungi and yeast species. Environ. Pollut. 117: 159 - 163.

Sack U, Günther T (1993). Metabolism of PAH by fungi and correlation with extracellular enzymatic activities. J. Basic Microbiol. 33, 1-9.

Sack U, Heinze TM, Deck J, Cerniglia CE, Martens R, Zadrazil F, Fritsche W (1997). Comparison of phenantrene and pyrene degradation by different wood-decaying fungi. Appl. Environ. Microbiol. 63: 3919-3925.

Steffen KT, Hatakka A, Hofrichter M (2002). Degradation of humic acids by the litter-decomposing basidiomycete Collybia dryophila. Appl. Environ. Microbiol. 68: 3442-3448.

Sutherland JB, Rafii F, Khan AA, Cerniglia CE (1995). Mechanisms of polycyclic aromatic hydrocarbon degradation. In: Microbial transfor-mation and degradation of toxic organic chemicals L Young and CE Cerniglia (eds). Wiley-Liss., New York, N.Y. pp 269-306.

Tortella GR Diez MC, Durán N ((2005). Fungal diversity and use in decomposition of environmental pollutants. Crit. Rev. Microbiol. 31: 197-212.

Zheng ZM, Obbard JP (2002). Oxidation of polycyclic aromatic hydrocarbons (PAH) by the white rot fungus Phanerochaete chrysosporium. Enzym. Microb. Technol. 31: 3-9.

17

Studies on biodiversity of some mushrooms collected in Lagos State, Nigeria using biotechnological methods

Bankole, P. O.*, and Adekunle, A. A.

Department of Botany, University of Lagos, Akoka, Lagos State.

The biodiversity of mushrooms in Lagos State, Nigeria was studied using modern biotechnological method of DNA sequence analyses. Sixteen mushrooms were collected in Lagos State for 12 months. The mushrooms collected include: *Agaricus campestris, Coprinus comatus, Daldinia concetrica, Ganoderma adspersum, Ganoderma applanatum, Ganoderma lucidum, Mycena haematopus, Mycena* sp., *Pleurotus ostreatus, Pleurotus tuber-regium, Polyporus* sp., *Polyporus squamosus, Polyporus sulphureus, Trametes versicolor, Xylaria polymorpha,* and *Xylaria* sp. Subsequently, eight different pure mushroom mycelia were isolated using potato dextrose agar medium (PDA) from sterile tissues of *A. campestris, C. comatus, G. lucidum, P. ostreatus, P. tuber-regium, P. sulphureus, X. polymorpha,* and *Xylaria* sp. A mycelia spawn of each of the isolated mushrooms was made and kept in duplicates for DNA sequence analyses. The DNA sequence analyses of each of the eight mushroom mycelia pure isolates were carried out. No similarity was observed when a BLAST query of the mushroom DNA sequences was carried out to determine the similarity with the GenBank database previous submissions. The sequences were then submitted to the GenBank database for reference purpose. Comprehensive description given by the GenBank provides a detailed, reliable and accurate identification than visual characteristics and phenotypic properties of the mushrooms.

Key words: Biodiversity, mushroom, deoxyribonucleic acid (DNA) sequence, phylogenetic tree.

INTRODUCTION

Mushrooms are abundant in various parts of Lagos environs. While some grow in singleton like the *Ganoderma lucidum*, others grow in clusters like *Coprinus comatus and Mycena haematopus* and sometimes in layers like *Trametes versicolor*. The mushrooms showed to a large extent diverse colour variations especially of their pileus ranging from cloud/milk white; *Pleurotus ostreatus* to yellow; *Polyporus sulphureus* to red *Mycena haematopus* and some possess two or more colours; *G. lucidum*. Some grow stipes; *Agaricus campestris* while others do not; *Ganoderma adspersum* and *Daldinia concentrica*. Most of the studied mushrooms grow in humid conditions on dead logs and trees like *G. lucidum, G. adspersum, and M. haematopus* and in the soil like *A. campestris*. Basically, a mushroom is the fleshy, spore-bearing fruiting body of a fungus, typically produced above ground on soil or on its food source (substratum). Like all fungi, mushrooms are not plants and do not undergo photosynthesis. Some of the mushrooms grow fleshy fruiting bodies; *A. campestris, C. comatus*, and *Mycena* spp. while others grow woody or leathery fruiting bodies; *G. lucidum, T. versicolor* and *Xylaria polymorpha*. They showed varying growth patterns ranging from umbrella-like to ball-like and semi-circular forms.

Deoxyribonucleic acid (DNA) is the component containing the blueprint that specifies the biological development and composition of every form of life with no exception of mushrooms. In living organisms, DNA does not usually exist as a single molecule, but instead as a

*Corresponding author. E-mail: pbank54@yahoo.co.uk.

pair of molecules that are held tightly together (Watson and Crick, 1953; Berg et al., 2002). These two long strands entwine like vines, in the shape of a double helix. The nucleotide repeats contain both the segment of the backbone of the molecule, which holds the chain together, and a nucleobase, which interacts with the other DNA strand in the helix. A nucleobase linked to a sugar is called a nucleoside and a base linked to a sugar and one or more phosphate groups is called a nucleotide. Polymers comprising multiple linked nucleotides (as in DNA) are called a polynucleotide.

The backbone of the DNA strand is made from alternating phosphate and sugar residues (Ghosh and Bansal, 2003). One major difference between DNA and RNA is the sugar, with the 2-deoxyribose in DNA being replaced by the alternative pentose sugar ribose in RNA (Berg et al., 2002). DNA sequencing of mushroom is the process of determining the exact order of the nucleotide bases in them; Adenine (A), Thymine (T), Cytosine (C) and Guanine (G) in a piece of DNA oligonucleotide. Generating a DNA sequence for a mushroom is to determine the patterns and make up genetic traits, diversity and growth forms.

Fungal genomes contain 50 to 150 identical copies of the ribosomal DNA (rDNA), a higher copy number than in bacterial genomes. Until now, phylogenetic typing of fungi from both single cultures and mixed populations involved DNA extraction, amplification of regions of rDNA or internal transcribed sequences (ITS), and purification (or cloning) of the polymerase chain reaction (PCR) product.

Aim of the research work

The most correct and accurate method of identifying an organism is through DNA sequence analyses. The need to collect diverse mushrooms from Lagos State, Nigeria and study their biodiversity through DNA sequence analyses was the pulling force behind this research. Therefore, the aim of this research is to deposit and document the Nigerian mushroom genomic DNA sequence data in the GenBank for reference purpose.

MATERIALS AND METHODS

Sources of mushrooms

Sixteen (16) fresh and healthy mushroom samples were collected for 12 months in Lagos State ranging from Araga-poka in Epe, Lagoon front, gate and Staff Quarters all in University of Lagos (Main Campus), Owodunni suburbs of Alagbado, Ijede in Ikorodu and old Ota road in Ile-Epo. They were all collected from January to December, 2011 (Table 1).

Mushroom collection

The mushrooms were collected for twelve months; both on ground

and on substratum such as trees and decaying tree logs. Their photograph and the Global positioning system (GPS) caption were taken with a Nokia device *in situ* before collection. Detailed notes of their morphological and ecological features were also recorded. A sharp sterilized knife was used to collect the whole and parts of mushrooms growing on trees and on the ground. Mushroom samples collected were free of infection or insect attack.

The earth was carefully removed from the lowest part of the stipe with knife or through dusting so as to save the fresh samples from getting too dirty. Care was also taken not to remove or damage any part of the mushroom sample. The non-fleshy mushrooms were collected, wrapped with tissue paper and kept inside a sterilized polythene bag. The heaviest and the lightest were placed on the bottom and top, respectively of the polythene bag in order to prevent damage of the samples.

Identification of the mushrooms

The mushrooms were identified based on their morphological, macroscopic, physiological and ecological features according to a previously published guide (Wei, 1979). Further identification was based on their structure, substratum they were attached to, spore growth, colour, shape, and mycology textbooks.

Mycelia isolation

Of all the sixteen mushroom samples, eight mushroom mycelia were isolated through random selection. A fresh, healthy part of each of the mushroom was cleaned with absolute alcohol to remove dirt and placed on a sterilized workbench in the ultra violet room. The mushroom was split into half longitudinally and some inside tissues at the upper part of the stipe were taken with the aid of forceps. The tissues were then sterilized in a 40% sodium hypochlorite solution after which they were rinsed thrice in separate beakers with distilled water. This is done to remove bacteria or fungal spores that might be on the mushroom tissues. The tissues were then seeded in two's with the use of a flamed forceps and inoculating needle on each Petri dish with previously prepared potato dextrose agar (PDA). Afterwards, the plates were dated and incubated at a temperature of 29 to 32°C for four days of full growth.

The mycelium growth culture was aseptically subcultured into freshly prepared PDA plates and incubated until the mycelium begins to grow. This was then followed by subsequent sub culturing for three times to ensure pure mycelia isolate. Some part of the mycelia isolated was then aseptically transferred into two sterile agar slants which had previously been prepared in McCartney bottles. The bottles were then incubated at 29 to 32°C till full mycelia growth is observed which serves as the stock culture. This procedure was repeated for all the eight mushroom mycelia isolated.

DNA extraction and analyses

The DNA extraction and analysis was carried out at the Biotechnology Centre of the Federal University of Agriculture, Abeokuta (FUNAAB), Ogun State, Nigeria. Total genomic DNA was extracted with cetyl trimethyl ammonium bromide (CTAB) buffer as described by Graham et al. (1994).

Procedure for DNA extraction

Some of the isolated mycelia of respective mushrooms were taken from the incubated stock in the McCartney bottles and transferred

Table 1. Geographical points of different locations in Lagos State, Nigeria where the mushrooms were collected.

Location	Points	Mushroom samples	Latitude	Longitude	Altitude (Metres)	Altitude accuracy (Metres)	Accuracy (Metres)
Lagoon Front, University of Lagos	1	*Xylaria polymorpha*	6°31′06.08″N	3°24′06.42″E	19	55	25
	2	*Daldinia concentica*	6°31′07.04″N	3°24′06.21″E	44	150	38
	3	*Pleurotus ostrearus*	6°31′08.51″N	3°24′06.04″E	45	30	84
	4	*Ganoderma lucidum*	6°31′07.64″N	3°24′06.16″E	41	85	106
	5	*Xylaria* spp.	6°31′07.04″N	3°24′06.25″E	47	76	99
Ozolua road, University of Lagos	1	*Ganoderma applanatum*	6°30′49.07″N	3°23′50.49″E	53	56	34
	2	*Ganoderma adspersum*	6°30′53.25″N	3°23′43.16″E	62	50	70
Araga-poka, Epe	1	*Coprinus comatus*	6°35′19.08″N	3°28′54.78″E	63	32	30
	2	*Polyporus* spp.	6°35′24.05″N	3°28′41.82″E	77	54	61
Ijede, Ikorodu	1	*Agaricus campestris*	6°30′58.94″N	3°23′57.03″E	34	42	115
Old Ota road, Ile epo.	1	*Pleurotus tuber-regium*	6°33′51.09″N	3°25′42.54″E	55	60	75
	2	*Mycena haematopus*	6°33′35.74″N	3°25′73.25″E	57	69	88
	3	*Mycena* spp.	6°33′37.46″N	3°25′71.33″E	82	45	91
Owodunni, Alagbado	1	*Trametes versicolor*	6°38′03.67″N	3°27′20.09″E	58	38	77
	2	*Polyporus squamosus*	6°38′03.67″N	3°27′20.09″E	56	43	81
Gate (University of Lagos), Akoka	1	*Polyporus sulphurous*	6°31′33.29″N	3°23′20.09″E	74	47	83

into Eppendorf tubes. Five hundred microliter (500 µl) of cetyl trimethyl ammonium bromide (CTAB) buffer was added. It was later mixed for homogenisation. Afterwards, the mixture was incubated at 65°C for 15 min and allowed to cool. Ten microliter of Protenase K solution (an equal volume of a mixture of phenol: chloroform: isoamyl alcohol (25:24:1, v/v) was then added to the mixture and incubated at 37°C for 30 min. This is done in order to remove the RNA molecules in the mushroom samples.

On cooling, five hundred microliter (500 µl) of chloroform was added again to the mixture and mixed for 5 min. The resultant mixture was then centrifuged at 10,000 rpm for 10 min in order to separate each component in the homogenous mixture. The supernatant was collected into a fresh Eppendorf tube without taking the white phase. Five hundred microliter of both cold isopropanol and absolute ethanol were added to the supernatant. The mixture was kept at the temperature of -20°C inside the refrigerator for 1 h. It was later brought out for centrifugation at 10,000 rpm for 10 min in order to separate the mixture. Afterwards, the supernatant was decanted carefully. 500 µl of 70% ethanol was added to the decanted supernatant which was mixed very well to achieve a homogenous solution. This was centrifuged at 10,000 rpm for 10 min and the supernatant was discarded. The pellet left was air dried for 1 h and then re-suspended in 200 µl sterile water. This procedure was repeated for each of the eight mushroom samples for 2 days. The samples were kept in sterilized PCR tubes for subsequent DNA sequencing analysis.

Preparation of 1% agarose gel

One gram of agarose was poured into 100 ml Tris/Acetic Acid/EDTA (TAE) and was placed in the microwave oven at 50°C for 3 min. On cooling, 3 µl of ethidium bromide was added to 100 ml agarose gel. The mixture was allowed to solidify after 10 min. 3 µl of loading dye was added to the extracted DNA samples. This mixture and marker were then loaded on the electrophoresis gel machine which was allowed to run at 110 V for 1 h. Genomic DNA was visualized in a Gel Documentation System LG 2020 (Hangzhou Langqi, Inco., China) under ultra violet light source (Plate 2).

Amplification of regions of rDNA, PCR analysis and DNA sequencing

The voucher extracted genomic DNA samples in properly labelled eight different PCR tubes tagged BPO; 1 to 8 were sent to Macrogen Incorporation, Washington, U.S.A for subsequent DNA sequence analysis. Procedure – terms and conditions was applied (Macrogen Inc., Washington, U.S.A.).

Basic local alignment search (BLAST) sequence

The basic local alignment search tool (BLAST) of both the ITS 1 and ITS 4 sequences of each of the eight mushrooms was carried out to find the region of local similarity between sequences. The BLAST search was also carried out to compare a query sequence of the eight genomic DNA mushroom sequences with the library database of the GenBank (http://www.ncbi.nlm.nih.gov) and identify library sequences that resemble the query sequence above a certain threshold. This was achieved through the online analysis tool provided by the GenBank. The BLAST program was carried out as described by Altschul et al. (1990).

Submission of genomic DNA nucleotide of the eight mushrooms

Registration was made online on the GenBank portal prior to submission of the genomic DNA FASTA sequences of the eight mushrooms. The FASTA sequences of the eight mushrooms were submitted using BankIt facility provided by the GenBank (http://www.ncbi.nlm.nih.gov). This was done to authenticate the genomic DNA sequence of the eight mushrooms for reference purpose and also for proper identification.

RESULTS

Identity of the mushrooms

A total sixteen (16) mushrooms were collected. The mushrooms collected showed varying morphological characteristics (Plate 1).

FASTA sequences of the studied mushrooms genomic DNA

The blast sequence query showed that *Issatchenkia orientalis* (FM199965.1) and *Pichia kudriavzevii* (HM771638.1) has the maximum identity (of 90%) with the genomic DNA sequence of *A. campestris* at ITS 1 (Figure 1) while at ITS 4 (Figure 2) no similarity of the *A. campestris* sequence was discovered with that in the GenBank library database (htpp://www.ncbi.nlm.nih.gov).

The blast sequence query showed that at ITS 1 (Figure 3) and ITS 4 (Figure 4) no similarity of the *P. sulphureus* sequence was discovered with that in the GenBank library database (htpp://www.ncbi.nlm.nih.gov).

The blast sequence query showed that *Lasiodiplodia pseudotheobromae* (GQ471832.1) has the maximum identity (of 100%) with the genomic DNA sequence of *G. lucidum* at both ITS 1 (Figure 5) and ITS 4 (Figure 6) with that in the GenBank library database (htpp://www.ncbi.nlm.nih.gov).

The blast sequence query showed that at ITS 1 (Figure 7) and ITS 4 (Figure 8) no similarity of the *X. polymorpha* sequence was discovered with that in the GenBank library database (htpp://www.ncbi.nlm.nih.gov).

The blast sequence query showed that *Lasiodiplodia theobromae* (EF622075.1) and *Botryosphaeria rhodina* (AY343481.1) has the maximum identity (of 84%) with the genomic DNA sequence of *C. comatus* at ITS 1 (Figure 9) while at ITS 4 (Figure 10) no similarity of the *C. comatus* sequence was discovered with that in the GenBank library database (htpp://www.ncbi.nlm.nih.gov).

The blast sequence query showed that *Schizosaccharomyces pombe* (U41410.1) has the maximum identity (of 86%) with the genomic DNA sequence of *Xylaria* spp. at ITS 1 (Figure 11) while at ITS

4 (Figure 12) no similarity of the *Xylaria* spp. Genomic DNA sequence was discovered with that in the GenBank library database (htpp://www.ncbi.nlm.nih.gov).

The blast sequence query showed that at ITS 1 (Figure 13) and ITS 4 (Figure 14) no similarity of the *P. tuber-regium* genomic DNA sequence was discovered with that in the GenBank library database (htpp://www.ncbi.nlm.nih.gov).

The blast sequence query showed that *L. pseudotheobromae* (GQ471832.1) has the maximum identity (of 99%) with the genomic DNA sequence of *P. ostreatus* at both ITS 1 (Figure 15) and ITS 4 (Figure 16) with that in the GenBank library database (htpp://www.ncbi.nlm.nih.gov).

Therefore, of all the BLAST query of the eight mushroom genomic DNA sequences carried out, it was discovered that there were no similarity with the exact sequences that were queried with the GenBank library database (htpp://www.ncbi.nlm.nih.gov). This may be attributed to the virtual absence of the various genomic mushroom DNA sequences in the GenBank database (htpp://www.ncbi.nlm.nih.gov).

Further efforts were then made to submit the eight genomic DNA sequence in the GenBank library through the BankIt facility through documentation for reference purpose. The phylogenetic relationship of each of the eight mushrooms provided by the database (Figure 17) showed that the *X. polymorpha* and *Xylaria* spp. were of the division Ascomycotina while others (*A. campestris, G. lucidum, P. tuber-regium, P. ostreatus, C. comatus, P. sulphureus*) belong to the division Basidiomycotina. *A. campestris, C. comatus, P. tuber-regium* and *P. ostreatus* belong to the order *Agaricales* while *X. polymorpha* and *Xylaria* spp. belongs to the order *Xylariales*. *A. campestris* and *C. comatus* belong to the family *Agaricaceae* while the duo of *P. tuber-regium* and *P. ostreatus* belongs to the family *Pleurotaceae*. *G. lucidum* and *P. sulphureus* belong to the Order *Polyporales* and of the family *Ganodermataceae*.

DISCUSSION

The biodiversity of some mushrooms have been studied in Lagos State through DNA sequence analyses. The mushrooms were collected in and around the Lagos environs for the eight marked bands indicated the mushroom RAPD/PCR products when viewed under the ultraviolet light source as shown in Plate 2. DNA sequencing of eight isolated mushroom mycelia. Also, a blast query of the mushrooms genomic DNA sequence was carried out to detect if there would be any similarity with the sequences that might have been submitted by any researcher previously in the GenBank database.

It was discovered that some of the mushroom genomic DNA sequences showed maximum identity with other fungi that were not mushrooms while others did not show any similarity with previous submission in the GenBank

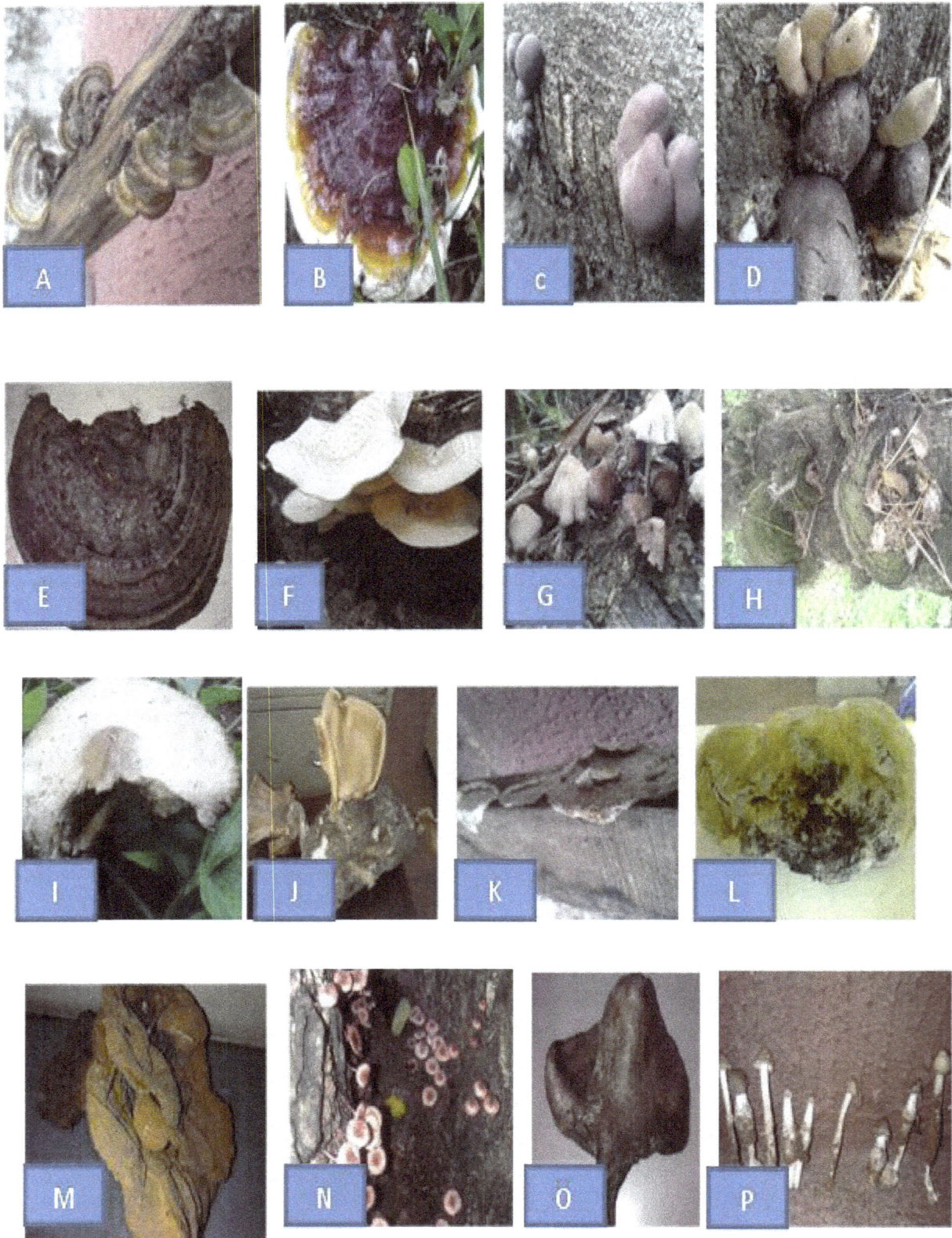

Plate 1. *Mushroom samples collected in Lagos State, Nigeria. (a) Polyporus* spp. *(B) Ganoderma lucidum (C) Daldinia concentric (D) Xylaria polymorpha (E) Ganoderma applanatum. (F) Pleurotus ostreatus. (G) Coprinus comatus (H) Trametes versicolor (I) Agaricus campestris (J) Pleurotus tuber-regium (K) Ganoderma adspersum (L) Polyporus sulphurous (M) Polyporus squamosus* (N) Mycena *haematopus* (O) Xylaria spp.(P) *Mycena* spp.

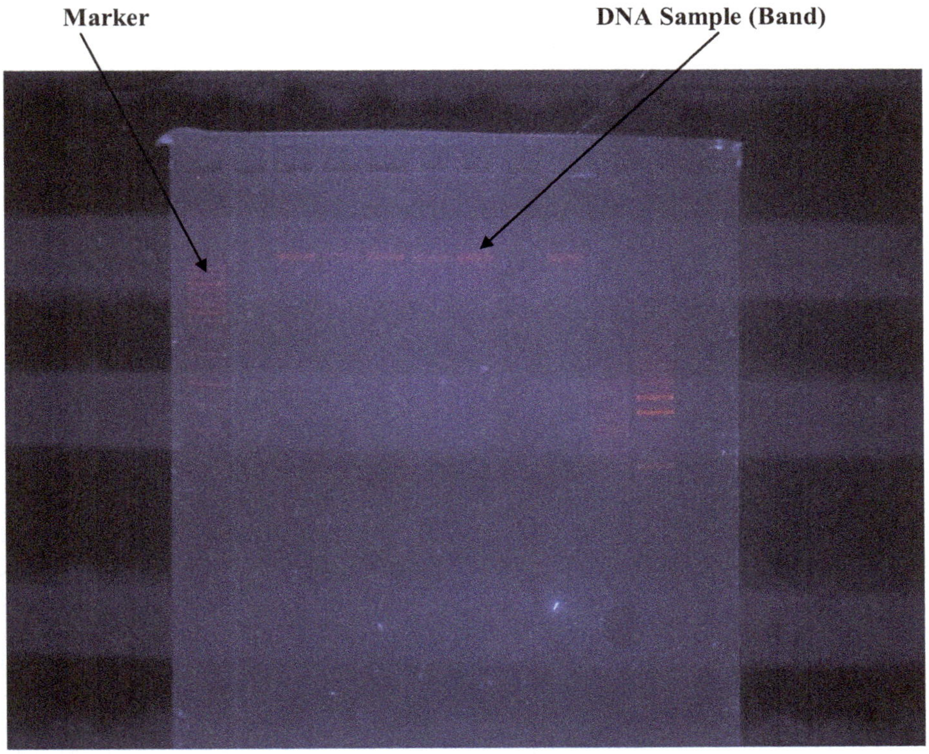

Plate 2. Agarose gel electrophoresis of RAPD/PCR products with eight identifying mushroom DNA bands viewed under the ultra-violet light.

>110810-11_E09_BPO-1-ITS1.ab1   488

TGAGTCGTTTTTGTTTCTCGGTGGAGGGGTGGGGCGGCCAAGCGGCCCCGGAAAAAAGTCTACTTCGCTCGGCCACCTTCT
CGCCCTTTCAGGGAAGTCCAGCTCCCACCTCTTTACACGTCCTCCGCTCGGCTCCCCCAACTCTGCGCACGCGCGAGAGGGA
AACGACGCTCAGACGAGCATGCCCCCCGGAAGTGGAAGGCGCAATGTGCGTTCAAGAACTCGATGATTCACATGGCTGCA
ATTCACACTAGGTATCGCATTTCGCTGCGCTCTTCATCGATGCGAGAACCAAAGATCCGTTGTTGAAAGTGTTGTTTGTGTT
TTCCGAGATTTCTCTTGTCCACTATATGCTATATTCCACATTTTAAGTGTTGTTGGTTTCCTTCCGCTCACCCCGTGTAGAAT
AAATCACAATAATGATCCTTCCCGCGGCCCTTCGGAAA

Figure 1. FASTA sequence of *Agaricus campestris* (90% homologous maximum identity with *Issatchenkia orientalis*).

>110810-11_G13_BPO-1-ITS4.ab1   1131

ACTTGTTGCCCGAGATAATCAGTGGTGGTGGTGGTGGTGCTCGAGTGCGGCCGCAGAGTCGACACCACTACCGCTACCGCC
CTGCTTGTCGGCCATGATATAACGTTGTGGCTGTTGTAGTTGTACTCCAGCTTGTGCCCCATGATGTTGCCGTCCTCCTTGAA
GTCGATGCCCTTCAGCTCTGTGCGGTTCACCAGGTGTCGCCCTCGAACTTCACCTCGGCCCGGGTCTTGTATTGCCGTCGTC
CTTGAAAAAATGGTGCGCTCCTGGACGTAACCTTCGGGCATGGCGGACTTGAAGAAGTCGTGCTGCTTCATGTGGTCGGGG
TAGCGGCTGAAGCACTGCACGCCGTAGGTCAGGGTGGTCACGAGGGTGGGCCAGGGCACGGGCAGCTTGCCGGTGGTGCA
GATGAACTTCAGGGTCAGCTTGCCGTATGTGGCATCGCCCTCGCCCTCGCCGGACACGCTGAACTTGTGGCCGTTTACGTCG
CCGTCCAGCTCGACCAGGATGGGCACCACCCCGGTGAACAGCTCCTCCCCTTGCTCACATGGTGATGATGGTGGATGGGTG
ATGTTCTTTCATATGCTATTATGTATATCTCCTTCTTAAAGTTAGACAAAATTATTTCTACAGGGGGAATTGTTATCCGCTCA
CAATTCCCCTATAGTGAGTCGTATTAATTTCGCGGGGATCAGATCGATCTCGATCCTCTACCCCGGACGCATCTGGCCGGCA
TCACCGGGCCCAGGTGCGTGCTGGCCCTATATCCCAATCCCAGGAATCAGCTCGCCACTTTCGGCTCCTGACGCTTGTTCGC
GGGGTGGTGGGCCCGGCCGGGAAGAACTGTTGAGCATCGCTGGATGCCAATCTGGGGGACGGAAAGGCCTTCACTCACGCT
GGTTCAAGGAAAGATCTCGAAGGAAAGCAAGTATCCTGACATCCAACCTTGTGAAACCCAAAAATATTTTTTGGCAAGCTT
CAAAGTCCTGCTTC

Figure 2. FASTA sequence of *Agaricus campestris* (90% homologous maximum identity with *Issatchenkia orientalis*)

>110810-11_G09_BPO-2-ITS1.ab1   490

GGCTGAGGCGGCGCTCCCTTTTTTTGTGTCGTGCAGTCGGTGTTGGCCGCCAATCCCCCTCAAAAAGGACTTCTTTCGCCCC
CCCCTGCTCCTCCCTTTTGACAGATTCCTAGTTCCACCCTTTTTTACCGTCCGTCTCTTCCCTCCCCAACTTCTCGGAAGGGC
GAGAGGAAAAGCGCGACTCGTGTGATGGCCAACTTAGGGACCCTACAAGCCTGTGAGTCCAAAACCCCCGAATGCCCAAC
GGGGGCAATTAAAACTAGGAACCGCATTTCTTTGCCCTCTTCAATTCCGAAAACCAAAAAACCCCCTGTGGAAGTTTTTTTT
GGTTTTTCCGAAATTTTCTCTTGTCCATAATTTCCAATTTTCCCCATTTAAGTTTTTGTTGTTTTCTTCCCCCCCCACCCGGGGA
GTATTAAATAAAAATAATGATCCTTCCGCGTGTACGTACCCATAAAG

Figure 3. FASTA sequence of *Polyporus sulphureus* (no homologous identity with previous submissions in the GenBank).

>110810-11_I13_BPO-2-ITS4.ab1   1138

CGTCTGTTAGCACGGATCTCACTGCTGGTGGTGGTGGTGCTGCACTGCCGCCCCACACTCGACACCACTACCGCCACCTACC
TGCTTGTCGGCCATGATATATACGTTGTGGCTGTTGTACTTGTACTCCATCTTGTGCCCCACCATGTTCCCCTCCTCCTTGAA
GGCTATGCCCTTCCTCGCAGTGCCGTTCCCCCCGGTGTCCCCCTCGGACTTCTGTTCGTCCCAAATTTGAAGATTCACTGAAT
TTTGAATTAATTTCACTTAGCCAATTTCGATGCTTCTTCTGGCATGCTTTCCTTAAAGCCTTTTTGTGGTCGGGGTTTATCCTT
GAAGCTTTAGACGTCAGGGTATGCTGACCGAATTTGGGGGCCCTTTGGCATTTTGCCGTGGTGCAAATGCTTCATGCTCAAC
GGGCCCCCGGCCTCGTCCTCACCCTCAACGAACACATAATCCTGCCGATGTTTTACGTCGGAGTCCACCTGGACCAGGATG
GCCACCACCCCGGTGAACAGTTCCTCCCCTTGCTCACATGACTTCACCGATGGTGATGATCGTTCTTCGCCATTTCGGGATC
TAGATCTCCTTCTTCGGTTAGAAAAAAGTAGCTCTACAGGGAGAATTGATACGGTTGGGGTTCCCCTATATTGAGGCGTATT
TCCTTGCGGGGATGGAGCTCGATCGCATCCTCTACTGGACGCAGCTTTCCACGGCTCGACTCCATCAGGGAGAGCGTGCGA
GTTCCCAATCCCTAATTGCCCAAGCGATTATCAGTATGCCATGATAGGCCTGAACGTCTCCTCGGTGATGAGAGCCAGCCC
CGCGCGTATAACAGGTCGCACTTTGCCAGGGTGTATGCTCGGTTCGGAGGAAGCCATCAACTCGTTGGGGGGGATAGGGGA
AAAAATCGAAACGAGAGAATAATTACCCACTGGCGCACGATACTGTGGGAAACGCCGCAAAATAGCTTCCCTCCCTGGTC
CCCCCGCCTCCATGGTCACTTATTCTCCTC

Figure 4. FASTA sequence of *Polyporus sulphureus* (no homologous identity with previous submissions in the GenBank).

>110810-11_I09_BPO-3-ITS1.ab1   519

ATTCGGGCTTCGGCTCGACTCTCCCACCCTTTGTGAACGTACCTCTGTTGCTTTGGCGGCTCCGGCCGCCAAAGGACCTCCA
AACTCCAGTCAGTAAACGCAGACGTCTGATAAACAAGTTAATAAACTAAAACTTTCAACAACGGATCTCTTGGTTCTGGCA
TCGATGAAGAACGCAGCGAAATGCGATAAGTAATGTGAATTGCAGAATTCAGTGAATCATCGAATCTTTGAACGCACATTG
CGCCCCTTGGTATTCCGGGGGGGCATGCCTGTTCGAGCGTCATTACAACCCTCAAGCTCTGCTTGGAATTGGGCACCGTCCTC
ACTGCGGACGCGCCTCAAAGACCTCGGCGGTGGCTGTTCAGCCCTCAAGCGTAGTAGAATACACCTCGCTTTGGAGTGGTT
GGCGTCGCCCGCCGGACGAACCTTCTGAACTTTTCTCAAGGTTGACCTCGGATCAGGTAGGGATACCCGCTGAACTTAAGC
ATATCAGGGAGGAA

Figure 5. FASTA sequence of *Ganoderma lucidum* (100% homologous maximum identity with *Lasiodiplodia pseudotheobromae*).

>110810-11_K13_BPO-3-ITS4.ab1   517

CAGATCCGAGGTCACCTTGAGAAAAGTTCAGAAGGTTCGTCCGGCGGGCGACGCCAACCACTCCAAAGCGAGGTGTATTCT
ACTACGCTTGAGGGCTGAACAGCCACCGCCGAGGTCTTTGAGGCGCGTCCGCAGTGAGGACGGTGCCCAATTCCAAGCAG
AGCTTGAGGGTTGTAATGACGCTCGAACAGGCATGCCCCCCGGAATACCAAGGGGCGCAATGTGCGTTCAAAGATTCGAT
GATTCACTGAATTCTGCAATTCACATTACTTATCGCATTTCGCTGCGTTCTTCATCGATGCCAGAACCAAGAGATCCGTTGT
TGAAAGTTTTAGTTTATTAACTTGTTTATCAGACGTCTGCGTTTACTGACTGGAGTTTGGAGGTCCTTTGGCGGCCGGAGCC
GCCAAAGCAACAGAGGTACGTTCACAAAGGGTGGGAGAGTCGAGCCGAAGCCCGAAAACTCGGTAATGATCCTTCCGCAA
GGCACCTATACGAAA

Figure 6. FASTA sequence of *Ganoderma lucidum* (100% homologous maximum identity with *Lasiodiplodia pseudotheobromae*).

CGCGCCCTTTTGTGTCCGCACGCGGTGGCGCCCCCTGGATCGGAAAAAAACCCTTTTCGATCGCAGCTTCCCCCTTGTTGGA
GTTAGCCGCTCCCCCTCTCTTATCAACTCCTCATCTTCTCTTCCCCCCCCTGCTCGGCAACGCGGGAGAAGGAAACCACGCAA
AGACAGGTGTGCCCCCGAGAAATCCGAGGGGCTGTATGTGCACTTTAAAACCCATTGACTCCCGACGCCTGCCATTGATCT
ATCTTCTTATTTACCAAAAGTTAGGTTCGTGAAAACCTAAATCCCCCGTGGGAAGAAAATTCAGAGATTTCCGAAATATCC
CTTTGGCCCTGAATTCTAAGGTCCCCCCTTTATGTTTTGGTTGTTTTCTTCCGCCCCCCCCCCGGGATGAATAAATAATCCAC
ATGATTATCCTGCGGCAACGTACCGATACAGAATAATTAGCATTTATTAATGTCTCTGGAAGCTTTGGAACTAGCTCACGCA
ACAGACGTATATCTTTCGGGTGCTTAAACAATCGTTTCGGTAGGAGTATGACATCCGAGTTTCACTTGTATAAGAAACATG
AGGTTACTACAGGTAGACATGCTTAATCTCTGGGAAGATGAAGTAAAGAAACTATTCACTCTATTGCGACGCCCATGCCCC
TAGCTGTCTGGACGGGGATGCCAACATATTAATACAATGACCTCGTAGGCCGTGTAGGAATTCCAGCAATTCCAGATAACA
CTATATAATAAAAACGCTGAGCTAATAAATAAGTAATAGATGTTAGCTTAACCGGAGTCCTTATATAATAGCGAGACTATT
AGCTAGGTATATAGGTAAATAGTAAACGAGACTTTCTCTCCTCTGCCACAT

Figure 7. FASTA sequence of *Xylaria polymorpha* (no homologous similarity with previous submissions in the GenBank).

>110810-11_M13_BPO-4-ITS4.abl   511

TTACTTGATAATAGTGCTGGTGGTGGTGGTGCTCCCTGCTGCCCCACACAAACCCACGGCCTCCACCGACCTGCTTGTCCGT
GCTGAGCTCGTTCAGGCTGACGCGAACTCTCAATGGAGAGCCCCCCCAAGTTTCTGGCTAGCGTGAGGGCTATGCCCACAC
TGGATGTGCCCTTCCCCCCTGAGTGGCCCCAGGACCCCTGTGCCTCCAAAAACTTGATGATTCACGGAATTCTGAAATTAAT
TTTACTTAGCCCAATTCGATGTTGTCTTCTACCATGATAACCCAGAGATCCATTGTTAAAGCTTTTAGTTTATTAACTTGTGA
TCTGACGTCTGCCTTTGCTGACTCTAATTTGGAGGTGGTTTGGCATCCGTTGCCTTCCCGGACTAGGGTTCCTTCTCAAAGTG
TGTGTGGTCATACTCACTAAAACGGAAGATAATGCTCCCAATTGCCGCCGCCGCCGGCT

Figure 8. FASTA sequence of *Xylaria polymorpha* (no homologous similarity with previous submissions in the GenBank).

>110810-11_M09_BPO-5-ITS1.abl   908

CTGCTCCCCATTTTGTGCCGCACTCTGTCGCTTTGGCGCCTCCTGCCGCCGGAGGACCTCCGACTCCCCACCCCAACTGCCC
CTCTGATAAACGACTTAACTTAATATAACATTAAACAATTGATTTTTTGGTTCTGGCATCGATGAACAAGGCACGAAATGC
GATAAGTTATGAGAATTGCACACCCCGAAAAATCATCGAATCTATGAACGCACAAAGAGCCCATTGATATTCCAGGGGGC
CTGCCTGCTCGAGCGTCATTACAACCCTCAGGCTCTGCTTGGAATTGGGTACCGCCCTCACTGCGGACGCGCCTCAAAGAC
CCCGACGGAGGCCGGTCATCCCTCAAGCGTGCTCCAATACCCCTTTTTTTGTAGTGGCTGGCGTCCCCGCCGGACGAACCT
TCTGAACTATCCTCAATGTTGACCTCGGATCAGGTAGGGAAACCCGATGAACTTAAGCATATAATGTGCCGGAGGACCGAC
GAAGTTAGTTCACGAATTGATATTTTATCGTCTTCGGTTTCAGAATCAATCTCATCGGCAAGTTCCTGTTCTTTGACGTATCC
TTCAACAGATTTTGTCATTTTTGGAAAGAAGAGCTTGCTTAAAATATTTGAATGGAGATAGAAAGTTACATTACACAAATTT
ACTGTTATGAGCTTTTGAACCTGTCTTCCGTATGCCAATTTGATGTTATCCATGACAGTTCCCAAGTATGCTTCCTCTATACA
TAGATATAAGTAAGTTCAAATCTCTCGCTAACTAAATACGAAAGTATATGGAGATGTTGTGTTATGTATTAACAGTATGCAT
TCACAAGAGACAAACGAATCATCGCCTCAAACAAACAAT

Figure 9. FASTA sequence of *Coprinus comatus* (84% homologous maximum identity with *Botryosphaeria rhodina*).

>110810-11_O13_BPO-5-ITS4.abl   515

TTCTGTGTTTGAGAAGTTCAGAAGGTTCGTCCGGCGGGCGACGCCAACCACTCCAAAGCGAGGTGTATTCTACTACGCTTG
AGGGCTGAACAGCCACCGCCGAGGTCTTTGAGGCGCGTCCGCAGTGAGGACGGTGCCCAATTCCAAGCAGAGCTTGAGGG
TTGTAATGACGCTCGAACAGGCATGCCCCCCGGAATACCAAGGGGCGCAATGTGCGTTCAAAGATTCGATGATTCACTGAA
TTCTGCAATTCACATTACTTATCGCATTTCGCTGCGTTCTTCATCGATGCCAGAACCAAGAGATCCGTTGTTGAAAGTTTTAT
TTTATTAACTTGTTTATCAGACGTCTGCGTTTACTGACTGGAGTTTGGAGGTCCTTTGGCGGCCGGAGCCGCCAAAGCAACA
GAGGTACGTTCACAAAGGGTGGGAGAGTCAGCCGAAGCCCGAAAACTCTGTAATGATCCTTCCGCAAGGTTCACCCTACCG
GA

Figure 10. FASTA sequence of *Coprinus comatus* (84% homologous maximum identity with *Botryosphaeria rhodina*).

>110810-11_O09_BPO-6-ITS1.ab1   918

GTTGGCTTGCTCGTGCATATAAAATACTGCACTAGAAGGCTTGACCCCCAAGACACTTGGGCTTATCTATGTGGCTTACGTT
TGCCTGTGGCTTGGTGAAGTTCTCTGATCTGCATCTCCGTCTCTTCCATCCATGCCGCCACTCAATGGAGTGTTACCAGGTG
CACGGCCCTCCCTTGAAGCATCTATTGAAAACCAAAGTAATGGTATACGGAATTCTACTGTCACAAAAGACCGTGTAGAAA
TACTACTGCGGATGTCTCTCAATCCCAAACCATTGAAACATCTGGACCTTCTAACGAAGTTCCTACTGCTCAACCGGATGCT
TGTGCTGTAGCCTACATTTTTATTGGACTTTCTAACTGCCATTCCCAAGTTTTGATCGTTGAAAGACGATTTGTCCTCAGGAA
AAAATGATGAAAATCACGAATAATTTCCCTACCCTAGCCAATCTTTTTTTGATAATATTAGTTGTGAAAGTAAAGAGAAGG
GTATGGAAGCTGCAGATAGAAGAGCTCGTTTAAACTGGATGGCGGCGTTAGTATCGAATCGACAGCAGTATAGCGACCAG
CATTCACTTACGATTGACGCATGATATTACTTTCTGCGCACTTAACTTCCATCTGGGAAGGGACGATTCAGAAACTTACTAT
AATTTCGACGGCAATGTCCAAGCTGTCATACAGCGATGCCAAGATGTTGTGACAGCACCACCCAGGCCGAGGTTGATTTCC
GGCAATCACAGAGATAACATATATAATTAGAAACTCTCACTATTTAAAGTAAAGATAAAGAGAATGTTATAATTCCTTAAA
AGGAGCTCATTCAAAGATAAGAGAGACTATTTAATCCGACCCAAGGTAGATAAGTAA

Figure 11. FASTA sequence of *Xylaria* sp. (86% homologous maximum identity with *Schizosaccharomyces pombe*).

>110810-11_A15_BPO-6-ITS4.ab1   1229

CTGGACGGGCTTTCGCGAATGGAGGTGGTGGTGGGGGGGTTGCGGACTGCAATTGGTCATTGGCCCCCCCGGTAGCCGGGCA
TTTGGTGTGACGATCCACCAACGCTCACTGGTAAACTACCTGGCCGATCCCCCAATAATGAATTGAAAAAGTGCATGGTCT
CGTCGCGCATGCAGCCCCGAAAGTGGTTTGGCCCATGGGGATCTTGAGAACTCGGATGAGCACGAGGGCTGACCTCTGGTT
GATTTGGGTTCAAATGAGGCGCTCTCTGGAGGATTTGAACAAAACGGGGCGGTTCATAATGTTGAGGTTGCGTTTCCATTT
GTTGCTCATGGCGAAGCTGTGCTATCTCCCACGAGATGGAGGGTGTTTTCTTTTCGGACGTTCCACAATATACTCTCATAGT
AATGATCCGCCCTAAGAACATCCTCATCACTCACCAACATCACCATCGTACTCCCCAACGTCACCATCACAGTCACCTACAT
CACCATCGTTCCAACGTCGCCATCATATTCGCCAACGTCATGATGATAGTCGCCACGTCCCTCGTATTCTCCAAGGGGTCCA
TCGTATTCACCAACGGGCCTTCCTACTAACTCACGGCCCCAAGCAATCCCCCTACGGGGTCCTTCTTATTCTCCTACGGATC
CATCATATGGTCCTACGGCACCGAGTTACGATCCCAGGGGCACCAGTTGCGGCCCAACGGGGTCCACCCTATTCCCCATCA
CCCACCAAGTTACTAGTCCTAAAACGCCATCATCCTCTTCACGCTACTATCCCTTTTCCCCAACCACCCTTCCTACTCTCCCC
GGCGCCCAAAAAGCTCGACTGCATTGTAACGCCCAACTCGAGGTACGCGCAAGTCCAGTATTCGCAGAACAAACAAACAT
ATGAAAAGGAAAATTCCCAAGAATACCATTCCTCCTGACCTATTTCCGTTTAATAAAGTCCAAAATAAGTATGTCTCANGTT
ATTTTTGGATCCTAACAANACCGTCTACAAGAAACCCTTATTTGTACTCCATTAGGGGGGGGACGGGAACGCATACCTATTTT
TTCGCCAAAGAGTGGTAAAACCCGGGGAAAAT

Figure 12. FASTA sequence of *Xylaria* sp. (86% homologous maximum identity with *Schizosaccharomyces pombe*).

>110810-11_A11_BPO-7-ITS1.ab1   54

TATGCTAGGTCGATTTTTTTCACCCCG

Figure 13. FASTA sequence of Pleurotus tuber-regium (no similarity with previous submissions in the GenBank).

>110810-11_C15_BPO-7-ITS4.ab1   5

GATCGATTTTCAGCTTAGTAAAGCAAA

Figure 14. FASTA sequence of Pleurotus tuber-regium (no similarity with previous submissions in the GenBank).

Database (http://www.ncbi.nlm.nih.gov). This might have been as a result of no previous submissions of the mushroom genomic DNA sequences by any researcher in the GenBank database. Furthermore, the genomic NA

>110810-11_C11_BPO-8-ITS1.ab1   518

TCGTTCGGCTCGACTCTCCCACCCTTTGTGAACGTACCTCTGTTGCTTTGGCGGCTCCGGCCGCCAAAGGACCTCCAAACTC
CAGTCAGTAAACGCAGACGTCTGATAAACAAGTTAATAAACTAAAACTTTCAACAACGGATCTCTTGGTTCTGGCATCGAT
GAAGAACGCAGCGAAATGCGATAAGTAATGTGAATTGCAGAATTCAGTGAATCATCGAATCTTTGAACGCACATTGCGCCC
CTTGGTATTCCGGGGGGGCATGCCTGTTCGAGCGTCATTACAACCCTCAAGCTCTGCTTGGAATTGGGCACCGTCCTCACTGC
GGACGCGCCTCAAAGACCTCGGCGGTGGCTGTTCAGCCCTCAAGCGTAGTAGAATACACCTCGCTTTGGGAGTGGTTGGCG
TCGCCCGCCGGACGAACCTTCTGAACTTTTCTCAAGGTTGACCTCGGATCAGGTAGGGATACCCGCTGAACTTAAGCATAT
CATGGAAGGAAA

Figure 15. FASTA sequence of *Pleurotus ostreatus* (99% homologous identity with *Lasiodiplodia pseudotheobromae*).

>110810-11_E15_BPO-8-ITS4.ab1   517

GCCCTGATCAGGTCACCTTGAGAAAAGTTCAGAAGGTTCGTCCGGCGGGCGACGCCAACCACTCCAAAGCGAGGTGTATTC
TACTACGCTTGAGGGCTGAACAGCCACCGCCGAGGTCTTTGAGGCGCGTCCGCAGTGAGGACGGTGCCCAATTCCAAGCAG
AGCTTGAGGGTTGTAATGACGCTCGAACAGGCATGCCCCCCGGAATACCAAGGGGCGCAATGTGCGTTCAAAGATTCGAT
GATTCACTGAATTCTGCAATTCACATTACTTATCGCATTTCGCTGCGTTCTTCATCGATGCCAGAACCAAGAGATCCGTTGT
TGAAAGTTTTAGTTTATTAACTTGTTTATCAGACGTCTGCGTTTACTGACTGGAGTTTGGAGGTCCTTTGGCGGCCGGAGCC
GCCAAAGCAACAGAGGTACGTTCACAAAGGGTGGGAGAGTCGAGCCGAAGCCCGAAAACTCGGTAATGATCCTTCCGCAA
GGTCCTTACGGAAG.

Figure 16. FASTA sequence of *Pleurotus ostreatus* (99% homologous identity with *Lasiodiplodia pseudotheobromae*).

sequences (Figures 1 to 17) were deposited in the GenBank for academic and reference purpose through D the BankIt Online tool provided by the web portal. This provides a detailed and comprehensive identification of the mushrooms through the description given by the GenBank in form of phylogenetic relationship/tree (Figure 17).

The phylogenetic tree (Figure 17) given by the GenBank further confirms the assertion that mushrooms falls into two fungal divisions of Basidiomycotina and Ascomycotina. The result of this research shows the genome rDNA sequences of the mushroom samples which is more accurate and reliable in phylogenetic typing and identification of mushrooms than the conventional means. Identification of mushrooms is mainly by morphological description of the fruiting bodies, host specificity and geographical distribution (Seo and Kirk, 2000). In most cases, morphological characteristics have their limitation in allowing a reliable distinction of intraspecific characteristics. It has been noted that the genus *Ganoderma* presently represents taxonomic chaos (Ryvarden, 1991). Molecular techniques could be used to adequately characterize and identify intra and inter species (Zakaria et al., 2009).

The ability to accurately and reproducibility identify fungi; both yeasts and molds has been greatly enhanced through comparative DNA sequencing (Hall et al., 2003). Fungal taxonomists have used DNA sequences for many years as a basis for reclassification of all fungal taxa and have more recently moved to ITS sequencing.

In the past three to four decades, research findings had pointed to mushrooms as important sources of pharmacologically important bioactive compounds that can improve health. Hence, correct identification procedure of medicinal mushrooms is required for quality control of functional health-aid preparations as well as nutritional supplements (Lee et al., 2006). Gene sequences also serve as a basis of molecular tools for sensitive and incisive identification of mushrooms. Highly conserved genes such as ribosomal RNA (rRNA) genes provide information on the general properties of the organism based on the properties of their known relatives. In addition, analysis of the rDNA sequences has the advantage that it not only enables species identification but also permits phylogenetic analysis.

In conclusion, it was demonstrated that DNA sequencing is an accurate identification measure of mushroom samples than the conventional macroscopic, ecological and geographical means. More research should be carried out on sequencing the DNA of mushroom samples of specific medicinal, edible, toxic, psychoactive properties.

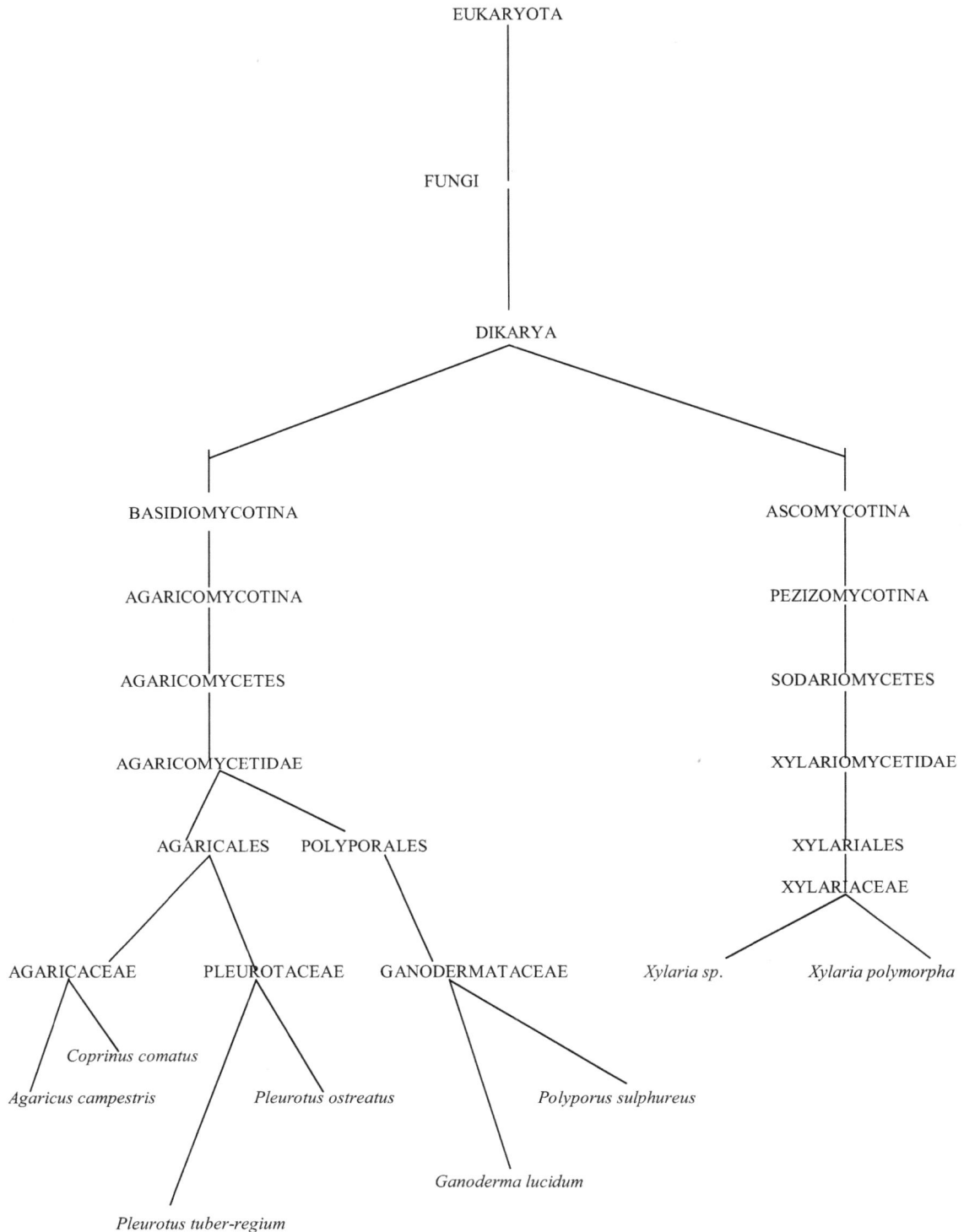

Figure 17. Phylogenetic relationship/tree of the studied mushrooms.

REFERENCES

Altschul SF, Gish W, Miller W, Myers EW, Lipman DJ (1990). "Basic local alignment search tool". J. Mol. Biol. 215(3):403–410.
Berg J, Tymoczko J, Stryer L (2002). Biochemistry. W. H. Freeman and

Company, London. pp. 4955-4956.
Ghosh A, Bansal M (2003). "A glossary of DNA structures from A to Z". Acta Crystallogr. D. pp. 620–626.
Graham GC, Mayer P, Henry RJ (1994). A simplified method for the preparation of fungal genomic DNA for PCR and RAPD analysis.

Biotechniques 16(1):48-50.

Hall L, Wohlfiel S, Roberts GD (2003). "Experience with the MicroSeq D2 Large-Subunit Ribosomal DNA Sequencing Kit for Identification of Filamentous Fungi Encountered in the Clinical Laboratory". J. Clin. Microbiol. 42(2):622–626.

Lee JS, Linz MO, Cho KY, Cho JH, Chang SY, Nam DH (2006). Identification of Medicinal Mushroom Species Based on Nuclear Large Subunit rDNA Sequences. J. Microbiol. 44(1):29-34.

Ryvarden L (1991). Genera of Polypores. Nomenclature and Taxonomy. Synopsis fungorum 5. Fungi flora, Oslo, Norway. p. 363.

Seo GS, Kirk PM (2000). Ganodermataceae: Nomenclature and classification, In: *Ganoderma* Disease of Perennial Crops. Flood, J. P.D. Bridge and P. Holderness (Eds.). CABI Publishing, Wallingford, UK. pp. 3–22.

Watson JD, Crick FHC (1953). "A Structure for Deoxyribose Nucleic Acid". Nature 171(4356):737–738.

Wei JC (1979). Fungi identification manual (in Chinese). Shanghai Scientific and Technological Press, Shanghai. p. 780.

Zakaria L, Ali NS, Salleh B, Zakaria M (2009). Molecular analysis of *Ganoderma* spp. from different host in Peninsula Malaysia. J. Biol. Sci. 9(1):12–20.

18

Effect of environmental conditions on the growth of *Cryptosporiopsis* spp. causing leaf and nut blight on cashew (*Anacardium occidentale Linn.*)

Dominic Menge[1,2,3]*, Martha Makobe[1], Shamte Shomari[2] and Andreas. V. Tiedemann[3]

[1]Jomo Kenyatta University of Agriculture and Technology (JKUAT), P. O. Box 62000-00100 Nairobi, Kenya.
[2]Cashew Research Programme, Naliendele Agricultural Research Institute (NARI), P. O. Box 509, Mtwara, Tanzania.
[3]University of Göttingen, Grisebachstrasse 6, 37077 Göttingen, Germany.

A new disease (cashew leaf and nut blight) in Tanzania caused by a fungus related to the genus *Cryptosporiopsis* was identified in 2006. The present work investigated the effects of environmental factors on the growth of *Cryptosporiopsis* spp. causing blight on cashew. The mycelial growth, colony character and sporulation pattern of 10 fungal isolates, grown on seven different culture media namely, corn meal agar (CMA), malt extract agar (MEA), tryptone dextrose agar (TDA), potato carrot agar (PCA), water agar (WA), potato dextrose agar (PDA) and host leaf agar were observed after 10 days of incubation at 25±2°C. The colony diameter, culture characteristics and sporulation of the 10 isolates were greatly influenced by the type of growth medium used. The best mycelial extension was recorded in 12 h alternating light/dark followed by total light and total dark conditions, respectively. Seven media were evaluated for best growth of the fungi that is, *Cryptosporiopsis* spp. grew maximum on WA followed by host leaf extract media and PDA, respectively but least grew on the TDA medium. The growth of *Cryptosporiopsis* spp. was maximum in temperature range of 25 to 30°C. The most suitable pH level for growth of fungus was 7.0 and 6.0. These results will be useful for fungal taxonomic studies.

Key words: *Cryptosporiopsis* spp., culture media, light, temperature.

INTRODUCTION

Cashew (*Anacardium occidentale L.*) is a tropical nut crop that belongs to the family Anacardiaceae, which consist about 75 genera and 700 species (Nakasone and Paull, 1998). *Anacardium* contains eight species all of which are native to the coastal parts of north eastern Brazil (Azam-Ali and Judge, 2000). Cashew is an important cash crop traded worldwide that originated from South American countries like Brazil, Bolivia, Ecuador and Kenya. Cashew world production is about 400,000 tonnes. More than 50% of this production comes from

South Asia and South Africa, especially India and and Peru (Behrens, 1998). Cashew major producers are India, Tanzania, Mozambique, Nigeria, Guinea-Bissau Tanzania (Opeke, 2005). It was introduced in India and Africa in the 16th century, by the Portuguese initially to protect the soil from erosion (Azam-Ali et al., 2001). The global area under cashew cultivation has risen tremendously, from about half a million hectares to four million hectares from 1961 to 2008, respectively (FAOSTAT, 2008). Cashew trees are genuinely tropical and very frost sensitive. The trees grow in a wide spectrum of climatic regions between the 25°N and S latitudes. Although the cashew can withstand high temperatures, a monthly mean of 25°C is regarded as optimal. Yearly, rainfall of 1000 mm is sufficient for production but 1500 to 2000 mm can be regarded as optimal. Diseases constitute limiting factors in production of cashew in cashew producing regions of Tanzania

*Corresponding author. E-mail: dominicmenge@yahoo.co.uk.

Abbreviations: CMA, Corn meal agar; MEA, malt extract agar; TDA, tryptone dextrose agar; PCA, potato carrot agar; WA, water agar; PDA, potato dextrose agar.

Table 1. Isolates obtained from sampled locations.

Isolate	Infected organ	Place of origin
AA1	Leaf	Naliendele
AA2	Leaf	Madangwa
AA3	Leaf	Mnazi mmoja
AA4	Pseudo fruit	Malamba
AA5	Pseudo fruit	Lyenje
AA6	Nut	Namiyonga
AA7	Nut	Chiola
AA8	Nut	Nandagala
AA9	Pseudo fruit	Nachingwea
AA10	Leaf	Newala

because the environment is conducive to the growth and multiplication of disease pathogens. In 2003, during a survey carried out by Naliendele Agricultural Research Institute in Tanzania, cashew leaves with spots were collected from cashew trees. According to Sijaona et al. (2006) the causative agent was identified as *Cryptosporiopsis* spp. Leaf and nut blight caused by *Cryptosporiopsis* spp. is a major limiting factor affecting cashew nut production in Tanzania, causing 48.4% crop loss annually (ACRR, 2006). The cashew blight samples were forwarded to Global Plant Clinic where they were deposited as Herbarium specimen (IMI 391611). The pathogen is being characterized further into its taxonomical nomenclature (GPC, 2010). An understanding of the role of environmental conditions and its effect on infection and survival of the pathogen is necessary to develop cultural disease management practices. The objective of this study was to provide information on effects of culture media and various environmental factors including temperature, pH and light on mycelial growth and conidia production of *Cryptosporiopsis* spp. causing blight disease in cashew (*Anacardium occidentale L*).

MATERIALS AND METHODS

Sample collection

The present investigations were carried out both in the field and laboratory during the period 2010-2011. All the experiments were conducted at the Agricultural Research Institute (Naliendele), Mtwara, Tanzania. One cashew leaf and nut blight sample was forwarded to the Naliendele Agricultural Research Institute Plant Pathology Herbarium, Tanzania where they were deposited as herbarium specimen *Cryptosporiopsis* spp. IMI396316 for reference. *Cryptosporiopsis* spp. was found in all of the lesions and was identified based on Sijaona et al. (2006) well illustrated and detailed description of the fungus, which is reproduced here. The typical cashew leaf blight diseased leaf and nut samples were collected from farmers' fields on commercially cultivated clones at 10 locations comprised of different agroclimatic zones in southern Tanzania (Table 1). These and brought to the laboratory for isolation of disease causing fungi.

The pathogen was isolated by direct conidial transfer method on potato dextrose agar (PDA) medium. Cashew leaves showing leaf blight symptoms were cut into small pieces of 1.2 cm, surface sterilized by sodium hypochloride for 1 min and washed in sterilized distilled water three times. The leaf bits were placed in Petri plates containing moist filter paper and incubated for four days at 25±2°C. Sporulated leaf bits were shaken onto new PDA medium to release spores thereafter the plates were incubated for four days at 25±2°C. The fungus was purified hyphal tip isolation technique (Harlapur, 2005) and the isolates were maintained on PDA slants. *Cryptosporiopsis* spp. (IMI396316) was obtained from CABI, UK for reference.

Pathogenicity test

Pathogenicity test was performed on cashew seedlings by spraying conidial suspensions (10^6 spores mL^{-1}) of the 10 isolates selected randomly on young tender leaves of nine-month-old plants. Inoculated plants were enclosed in wet plastic bags. Control plants were sprayed with sterilized deionised water. After 48 h, the plastic bags were removed and the plants observed daily for 10 days. The symptoms were observed. The inoculated cashew seedlings were observed for symptom expression on the 1st, 2nd and 3rd leaves of the seedlings. Koch's postulates were observed by consistently re-isolating the pathogen from the inoculated plants.

Maintenance of the culture

The pure cultures of the fungus were sub-cultured on potato dextrose agar slants and kept in laboratory at 25±2°C for 10 days. Such mother culture slants were preserved at 5°C in refrigerator. Further, these cultures were sub-cultured once in a month and used for future studies.

Effect of culture media on mycelial growth

Following the culture media, seven were used to find out the most suitable one for the mycelial growth of the fungus. Cultural characters of eleven isolates of *Cryptosporiopsis* spp. from different geographical regions were studied on seven different media. The growth characters of *Cryptosporiopsis* spp. were studied on seven solid media namely leaf extract, corn meal agar, malt extract agar, potato dextrose agar, dextrose tryptone agar, water agar and potato carrot agar. All the media were sterilized at 1.1 kg/cm² pressure for 15 min. To carry out the study, 20 ml of each of the medium was poured in 90 mm Petri plates. Such Petri plates were inoculated with 5 mm disc cut from periphery of actively growing culture and incubated at 25±2°C for 10 days. Each treatment was replicated thrice. Observations were taken when the fungus had completely covered the Petri plate in any one of the media. The colony diameter was recorded. The fungus colony colour, margin and sporulation were also recorded. The data on radial growth was analyzed statistically. Petri plates containing 20 ml of each of following media were inoculated with 9 mm diameter disc from 10 day old cultures of different isolates. *Cryptosporiopsis* spp. (IMI396316) was obtained from CABI, UK for reference.

i. Corn meal agar (CMA; Unipath LTD, Hampshire, England) with the following composition (g L^{-1}); corn meal extract from 50 g whole maize 2.0, agar 15 and distilled water 1000 mL.
ii. Potato Dextrose agar (PDA; Unipath LTD, Hampshire, England) – Potato infusion 200, dextrose 20.0, agar 15 and distilled water 1000 mL.
iii. Dextrose Tryptone Agar (DTA; Unipath LTD, Hampshire, England)- Tryptone 10.0, Glucose 5.0, Bromocresol purple 0.04, agar 12.0 and distilled water 1000 mL.

iv. Malt extract agar (MEA; malt extract 30.0, peptone from soymeal 3.0, agar 15.0 and distilled water 1000 mL.

v. Potato carrot agar (PCA); grated potato 20g, grated carrot 20g, agar 20g and distilled water 1000 mL prepared according to Tuite, 1969.

vi. Water agar (WA) – agar 20 g in 1000 mL distilled water and

vii. Host leaf agar (young cashew leaves 200 g, agar 20 g and distilled water 1000 mL.

Morphological studies of the pathogen

Spores of leaf and nut blight were taken from infected host tissue and mounted on a clean glass slide. To characterize isolates by colony morphology, single germinating conidia were transferred to Petri dishes containing PDA. Dishes were incubated at 25±2°C in 12 h alternate light and darkness for six days. After incubation, cultures were examined for colony color, colony margin, colony texture, and the development of pigments or crystals in the agar medium. To characterize isolates by sporulation habit, single germinating conidia were transferred to Petri dishes containing PDA. Dishes were incubated for six days. After incubation, cultures were examined at ×40 to ×100 magnifications with a dissecting microscope and sub-stage illumination for characteristics of the sporulation apparatus, presence of conidiophores, and branching of conidial chains.

Effect of different pH levels on mycelial growth

After preparation of the PDA broth, their suitable volumes were adjusted at different pH 4, 5, 5, 6, 7, 8 and 9 using 1 N HCl or 1 N NaOH. The sterilized media of different pH levels was poured in the sterilized Petri plates in about 20 ml quantities and allowed to solidify. 9 mm discs from the actively growing 10 day old cultures of different isolates were placed on the centre of the Petri plates. The plates were incubated at 25±2°C for six days after which the mycelia growth diameter was measured. Three replications were maintained for each treatment.

Effect of temperature on mycelial growth

Temperature tolerance by cultivation of the isolated fungi was determined. Petri plates containing 20 ml of PDA medium were inoculated with 9 mm mycelia disc from 10 day old culture of different isolates. Disks of mycelium were cut with a flamed cork-borer and transferred to Petri dishes containing PDA media. These plates in triplicate were incubated at 5, 10, 15, 25, 30, 35 and 40°C for 10 days. The diameter of the growing colony was measured crosswise in two directions in 10-day old cultures. The average of these two readings was taken as diameter of the colony. The experiment was conducted in completely randomized design (CRD) and the data were statistically analyzed. Cryptosporiopsis spp. (IMI396316) was obtained from CABI, UK for reference.

Effect of light on mycelial growth

Effects of light on mycelial extension were determined by measuring the radial growth of the colony. PDA medium was autoclaved at 121°C for 15 min, and 20 ml of it was poured into Petri dishes. Mycelial disc of 9 mm of each isolate was used to inoculate Petri plates. After cooling, a 9 mm diameter disk of actively-growing mycelia from PDA was transferred into the medium, and then incubated at 25±2°C for six days in three different light conditions in the incubator (fluorescent light). Carbon paper was used to wrap the Petri dishes for darkness. Fluorescent lamp was used for light exposure. The light conditions were: a) first group was incubated in total darkness, b) the second group was in complete light, and c) the third group was in 12 h alternating shifts of total darkness and light. Colony diameter was recorded after 10 days of incubation. Cryptosporiopsis spp. (IMI396316) was obtained from CABI, UK for reference.

Statistical analysis

All the experiments were repeated once with similar results. The data were statistically analyzed according to Gomez and Gomez (1984). The package used for analysis was SAS ver 9.2 developed by SAS Institute, (1999). "SAS/Stat user's Guide". SAS Institute Inc. Cary, N.C.

RESULTS AND DISCUSSION

Pathogenicity test

Cashew seedlings inoculated with Cryptosporiopsis spp conidial suspensions exhibited small brown spots on multiple leaves. Spots enlarged over time and closely resembled spots observed in the field, although disease severity appeared lower than for field plants. Sterile water control did not display any disease symptoms. After 72 h, leaves sprayed with Cryptosporiopsis spp isolate began curling thereafter to developed dark, irregularly shaped brown spots. The younger first leaves of cashew seedlings were more susceptible than the older second leaves.

Effect of temperature on mycelial growth

The temperature range indicates that the pathogen Cryptosporiopsis spp. causing blight disease in cashew (A. occidentale L) can survive and be distributed in environments that are within the range (Table 2). The 11 isolates grew well at temperatures of 30°C (88.83 mm) followed by 25°C (82.40 mm) and 35°C (72.04 mm). The temperature better suited for mycelial growth ranged from 25 to 30°C. As the temperature increased, the mycelial growth increased but at 35°C the growth started to decline. This could be attributed to increase in enzymatic activity of Cryptosporiopsis spp. The least growth was produced at 5°C (9.19 mm). The fungus failed to grow at 40°C probably due to the inactivation of enzymes with a resulting effect on metabolism that affects growth. There were significant differences in mycelial growth of isolates (Table 2) in different range of temperatures, F (76.233) = 2664.7, P<0.0001.

Effect of pH on mycelial growth

There were significant differences in pH levels in the 11 isolates, F (5.197) = 3372.7, P<0.0001. pH 7 was found to be ideal and produced the maximum mycelial growth of 67.79 mm followed by pH 6.0 (62.94 mm) and pH 8.0

Table 2. Effect of temperature on growth of isolates.

Isolate	Mean mycelial growth (mm) of the isolate						
	5°C	10°C	15°C	20°C	25°C	30°C	35°C
AA1	9.43[t]	17.23[s]	37.57[q]	63.45[k]	79.37[e]	88.50[ab]	67.53[j]
AA2	9.87[t]	17.53[s]	43.43[no]	67.73[j]	83.57[d]	89.37[a]	75.93[f]
AA3	8.55[t]	19.03[rs]	44.43[mn]	70.13[i]	87.50[ab]	90.00[a]	75.50[fg]
AA4	9.00[t]	20.10[r]	47.37[l]	73.50[gh]	87.67[ab]	89.33[a]	74.33[fgh]
AA5	8.53[t]	19.20[rs]	45.43[m]	76.07[f]	87.90[ab]	89.33[a]	68.13[ij]
AA6	9.23[t]	17.53[s]	38.20[pq]	63.63[k]	76.40[f]	88.10[ab]	69.10[ij]
AA7	10.47[t]	18.23[rs]	40.27[p]	72.47[fgh]	85.33[c]	89.33[a]	70.13[i]
AA8	9.00[t]	17.43[s]	37.63[q]	64.67[k]	78.43[e]	87.67[ab]	68.37[ij]
AA9	8.60[t]	18.37[rs]	39.17[pq]	67.13[j]	76.37[f]	86.33[bc]	75.77[fg]
AA10	9.17[t]	18.17r[s]	42.33[o]	64.43[k]	82.10[d]	89.67[a]	73.47[gh]
IMI396316	9.27[t]	17.57[s]	38.30[pq]	65.40[k]	81.77[d]	89.47[a]	74.13[fgh]
Grand mean	9.19	18.21	41.28	68.06	82.4	88.83	72.04

*Means separated using least significant difference test (LSD) by the same letter are not significantly different ($P < 0.05$) from each other.

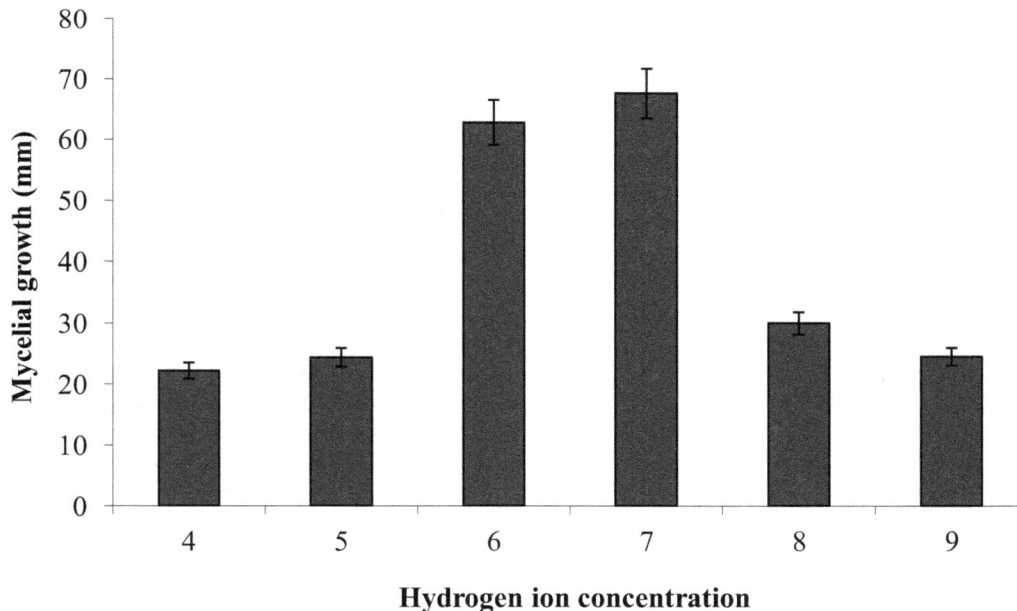

Figure 1. Effect of different pH levels on the mycelial growth of *Cryptosporiopsis* spp.

(30.09 mm). There was an observed decrease in growth when pH increased from 7.0 to 9.0. pH 4 recorded the lowest mean mycelial growth at 22.24 mm (Figure 1). The pH below 6.0 and above 7.0 produced inhibitory mycelial growth of the eleven isolates. The *Cryptosporiopsis* spp. isolates prefers pH range of 6.0 to 7.0. The fungus isolates preferred slightly acidic pH for the growth (Figure

1). The effect of pH on mycelial growth and conidial germination was not significant from pH 4 to 10.

Effects of light on mycelial growth

Photoperiod showed significant effect (F (29.89) = 750.5,

Table 3. Effect of light intensities on mycelial growth of *Cryptosporiopsis* spp. causing blight.

Isolate	Mean mycelial growth (mm) of the isolates		
	12 h photoperiod	Complete darkness	24 h photoperiod
AA1	*58.00±0.5^d	24.67±0.3^n	34.67±0.8^{hi}
AA2	52.00±0.5^f	25.33±0.3^n	32.00±0.5^j
AA3	48.00±0.5^g	29.00±0.5^k	31.00±0.5^j
AA4	52.33±0.3^f	27.67±0.3^{klm}	35.67±0.3^h
AA5	58.67±0.3^{cd}	17.33±0.3^o	27.00±0.5^m
AA6	54.00±0.5^e	31.00±1.0^j	31.00±0.5^j
AA7	63.67±0.3^a	17.67±0.3^o	35.67±0.3^h
AA8	63.67±0.3^a	27.33±0.6^{lm}	34.00±0.5^i
AA9	60.67±0.8^b	28.67±0.3^{kl}	34.67±0.3^{hi}
AA10	60.00±0.5^{bc}	18.67±0.3^o	28.33±0.6^{klm}
IM1396316	60.67±0.8^b	25.33±0.3^n	27.00±0.5^m
Grand mean	57.43	24.70	31.91

*Means separated using Duncan's Multiple Range Test (DMRT) by the same letter are not significantly different (P<0.05) from each other.

P<0.0001) on the growth of fungal mycelium (Table 3). The highest mean mycelial growth (57.43 mm) was observed in 12 h photoperiod, followed by 24 h photoperiod (31.91 mm). The lowest mean mycelial growth (24.70 mm) was found in complete darkness. This indicates that the fungus isolates require a period of dark followed by a light period for conidiophore formation and sporogenesis.

The fungus probably uses light to trigger the development of fruiting bodies and phototropic responses of reproductive structures. *Phycomes* and *Pilobolus* have been shown to use light in the formationP<0.0001) on the growth of fungal mycelium (Table 3). The highest mean mycelial growth (57.43 mm) was observed in 12 h photoperiod, followed by 24 h photoperiod (31.91 mm). The lowest mean mycelial growth (24.70 mm) was found in complete darkness. This indicates that the fungus isolates require a period of dark followed by a light period for conidiophore formation and sporogenesis.

The fungus probably uses light to trigger the development of fruiting bodies and phototropic responses of reproductive structures. *Phycomes* and *Pilobolus* have been shown to use light in the formation of reproductive structures (Alexopoulus et al., 1996). Most light sensitive fungi sporulate when exposed to continuous light, but some called diurnal sporulators, require a period of darkness followed by a light period (Leach, 1967). Mycelial colony was relatively dense when it was incubated in alternating shifts of dark/light conditions; this result shows that light is also an important factor in the growth of this fungus. There were significant differences in light duration among the isolates (P<0.0001). Isolates AA7 (63.67 mm) and AA8 (63.67 mm) had the highest mean mycelial growth in 12 h alternate light and dark conditions (Table 3). The lowest mean mycelial growth

was recorded in complete darkness by isolates AA7 (17.67 mm) and AA5 (17.33 mm). Adeniyi et al. (2011) reported that light was found to be suitable for maximum growth of *Pestalotia* species which causes leaf spots in cashew.

Cultural study

Growth characters of *Cryptosporiopsis* spp. studied on different solid media indicated that the growth was maximum on WA followed by host leaf medium and PDA supported maximum growth of fungal colony (Table 4). Growth behavior of 11 isolates on seven different media showed significant difference in color, morphology, margin, topography and pigmentation along with sporulation in PDA (Table 5, Figure 2). There were variations among the colony characters of the isolates collected from different locations. Most of the isolates (AA2, AA1, AA4, AA6, AA7, AA9 and AA10) had white colonies whereas AA3 and AA8 produced white brown colored colonies. AA5 had dark brown colored colonies. Pale brown exudates droplets occurred in the centre. The media became brownish in all isolates with time. The substrate color for most isolates ranged from light grey to grayish in most isolates 10 days after incubation. Isolates AA3, AA5, AA7 and AA8 had grayish substrate color while the rest were light grayish. Most isolates (AA1, AA2, AA4, AA5, AA6, AA7, AA9, IMI396316 and AA10) produced smooth margins except for isolates AA8 and AA3 that showed irregular margins.

The colony topography varied from medium to raised fluffy growth in all isolates after 10 days of incubation. Isolates AA4 and AA8 showed medium raised colony topography. AA6 and AA7 produced medium fluffy growth

Table 4. Cultural characters of *Cryptosporiopsis* spp. isolates in PDA medium.

Isolate	Colony color	Substrate color	Margin	Topography	Pigmentation	*colony diameter	Sporulation
AA1	White	Light greyish	Smooth	Raised fluffy growth	White	84.33 ± 0.3^{bcd}	++
AA2	White	Light greyish	Smooth	Raised fluffy growth	White	85.00 ± 0.6^{bcd}	+++
AA3	White brown	Greyish	Irregular	Raised fluffy growth	Light brown	86.33 ± 1.3^{abc}	++
AA4	White	Light greyish	Smooth	Medium raised	White	88.00 ± 0.6^{ab}	++
AA5	Dark brown	Greyish	Smooth	Raised fluffy growth	Dark brown	90.00 ± 0.0^{a}	+++
AA6	White	Light greyish	Smooth	Medium fluffy growth	White	88.00 ± 0.6^{ab}	++
AA7	White	Greyish	Smooth	Medium fluffy growth	White	87.67 ± 0.8^{ab}	++
AA8	White brown	Greyish	Irregular	Medium raised	Light brown	86.33 ± 3.7^{abc}	++
AA9	White	Light greyish	Smooth	Raised fluffy growth	White	82.00 ± 0.6^{d}	+++
AA10	White	Light greyish	Smooth	Raised fluffy growth	White	87.00 ± 1.5^{ab}	+++
IMI396316	White	Light greyish	Smooth	Raised fluffy growth	White	83.00 ± 0.6^{cd}	++

[+]Poor sporulation: 1-10 spore/ microscopic field (100x); [++]medium sporulation: 11-50 spores/microscopic field (100x); [+++] good sporulation: more than 100 spores/ microscopic field (100x); *mean of three replications, In a column, means followed by a common letter are not significantly different at the 5% level by DMRT.

Table 5. Cultural characters of *Cryptosporiopsis* spp. isolates in different media.

Isolate	PDA	CMA	Host Leaf Agar	MEA	TDA	PCA	WA
AA1	White	White	White brown	White brown	Reddish purple	White	White brown
AA2	White	White	White brown	White brown	Purple	Creamish white	Light brown
AA3	White brown	Creamish white	Dark brown	brown	Reddish purple	White	White brown
AA4	White	White	White brown	White brown	Reddish purple	White	White brown
AA5	Dark brown	White brown	White brown	Brown	Purple	Creamish white	White
AA6	White	White	White brown	White brown	Reddish purple	White	White brown
AA7	White	White	White brown	White brown	Reddish purple	White	Light brown
AA8	White brown	White	Dark brown	Brown	Reddish purple	White	White brown
AA9	White	White	White brown	White brown	Reddish purple	White	White brown
AA10	White	White	White brown	White brown	Purple	Creamish white	White brown
IMI396316	White	White	Creamish brown	White brown	Reddish purple	White	White brown
Sporulation	Good	Poor	Good	Good	Medium	Medium	Poor

PDA, Potato dextrose agar; **CMA**, corn meal agar; **MEA**, malt extract agar; **TDA**, tryptone dextrose agar; **PCA**, potato dextrose agar; **WA**, water agar. [+]Poor sporulation: 1-10 spore/ microscopic field (100x); [++]medium sporulation: 11-50 spores/microscopic field (100x); [+++] good sporulation: more than 100 spores/ microscopic field (100x).

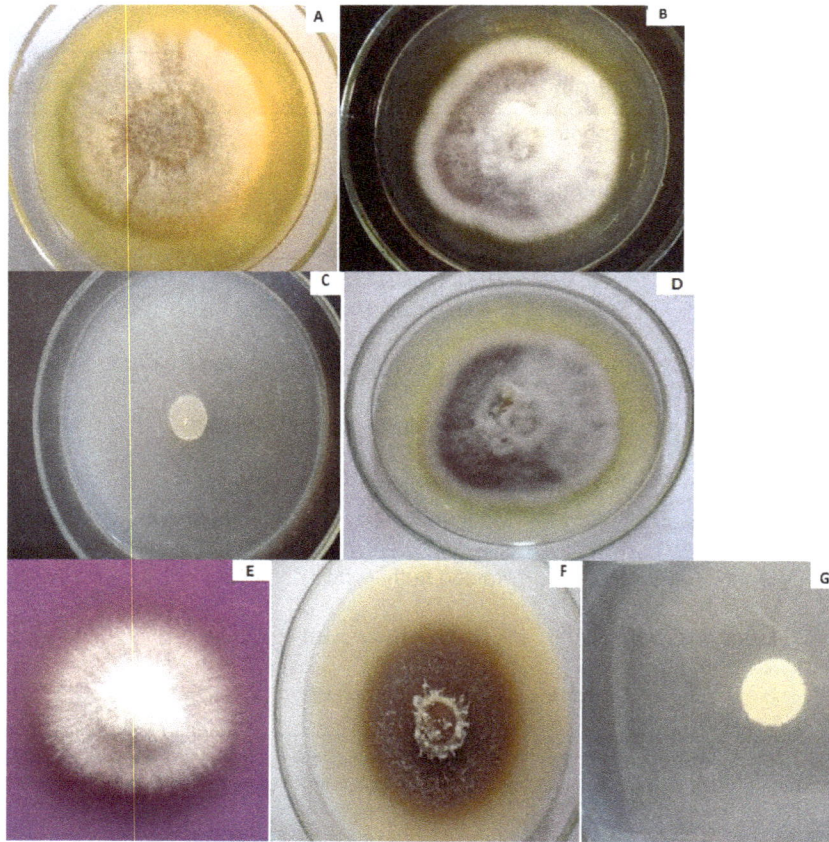

Figure 2. Growth behavior of isolate AA3 on seven different media: **A.** PDA; **B.** WA; **C.** PCA **D.** MEA; **E.** TDA; **F.** host leaf agar; **G.** CMA.

whereas the rest (AA1, AA2, AA3, AA5, AA9, AA10 and IMI396316) showed raised fluffy growth. There were variations in pigmentation of the 11 isolates. Isolates AA1, AA2, AA4, AA6, AA7, AA9, AA10 and IMI396316 produced white pigmentation. AA3 and AA8 showed light brown pigmentation whereas AA5 produced dark brown pigmentation. Good sporulation was achieved by isolates AA2, AA5 and AA10 on PDA medium. Other isolates AA1, AA3, AA4, AA6, AA7, AA8 and IMI396316 showed medium sporulation on PDA medium 10 days after inoculation. All isolates showed variations in colony color on different media (Table 5). Isolate AA1 showed white colonies on PCA, PDA and CMA while in MEA, WA and host leaf extract agar it produced white brown colonies. In the case of TDA, AA1 isolate exhibited reddish purple colored colonies. Other isolates showed variations in colony color on different media as shown in Table 5. Colonies on PDA at 25±2°C after 10 days were white to dark brown. In MEA, the colonies were white brown to brown in color. Colonies on PDA at 25±2°C after 10 days were centrally dark brown, mid zone higher brown with fine droplets of brown exudates. Different concepts have been used by the mycologists to characterize the fungal species, out of which morphological (phenetic or

phenotypic) and reproductive stages are the classic approaches and baseline of fungal taxonomy and nomenclature that are still valid (Davis, 1995; Guarro et al., 1999; Diba et al., 2007; Zain et al., 2009). Colonies of *Cryptosporiopsis* spp. were white initially becoming pale to dark brown in age. Conidiomata were numerous over the surface and they are cream colored. Macroconidia were slightly curved, aseptate, and arose from single phialides on vegetative hyphae or in conidiomata (Figure 3a, b, c and d). Spores were brown to olive brown in color, elliptical with tear shaped ends. Physical and chemical factors have a pronounced effect on diagnostic characters of fungi (Sharma and Pandey, 2010). It is necessary to use several media while attempting to identify a fungus in culture since mycelial growth and sporulation on artificial media are important biological characteristics (St-Germain and Summerbell, 1996).

Growth of *Cryptosporiopsis* spp. on media

To check the best growth of fungi in *Cryptosporiopsis* spp., seven different media; PDA, WA, TDA, CMA, PCA, MEA and host leaf extract agar were selected and

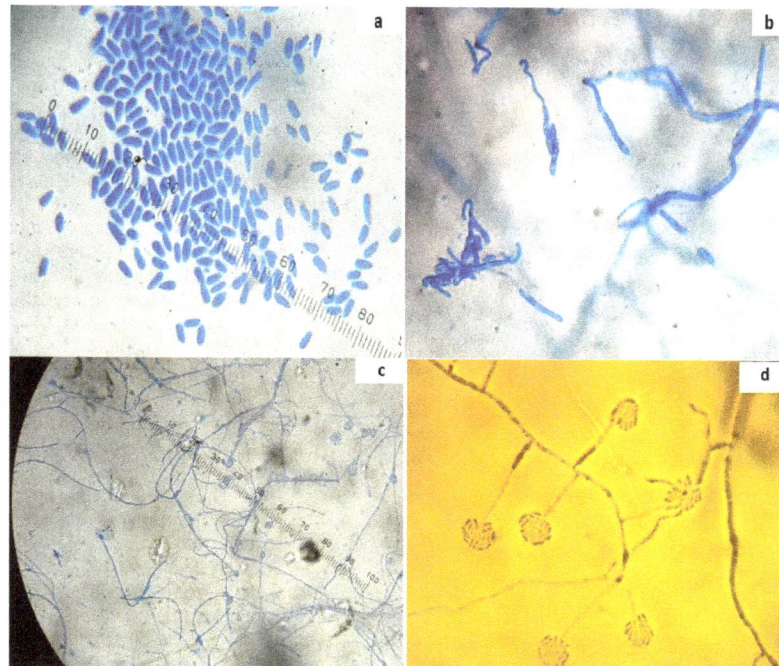

Figure 3. (a) Conidia of *Cryptosporiopsis* spp. causing leaf and nut blight; (b) Germinating Conidia (c) *Cryptosporiopsis* spp. hyphae; (d) *Cryptosporiopsis* spp. Conidiophores.

Table 6. Effect of different media on mycelial growth of *Cryptosporiopsis* spp. isolates.

Isolate	CMA	HOST LEAF	MEA	PCA	PDA	TDA	WA
AA1	*81.67±4.4^{fgh}	88.00±1.5^{abc}	73.00±1.5^{jklmn}	72.33±1.2^{klmno}	84.33±0.3^{bcdef}	69.33±1.2^{no}	84.33±4.7^{bcdef}
AA2	84.33±2.3^{bcdef}	90.00±0.0^{a}	67.67±5.9^{o}	77.33±1.5^{hijk}	85.00±0.6^{abcdef}	71.67±3.3^{lmno}	87.33±1.7^{abcd}
AA3	89.00±1.0^{ab}	86.67±3.3^{abcdef}	70.00±2.9^{mno}	89.00±0.6^{ab}	86.33±1.3^{abcdef}	75.33±2.7^{ijkl}	90.00±0.0^{a}
AA4	78.33±1.7^{ghi}	89.33±0.3^{ab}	85.00±2.9^{abcdef}	86.00±2.5^{abcdef}	88.00±0.6^{abc}	76.33±1.3^{ijkl}	90.00±0.0^{a}
AA5	86.00±3.1^{abcdef}	89.00±0.6^{ab}	87.00±2.5^{abcde}	86.00±0.6^{abcdef}	90.00±0.0^{a}	75.00±0.6^{ijklm}	89.00±0.6^{ab}
AA6	90.00±0.0^{a}	86.67±3.3^{abcdef}	88.00±0.5^{abc}	90.00±0.0^{a}	88.00±0.6^{abc}	76.33±2.3^{ijkl}	90.00±0.0^{a}
AA7	86.33±3.2^{abcdef}	89.67±0.3^{a}	88.67±1.3^{ab}	90.00±0.0^{a}	87.67±0.8^{abc}	74.00±0.6^{ijklmn}	90.00±0.0^{a}
AA8	86.67±2.8^{abcdef}	89.33±0.3^{ab}	89.00±1.0^{ab}	89.67±0.3^{a}	86.33±3.7^{abcdef}	75.67±0.8^{ijkl}	90.00±0.0^{a}
AA9	83.00±0.6^{cdefg}	89.67±0.3^{a}	89.00±0.6^{ab}	87.67±0.3^{abc}	82.00±0.6^{efgh}	73.33±2.7^{ijklmn}	89.00±0.6^{ab}
AA10	86.00±0.6^{abcdef}	89.67±0.3^{a}	85.00±2.9^{abcdef}	90.00±0.0^{a}	87.00±1.5^{abcde}	71.67±0.8^{lmno}	89.67±0.3^{a}
IMI396316	82.33±0.9^{defgh}	90.00±0.3^{a}	90.00±0.0^{a}	89.00±0.6^{ab}	83.00±0.6^{cdefg}	78.00±0.6^{ghij}	90.00±0.0^{a}
Grand Mean	84.8±0.8	88.9±0.4	82.9±1.5	86±1.0	86.2±0.5	74.2±0.6	89.0±0.5

*In a column, mean mycelial diameter (mm) followed by a common letter is not significantly different at the 5% level by DMRT. **PDA**, Potato dextrose agar; **CMA**, corn meal agar; **MEA**, malt extract agar; **TDA**, tryptone dextrose agar; **PCA**, potato dextrose agar; **WA**, water agar.

incubated for 10 days (Figure 3, Table 6). There were significant differences in radial growth of the isolates used in different solid media, $F_{(6, 231)} = 841.17$, $P<0.0001$. The results of the cultural studies of *Cryptosporiopsis* spp. on solid media indicated that the radial growth was maximum on water agar (89 mm) which was significantly superior over all other media. This was followed by host leaf medium (88.9 mm) and PDA (86.2 mm) which were on par. The least radial growth was obtained in TDA (74.2 mm). Several workers recognized the importance of reproductive structures for inoculums production and studies have been conducted on the effects of various media components along with important physiological parameters that lead to maximum sporulation (Kim et al., 2005; Saxena et al., 2001; Saha et al., 2008). Type of culture media and their chemical compositions significantly affected the mycelia growth rate and conidial production of *Phoma exigua* (Zhae and Simon, 2006).

Conclusion

Our findings reveal that culture media differentially influenced the growth, colony character and sporulation of the test fungi. Out of seven test media employed in the present study, PDA, MEA and host leaf agar were found to be the most suitable for good sporulation while PDA reproduced most visible colony morphology. Colonies grew rapidly on a variety of media. It is concluded that instead of using any single culture medium, a combination of two or more media will be more appropriate for routine cultural and morphological characterization of fungi to observe different colony features. This is the first report on physiological studies of cashew leaf and nut blight associated with *Cryptosporiopsis* spp. and this has implications for the epidemiology of the disease. Further studies will create bases for the morphological and molecular characterization of these organisms for better understanding of their biology, identification and classification.

ACKNOWLEDGEMENTS

We thank the German Federal Ministry for Economic Co-operation and Development (BMZ) for funding the cashew project, the International Centre of Insect Pathology and Ecology (ICIPE), Ministry of Agriculture and Food Security in Tanzania through Naliendele Agricultural Research Institute (NARI). We are grateful to CABI, UK, for providing a reference specimen.

REFERENCES

Adeniyi DO, Orisajo SB, Fademi OA, Adenuga OO, Dongo LN (2011). Physiological studies of fungi complexes associated with cashew diseases. ARPN J. Agric. Biol. Sci. 6(4).

Alexopoulus CJ, Mims CW, Blackwell M (1996) Introductory Mycology, 4th ed., John Wiley, New York.

Annual Cashew Research Report-ACRR (2006) Chapter 3 Pathology Report pp 98-99. Naliendele Agricultural Research Institute, Mtwara, Tanzania.

Behrens R (1998) About the spacing of cashew nut trees. Proceedings of International Cashew and Coconut Conference, 17-21 February 1997, Dar es salaam, Tanzania. BioHybrid International Limited, U.K., pp. 48-52.

Davis JI (1995) Species concepts and phylogenetic analysis. Introduction. Syst. Bot. 20:555-559.

Diba K, Kordbacheh P, Mirhendi SH, Rezaie S, Mahmoudi M (2007). Identification of *Asppergillus* species using morphological characteristics. Pak. J. Med. Sci. 23(6):867-872.

GPC (2010). Global Plant Clinic Issues. Volume 90.

Guarro J, Josepa G, Stchigel AM (1999) Developments in Fungal Taxonomy. Clin. Microbiol. Rev.12:454-500.

Harlapur, Sharanappa I, (2005) Phd Thesis: Epidemiology and Management of Turcicium Leaf Blight Of Maize Caused By *Exserohilum turcicium* (Pass.) Leonard and Suggs.

Kim YK, Xiao CL, Rogers JD (2005) Influence of culture media and environmental factors on mycelia growth and pycnidial production of *Sphaeropsis pyriputrescens*. Mycologia 97:25-32.

Opeke LK (2005). Tropical Commodity Tree Crops. Spectrum Books Limited, Ibadan, Nigeria. 360:371-373.

Saha A, Mandal P, Dasgupta S, Saha D (2008) Influence of culture media and environmental factors on mycelial growth and sporulation of *Lasiodiplodia theobromae* (Pat.) Griffon and Maubl. J. Environ. Biol. 29(3):407-410.

Saxena RK, Sangetha L, Vohra A, Gupta R, Gulati R (2001) Induction and mass sporulation in lignin degrading fungus *Ceriporiopsis subvermisppora* for its potential usage in pulp and paper industry Curr. Sci. 81:591-594.

Sharma G, Pandey RR (2010). Influence of culture media on growth colony character and sporulation of fungi isolated from decaying vegetable wastes. J. Yeast Fungal Res. 1(8):157-164.

Sijaona MER, Reeder RH, Waller JM (2006) Cashew leaf and nut blight – A new disease of cashew in Tanzania caused by *Cryptosporiopsis* spp. Plant Pathol. 55:576.

St-Germain G, Summerbell R (1996). Identifying Filamentous Fungi – A Clinical Laboratory Handbook, 1st Ed. Star Publishing Co., Belmont, California.

Tuite J (1969). Plant Pathological Methods: Fungi and Bacteria. Burgess Publishing Company, Minneapolis, MN. p. 239.

Zain ME, Razak AA, El-Sheikh HH, Soliman HG, Khalil AM (2009) Influence of growth medium on diagnostic characters of *Aspergillus* and *Penicillium* species Afr. J. Microbiol. Res. 3(5):280-286.

Zhae S, Simon FS (2006). Effect of culture media, temperature, pH and bio-herbicidal efficacy of *Phoma exigua*, a potential biological control for salal (*Gaultheria shallon*). Biocontrol Sci. Technol. 6:1043-1055.

19

The effects of culture parameters on the conidial germination and yields of *Ophiocordyceps sinensis*

Caiying Mei[1,2], Yi Zhang[2], Xiongmin Mao[2], Keqing Jiang[1,2], Li Cao[2], Xun Yan[2] and Richou Han[2]

[1]College of Life Sciences, Sun Yat-sen University, Guangzhou 510275, China.
[2]Guangdong Entomological Institute, Guangzhou 510260, China.

Ophiocordyceps sinensis is an insect-born fungus with various biological and pharmacological activities. Artificial cultivation of the fungus-insect complex with fruiting body is not successful for the market mainly due to the lack of optimal conditions for fungus-insect interaction. Mass production of the pathogenic conidia as fungal inocula for effective insect infection is one of the critical steps. In this paper, the effects of different culture parameters on the conidial germination (defined in this study as the fungal colonies developed from the conidia on agar plates) and yields of two *O. sinensis* isolates were investigated on solid media. The conidia of two fungal isolates were unable to grow at 4°C or over 25°C. The better conidial germination was recorded on a solid medium (Medium C) containing 2.5% malt extract, 1.0% soluble starch and 0.9% agar at pH 6.0 at 14°C. Medium C was also optimal for harvesting conidial yield. Soluble starch stimulated conidial germination and yields. More nitrogen sources appeared to inhibit the germination and yields of the conidia. Fe^{3+} addition to the medium facilitated the conidial germination but hindered the conidial yields. These results provide a basis for optimal mass production of the conidia for insect infection and artificial production of the fungus-insect complex with fruiting body.

Key words: *Ophiocordyceps sinensis*, conidia, conidial germination, culture parameter.

INTRODUCTION

Ophiocordyceps sinensisis is one of the entomopathogenic fungi used for medicinal treatment and health foods in many Asian Countries (Dong and Yao, 2008; Paterson and Russell, 2008). The fungus parasitizes larvae of moths (Lepidoptera), especially *Hepialus* spp., and converts each larva into a sclerotium, from which the fruiting body grows. This fungus-insect complex is endemic on the Tibetan Plateau and found only from above 3000 m in altitude up to the snow line, usually between 3600 and 5000 m (Li et al., 2011).

Natural production of the fungus-insect complex is very limited and the annual yield has been declining continually over recent years (Yang, 1997). It has been listed as an endangered species for protection (CITES Management Authority of China and China Customers, 2000). The market demand for *O. sinensis* is growing sharply in many Countries (Dong and Yao, 2008; Paterson and Russell, 2008). Due to over exploitation and habitat degradation, the yield of *O. sinensis* from the natural habitats decreases markedly (Liang et al., 2008). Since the mycelial fermentation products are demonstrated to have the same pharmacological efficacy as the wild *O. sinensis* (Li et al., 2001), the mass production of mycelia by submerged fermentation is considered as a promising alternative for natural products. Living strains isolated from natural *O. sinensis*, and *Ophiocordyceps*-based products from mycelial-submerged fermentation, are commercially supplied to meet the needs of human consumption and to reduce the pressure on natural resources of the species (Hamburger, 2007).

Different factors (e.g. temperature, humidity, light, pH, medium, aeration, etc.) influence the mycelial growth and production in submerged fermentation. Dong and Yao (2011) described the optimal parameters (temperature, pH) for mycelial growth on agar plates, showing that the best temperature and pH were 15 to 18°C and pH 5.5 to 6.0, respectively. The effect of light on the mycelial growth and conidial production was not significant (Dong and Yao, 2011; Li, 2002). Medium compositions were very important for mycelial growth. Different media used for the mycelial growth of *Ophiocordyceps* spp. in submerged cultures were reported (Dong and Yao, 2005; Kim and Yun, 2005; Xu et al., 2002). Those media contained ingredients of chemical composition, including beef extract, corn steep powder, potato, soya bean, etc. For cultures of *O. sinensis*, few reports on the mycelial production of *O. sinensis* on the agar plates were documented at present. Yin et al. (2009) found that modified PDA medium (potato 200 g, glucose 20 g, yeast extract 1 g, tryptone 1 g, agar 18 g, distilled water 1000 ml, pH = 6.5) supported quick growth of *O. sinensis*. Xie et al. (2010) reported an optimal medium (potato 100 g, sucrose 25 g, peptone 20 g, beef extract 5 g, yeast extract 2 g, lactoalbumin hydrolysate 5 g, KH_2PO_4 2 g, $MgSO_4$ 1.5 g, agar 18 g, distilled water 1000 ml, pH = 6.0) for *O. sinensis*. Ren et al. (2013) evaluated the influence of nutritional and physical stress on sporulation, conidial germination and vegetative biomass of *O. sinensis*, and demonstrated that the fungus had strong physiological adaptations to environmental stress. Although, mycelial mass production in the industrial fermentation systems had the potential advantage to replace the natural products, the fungus-insect complex with fruiting body is still strongly regarded as the traditional medicine by the growing market. In addition, modern research indicates that fungus-insect complex contains more active components with multiple pharmacological effects than the fermentated products (Hsu et al., 2002; Li et al., 2006). Unfortunately, until now, there have been no reports on the commercial cultivation of the fungus-insect complex with fruiting body in the low altitude area outside Tibetan Plateau.

For the successful commercial production of traditional fungus-insect complex, two steps are very important: artificial rearing of the *Hepialus* host insect, and production of the pathogenic conidia as fungal inocula for effective insect infection. As the initial crucial step, it is necessary to establish optimal media for mass production of the conidia, and the favorable conditions for conidia growth. However, there was only one present reference on the conidia production of *O. sinensis* (Li, 2002), which described the effects of several culture media on the conidia production of this fungus, and found that PDA medium supplemented with 1% peptone stimulated the mycelial growth but the media with corn powder and wheat flour were favorable for conidia production. In this study, th effects of temperature, pH and media on the

conidial germination (defined as the fungal colonies developed from the conidia on agar plates) and yields of two *O. sinensis* isolates were investigated, to provide cues for mass production of the conidia for insect infection and production of the traditional fungus-insect complex.

MATERIALS AND METHODS

Fungal isolates

Two isolates (N2009-1-1 and Qin) of *O. sinensis* fungus used for this study were isolated from Qinghai, China (by Guangdong Entomological Institute, Guangzhou), and from Sichuan, China (provided by Prof. Qilian Qin, Institute of Zoology, Chinese Academy of Sciences), respectively. They were maintained on PPDA agar (Table 1) plates at 4°C and subcultured at 14°C for 40 days as seed fungal cultures.

Molecular identification of the fungal isolates

To make sure of the correctness of these fungal isolates, molecular identification method was introduced (Jiang and Yao, 2005). Briefly, the internal transcribed spacer (ITS) of nuclear ribosomal DNA of the isolates was amplified and the sequences were compared with the dataset in GenBank, NCBI. The morphological characters were also observed on agar plates.

Observation of the fungal conidia

Strains N2009-1-1 and Qin were cultured in dark on Medium C (Table 1) for 40 days, and then the conidia were eluted from the plates according to the method of Zheng et al. (2011). Briefly, the conidia were harvested under sterile conditions by flooding the plate with sterile 0.05% Tween 80 and then scraping the colony with forceps. Conidia were stirred into suspension for 10 min in 5 ml of sterile distilled water, which was then filtered through cheesecloth to remove the mycelia. The filtrate was centrifuged at 12000 rpm at 10°C for 15 min. The harvested conidia were observed under phase-contrast microscope (Axioplan 2, ZEISS, Germany).

The conidia and mycelia were also observed by scanning electron microscopy according to the method (Anamika et al., 2012). Briefly, fungal colonies containing conidia and mycelia on Medium C were cut into 0.5 × 0.5 cm pieces with forceps and fixed in 2.5% glutaraldehyde and 2% paraformaldehyde in 0.1 M sodium phosphate, pH 7.2, for 12 h at 4°C. The glutaraldehyde solution was removed and replaced with 0.1 M sodium phosphate, pH 7.2. The samples were washed twice for 10 min at 4°C, postfixed in 1% osmium tetraoxide for 3 h at room temperature, replaced with 0.1 M sodium phosphate, pH 7.2, and then washed twice for 10 min. The resulting samples were dehydrated in a series of ethanol solutions (30, 50, 70, 80, 90, 100, 100 and 100%) for every 10 min, and replaced with tert-butyl alcohol, and further washed twice for 10 min. They were freeze dried and then coated with gold/palladium. The treated conidia and mycelia were observed using a JSM-6360LV scanning electron microscope (JSM-6360LV, Japan electronic Japan).

Media preparation

The following media (Table 1) were selected based on the published references or formulated by this study. The chemicals used were tryptone, yeast extract (Oxoid, England); malt extract, soluble starch

Table 1. The components of different solid media.

Component	Medium								
	C	D	E	F	PDA	M	PPDA	PPDA(Fe^{3+})	LA*
Malt extract (g)	25								
Soluble starch (g)	10								
Sucrose (g)		20		50					
Maltose (g)			15						
Glycerol (g)			10						
Mannose (g)		5							
Galactose (g)		1.5							
Glucose (g)					20	20	20	20	
Potato (g)					200	200	200	200	200
Tryptone (g)		3.5	10	10	10	1	10	10	10
Yeast extract (g)		1.5	10	3		1			
KH_2PO_4 (g)			1	1			3	3	
$MgSO_4·7H_2O$ (g)			0.2	0.5			1.5	1.5	
$CaCl_2$ (g)			0.5						
$Fe_2(SO_4)_3$ (g)								0.1	
Thiamine (VB_1) (g)				0.0005			0.02	0.02	
Gold theragran (pill)									2
Galleria mellonella larvae (g)									20
Agar (g)	15	15	15	15	15	15	15	15	15
H_2O (ml)	1000	1000	1000	1000	1000	1000	1000	1000	1000
References	In this study	Zhang et al. (2008)	Xiao et al. (2004)	Dong et al. (2005)	In this Study	Zhao et al. (2006)	In this study	In this study	In this Study

*In the preparation of Medium LA, living last instar G. mellonella larvae reared by this laboratory were placed at -20°C refrigerator for 30 min, then the frozen larvae were ground into powder with liquid nitrogen before us.

(Huankai Biotechnology, China); thiamine (VB_1) (Jiangmen Health & Pharmaceutical Company, China); gold theragran (Sino-American Shanghai Squibb Pharmaceutical Company, China); agar (Weijia Technology Company, China); reagents for preparation of scanning electron microscopic specimens (Guangzhou Chemical Reagent Company, China); other chemicals were from Guangzhou Shangbang Trade Company.

The initial pH of all the media was adjusted according to the experimental design, by using 1 N NaOH or HCl. The pH value was determined using a Delta-320-S pH meter (Mettler-Toledo, Shanghai, China). The media were autoclaved at 121°C for 30 min.

Effect of different media on the conidia yields

The fungal colonies with mycelia from seed cultures were cut into 0.5 × 0.5 cm pieces with forceps, and transferred onto different fresh agar plates (Table 1). The inoculated plates were sealed with parafilm and kept in dark at 14°C. The yield of the conidia from each medium was determined after 35 and 80 days according to the method of Zheng et al. (2011). The concentration of conidia in each suspension was determined with a hemocytometer (Weijia Technology Company, China). Three plates were sampled for each treatment.

Effect of culture parameters on the conidial germination

The effect of temperature, initial medium pH, agar content in the

medium and different media on the colony development from the conidia was determined, to establish the ideal conditions for conidial germination on agar plates. The conidia from 40-day old Medium C plates at 14°C were harvested under sterile conditions by the above method (Zheng et al., 2011).

Selective temperature effect was conducted by spreading 100 µl of the suspension containing 10^6 conidia onto the Medium M plates (d = 9 cm) with 1.5% agar and at pH 6.0, and placing these plates in the cold rooms or incubators at 4, 14, 25 and 30°C, based on the results from the reasonable screen tests and reference evaluation. For pH effect, 100 µl of the suspension containing 10^6 conidia were spread onto the Medium M with 1.5% agar and at different pH values (4.0, 5.0, 6.0, 7.0 and 8.0), and these plates were placed at 14°C.

To check the effect of agar percentage in the medium, 100 µl of the suspension containing 10^6 conidia were spread onto the Medium M plates with different agar percentages (0.9, 1.2, 1.5, 1.8 and 2.1%) and at pH 6.0 and 14°C. 100 µl of the suspension containing 10^6 conidia were also spread onto different media (Table 1) plates with 1.5% agar and at pH 6.0 and 14°C, to determine the medium effect. All these inoculated plates were checked after 60 days and the forming colonies from the spread conidia were calculated. Three replicates were established for each treatment.

Data analysis

The values were expressed as means ± SD. Data obtained from the

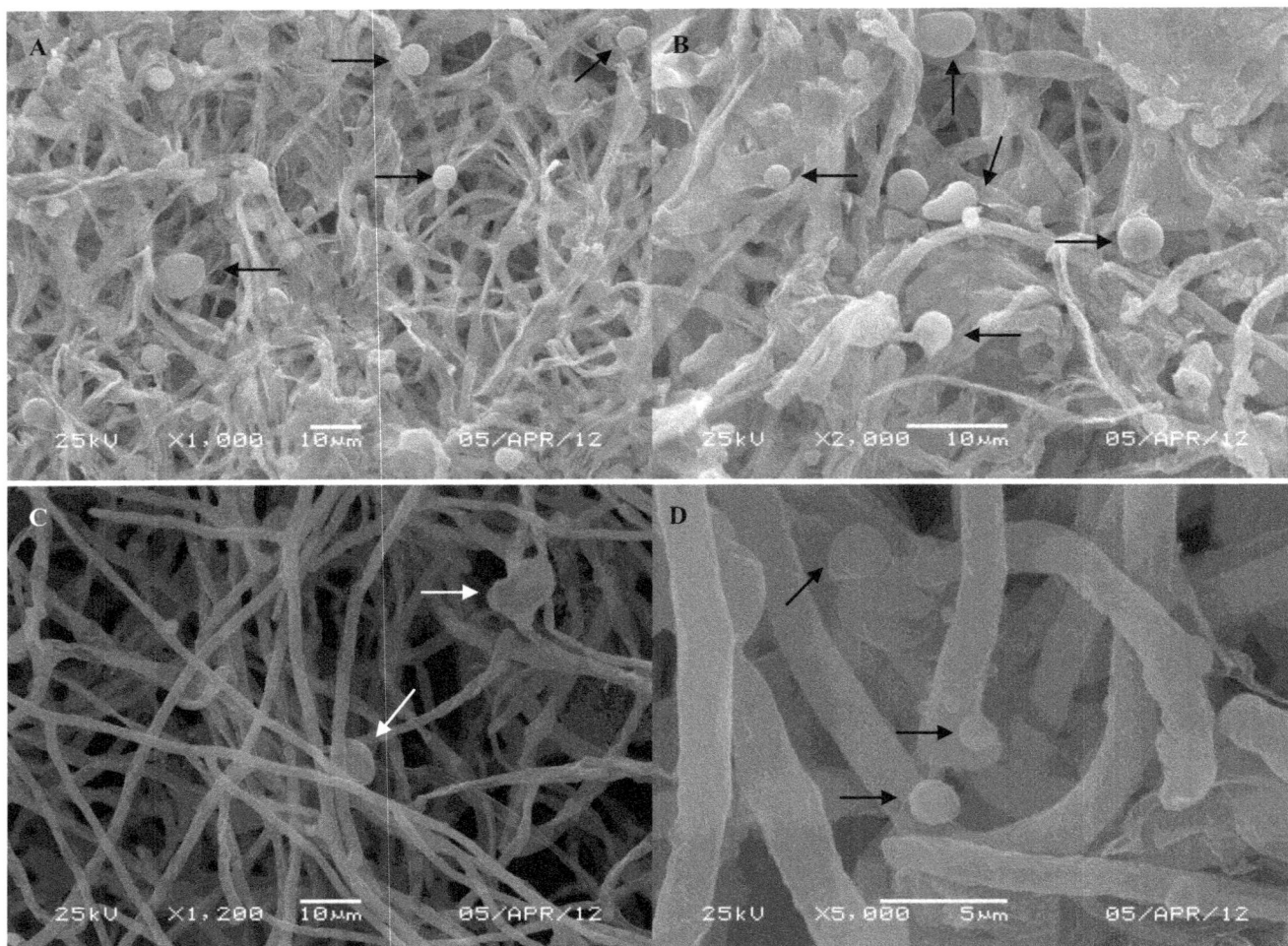

Figure 1. The micrograph of conidia of *O. sinensis* cultured in dark on Medium C at 14°C for 40 days by scanning electronic microscope. A-B: Scanning electron microscopy of conidia of strain N2009-1-1, C-D: Scanning electron microscopy of conidia of strain Qin. The arrows indicated the conidia of *O. sinensis* isolates, showing the subglobose, elliptical and kidney conidia with different sizes in the same *O. sinensis* isolates. Scale bar: A-C = 10 μm, D = 5 μm.

effect of culture parameters on the conidial germination or from the effect of culture media on the conidia yields were analyzed by normal one-way analysis of variance (ANOVA). Tests of significant differences were determined by Duncan's multiple range tests at P = 0.05 using SPSS 17.0 (SPSS Inc., Chicago, IL, USA).

RESULTS

Molecular identification of fungal isolates

The mycelia of two *O. sinensis* isolates were confirmed by rDNA ITS fragments, in addition to the morphological characters on agar plates. The amplified sequences including the complete ITS regions of N2009-1-1 and Qin isolates were 601 and 533 bp long, respectively, and have been submitted to GenBank (accession numbers JQ753314 and JQ753315, respectively). Two ITS sequences had 100% similarities with true *O. sinensis*

isolates of GU233806.1 and EU570958.1 from the GenBank, respectively.

The micrograph of conidia of *O. sinensis* isolates by scanning electronic microscope

The conidia of N2009-1-1 and Qin looked like subglobose, elliptical or kidney shapes (sizes about 2.6 - 10.2 × 2.1 - 5.2 μm), by scanning electronic microscope (Figure 1).

In the preparation of scanning electronic microscope samples, a double fixation was used to maintain the full appearance of the sample structure. As shown in Figure 1, the conidia were present inside or outside the mycelia in the asymmetric division; the conidial shapes and sizes were similar to those reported by other workers (Guo et al., 2010; Liu et al., 2003; He et al., 2011; Xiao et al., 2011). It appeared that different shapes and sizes of conidia were

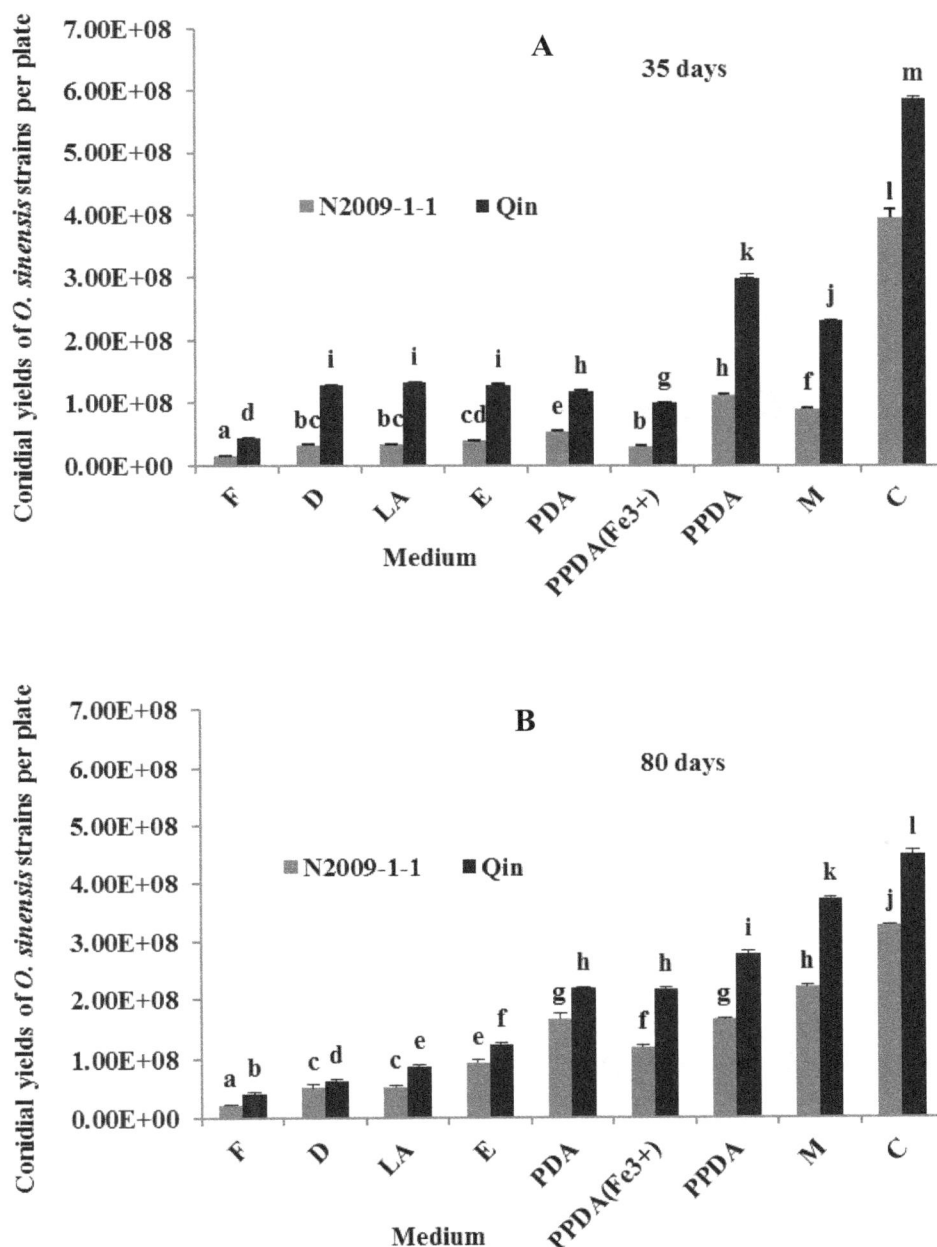

Figure 2. The conidial yields of two isolates (N2009-1-1 and Qin) of *O. sinensis* on 9 media (E, PDA, PPDA, PPDA(Fe^{3+}), F, LA, D, M and C) at different culture times (35 and 80 days) at 14°C. A. 35d; B. 80d. The columns with different letters indicated significant differences by Duncan's multiple-range test (P < 0.05).

observed in the same *O. sinensis* isolates.

Effect of different media on the conidial yields

There were significant differences in the conidial yields of two fungal isolates from different media, or from the same medium at different culture periods (Figure 2). The conidial yields of N2009-1-1 isolates were in order F < PPDA(Fe^{3+}) < D < LA < E < PDA < M < PPDA < C at 35 days (Figure 2A), or F < D < LA < E < PPDA(Fe^{3+}) < PPDA < PDA < M < C at 80 days (Figure 2B). The conidial yields of Qin isolates were in order F < PPDA(Fe^{3+}) < PDA < D < E < LA < M < PPDA < C at 35 days (Figure 2A) or F < D < LA < E < PPDA(Fe^{3+}) < PDA < PPDA < M < C at 80 days (Figure 2B), respectively.

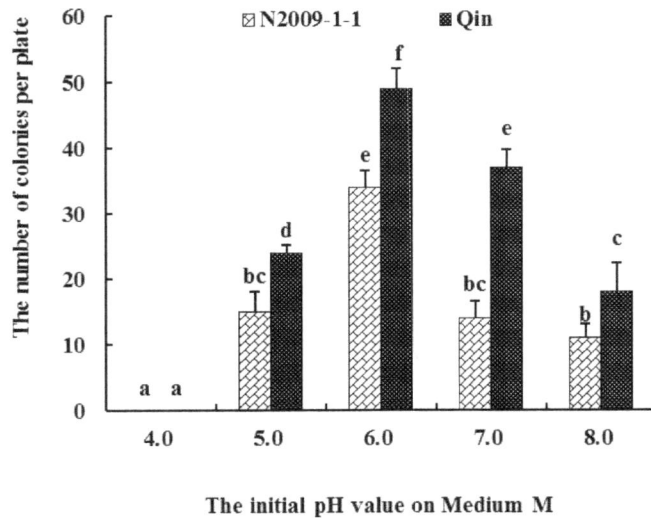

Figure 3. The number of the fungal colonies developed from the conidia of two *Ophiocordyceps* isolates (N2009-1-1 and Qin) on Medium M with different pH values (4.0, 5.0, 6.0, 7.0 and 8.0) at 14°C. The columns with different letters indicate significant differences by Duncan's multiple-range test (P < 0.05).

It seemed that Medium C was the ideal medium for conidial production of two fungal isolates, but Medium F was the worst. The conidial yields of Qin were much more than those of N2009-1-1, even from the same medium (P < 0.05). The maximum yields of two fungal isolates from different media were obtained in different culture times. N2009-1-1 and Qin reached their maximum yields both in 35 days from Medium C (Figure 2).

Effect of temperature on the conidial germination

In the tested temperature range, the conidia of N2009-1-1 and Qin isolates grew only at 14°C. No fungal colonies were observed on the Medium M plates after 60 days at 4°C or at 25 and 30°C. More colonies (49.0 ± 3.0 colonies per plate) of Qin isolate were found from the agar plates than those of N2009-1-1 (34.0 ± 2.5 colonies per plate).

Effect of medium pH on the conidial germination

The conidia were able to grow from the agar plates at pH 5.0-8.0, except those at pH 4.0 (Figure 3). The maximum colonies of two isolates were obtained from the plates at pH 6.0. An increase or decrease at pH beyond pH 6.0 resulted in decreased growth. No significant differences in the fungal colonies of N2009-1-1 were found from the plates at pH 5.0, 7.0 and 8.0 (P > 0.05); however, significant differences in the fungal colonies of Qin were found from the plates with different pH values (P < 0.05). Much more Qin colonies were observed from all the plates than those of N2009-1-1.

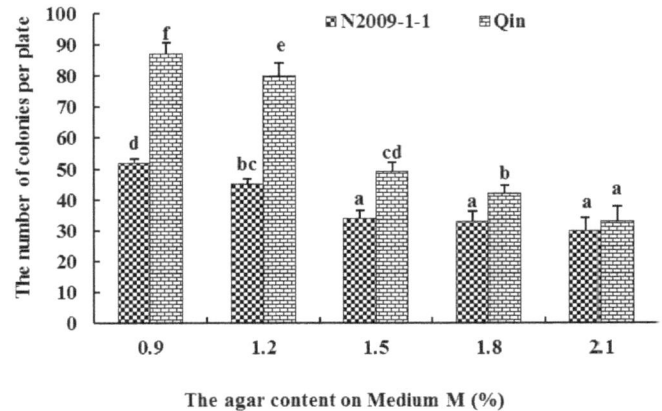

Figure 4. The number of the fungal colonies developed from the conidia of two *O. sinensis* isolates (N2009-1-1 and Qin) on Medium M with different agar content (0.9, 1.2, 1.5, 1.8 and 2.1%) at 14°C. The columns with different letters indicate significant differences by Duncan's multiple-range test (P < 0.05).

Effect of agar content on the conidial germination

The agar content in the solid medium reflected the hardness and free water of the medium, and significantly influenced the conidial growth. In the test range of agar percentages, the fungal colonies decreased with increasing agar content. The medium was too soft to culture the fungi if 0.8% agar was used (data not shown). N2009-1-1 colonies on the Medium M plates containing 1.5, 1.8 or 2.1% agar were not significantly different (P > 0.05), but significantly different on the plates with 0.9, 1.2 or 1.5% agar. For Qin isolate, the colonies on the plates with 0.9% or over agar were significantly different (P < 0.05). More colonies of Qin isolate were also present on all the plates than those of N2009-1-1 (Figure 4).

Effect of different media on the conidial germination

Different media also influenced markedly the fungal colonies developed from the spread conidia (Figure 5). The colonies from N2009-1-1 and Qin conidia on the media were in almost the same order: Medium C > M > D > LA > F > PPDA(Fe^{3+}) > PPDA > PDA > E. It seemed that the most colonies of two isolates were present on Medium C, but no colonies on Medium E. More colonies of Qin were found on all the plates than those of N2009-1-1.

DISCUSSION

In this study, culture parameters (temperature, initial medium pH, agar content in the medium, different media) were demonstrated to significantly influence the conidial germination and yields of two *O. sinensis* isolates in solid culture system.

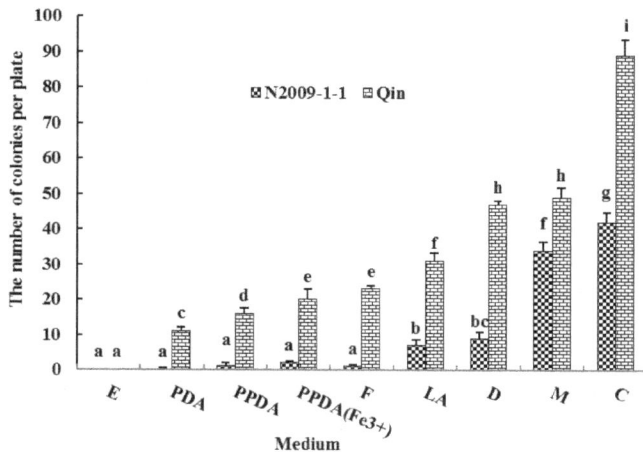

Figure 5. The number of the colonies developed from conidia of two *O. sinensis* isolates (N2009-1-1 and Qin) on different media (E, PDA, PPDA, PPDA(Fe^{3+}), F, LA, D, M and C) at 14°C. The columns with different letters indicate significant differences by Duncan's multiple-range test ($P < 0.05$).

The optimal temperature for the mycelia growth of these fungi was reported to be 15 to 18°C (Dong and Yao, 2011). So several typical or extreme temperatures (such as 4, 14, 25 and 30°C) were selected to test their effects on the conidial germination in this study. It was not surprising that the ideal temperature for conidial germination was 14°C, and that temperature over 25°C inhibited the conidial germination, because temperature at 25°C also stopped the mycelial growth (Dong and Yao, 2011; Yin et al., 2009). Although, the mycelia of two fungal isolates grew slowly at 4°C, it was interesting that their conidia were not able to grow at 4°C. It seems that conidia need higher temperature for germination than mycelia growth under the selective media.

The optimal initial medium pH for mycelial growth of *O. sinensis* was 5.5 to 6.0 (Dong and Yao, 2011), similar to that of other related fungi, e.g., *Cordyceps militaris* (Park et al., 2001) and *Cordyceps jiangxiensis* JXPJ0109 (Xiao et al., 2004). Dong and Yao (2011) also reported very weak growth of mycelia of *O. sinensis* at pH 4.3, however, for the first time from the present result, no colonies developed from the conidia were observed on the plates at pH 4.0, even with higher concentration of conidia (10^8/plate) (unpublished data). It appears that the temperature and pH are more crucial to the conidial germination than to the mycelial production. Free water content in the medium controlled by the agar content significantly influenced the conidial germination of *O. sinensis*, which was consistent with the other report on *O. sinensis* (Qiao et al., 2003). No agar content less than 0.9% was selected for this study because the plates with less than 0.9% agar were too soft to support the conidial germination.

Different media were developed to culture various *O. sinensis* isolates both in submerged (Dong and Yao, 2005;

Kim and Yun, 2005; Kim et al., 2003) and on solid cultures (Xie et al., 2010; Yin et al., 2009). However, all these media were utilized for the mycelial production of the fungi, except those published by only one reference (Li, 2002) which described the effect of media on the conidial production. In this study, different media referred from reliable papers and formulated by our laboratory showed their significant effects not only on the conidial germination but also on the conidial yields. Medium C containing malt extract and soluble starch was demonstrated to be the optimal medium both for conidial gemination and yields. The maximum colonies that grew from the conidia of two fungal isolates were obtained on Medium C. Maximum conidial yields of N2009-1-1 (400.00 ± 8.72 × 10^6/plate after 35 days) and Qin (588.00 ± 3.03 × 10^6/plate after 35 days) were recorded also on Medium C, which were at least 1000 times more than the conidial yields of a fungal isolate from Qinghai, China, reported by Li (2002). Except malt extract, no more nitrogen sources but carbon source (soluble starch) were composed in Medium C, indicating that soluble starch stimulated conidial germination and yields. Apparently, mycelial production was stimulated on the media containing more nitrogen sources than carbon sources, as reported by Kim et al. (2003) and Sehgal and Sagar (2006). But our results indicated that nitrogen source was not a key factor for conidial germination and yields. More nitrogen sources (such as tryptone) appeared to inhibit the germination and yields of the conidia.

Although, Medium F was selected as an optimal medium for submerged culture of an *O. sinensis* isolate (Dong and Yao, 2005), it did not improve the conidial yields of the present fungal isolates. More fungal colonies of Qin were obtained on PPDA(Fe^{3+}) medium than those on the PPDA but had an opposite result on conidial yields, indicating favorable Fe^{3+} addition for conidial germination but hindered the conidial yields. More conidial yields of both fungal isolates were harvested on the PPDA containing mineral elements and vitamins than on the PDA, showing the importance of mineral elements and vitamins for conidial production, as reported in other fungi by Kim et al. (2005) and Zhu et al. (2008). Medium E could not support the conidial germination of both isolates, probably due to the higher concentration of glycerol input in the medium. Conidial production of two isolates was poor but the conidial germination was much better on Medium LA containing the fresh powder of *Galleria mellonella* larvae. Maybe these fungi were adapted to the insect materials for the conidial germination.

Conidial germination is the first step for mycelia and conidial production. Ren et al. (2013) reported that stress conditions such as frozen-shock or heat-shock generally reduced the capacity for conidial germination. In the present result, temperature, pH and medium components are more crucial to the conidial germination than to the mycelial production. The optimal condition for germination was not the same as that for the optimal mycelia growth.

This result may provide information for selecting culture parameters for conidial germination and mycelia production.

ACKNOWLEDGEMENTS

This work was supported by the Research Project of Guangdong Province (Grant 2008B040200017, 2010A040301012) and by the Guangdong Province-Chinese Academy of Sciences Comprehensive Strategic Cooperation Project (Grant 2009B091300015).

REFERENCES

Anamika R, Sayali S, Patricia AC, Steven DH, Rolf P, Andrew JM (2012). *Phanerochaete chrysosporium* produces a diverse array of extracellular enzymes when grown on sorghum. Appl. Microbiol. Biotechnol. 93:2075-2089.

Dong CH, Yao YJ (2005). Nutritional requirements of mycelial growth of *Cordyceps sinensis* in submerged culture. J. Appl. Microbiol. 99:483-492.

Dong CH, Yao YJ (2008). *In vitro* evaluation of antioxidant activities of aqueous extracts from natural and cultured mycelia of *Cordyceps sinensis*. Lebensm. Wiss. Technol. 41:669-677.

Dong CH, Yao YJ (2011). On the reliability of fungal materials used in studies on *Ophiocordyceps sinensis*. J. Ind. Microbiol. Biotechnol. 38:1027-1035.

Guo YL, Xiao PG, Wei JC (2010). On the biology and sustainable utilization of the Chinese medicine treasure *Ophiocordyceps sinensis*. J. Chin. Modern Med. 12:3-8.

Hamburger M (2007). Comment on comparison of protective effects between cultured *Cordyceps militaris* and natural *Cordyceps sinensis* against oxidative damage. J. Agric. Food. Chem. 55:7213-7214.

He SQ, Wang SX, Luo JC, Li FQ, Jin XL, Wen ZH, Ma FQ, Zhou ZX, Tang DZ (2011). The re-study for morphology of *Ophiocordyceps sinensis* and *Hirsutella sinensis*. Microbiology 38:1730-1738.

Hsu TH, Shiao LH, Hsieh C, Chang DM (2002). A comparison of the chemical composition and bioactive ingredients of the Chinese medicinal mushroom Dong Chong Xia Cao, its counterfeit and mimic, and fermented mycelium of *Cordyceps sinensis*. Food Chem. 78:463-469.

Jiang Y, Yao YJ (2005). ITS sequence analysis and ascomatal development of *Pseudogymnoascus roseus*. Mycotaxon. 94:55-73.

Kim HO, Yun JW (2005). A comparative study on the production of exopolysaccharides between two entomopathogenic fungi *Cordyceps militaris* and *Cordyceps sinensis* in submerged mycelial cultures. J. Appl. Microbiol. 99:728-738.

Kim SW, Hwang HJ, Xu CP, Sung JM, Choi JW, Yun JW (2003). Optimization of submerged culture process for the production of mycelial biomass and exopolysaccharides by *Cordyceps militaris* C738. J. Appl. Microbiol. 94:120-126.

Kim YK, Xiao CL, Rogers JD (2005). Influence of culture media and environmental factors on mycelia growth and pycnidial production of *Sphaeropsis pyriputrescens*. Mycologia . 97:25-32.

Li SP, Li P, Dong TT, Tsim KWK (2001). Anti-oxidation activity of different types of natural *Cordyceps sinensis* and cultured *Cordyceps* mycelia. Phytomedicine. 8:207-212.

Li SP, Yang FQ, Tsimb KWK (2006). Quality control of *Cordyceps sinensis*, a valued traditional Chinese medicine. J. Pharm. Biomed. Anal. 41:1571-1584.

Li YL (2002). Study on the effect of different factors on conidiophores production by *Cordyceps sinensis*. Edible Fungi 5:8.

Liang HH, Cheng Z, Yang XL, Li S, Ding ZQ, Zhou TS, Zhang WJ, Chen JK (2008). Genetic diversity and structure of *Cordyceps sinensis* populations from extensive geographical regions in China as revealed by inter-simple sequence repeat markers. J. Microbiol. 46:549-556.

Li Y, Wang XL, Jiao L, Jiang Y, Li H, Jiang SP, Lhosumtseiring N, Fu SZ, Dong CH, Zhan Y, Yao YJ (2011). A survey of the geographic distribution of *Ophiocordyceps sinensis*. J. Microbiol. 49:913-919.

Liu ZY, Liang ZQ, Liu AY (2003). Investigation on microcycle conidiation of ascospores and conidiogenous structures of anamorph of *Cordyceps sinensis*. J. Guizhou Agric. Sci. 31:3-5.

Park JP, Kim SW, Hwang HJ, Yun JW (2001). Optimization of submerged culture conditions for the mycelial growth and exo-biopolymer production by *Cordyceps militaris*. Lett. Appl. Microbiol. 33:76-81.

Paterson M, Russell R (2008). *Cordyceps* - a traditional Chinese medicine and another fungal therapeutic biofactory? Phytochemistry 69:1469-1495.

Qiao ZQ, Zhang ZG, Yang D, Zhuo MC (2003). Report of germination of ascospores of *Cordyceps sinensis*. Gansu. Agric. Sci. Techno. 4:51-52.

Ren SY, Yao YJ (2013). Evaluation of nutritional and physical stress \ conditions during vegetative growth on conidial production and germination in *Ophiocordyceps sinensis*. FEMS Microbiol. Lett. doi: 10.1111/1574-6968.12190.

Sehgal AK, Sagar A (2006). In vitro isolation and influence of nutritional conditions on the mycelial growth of the entomopathogenic and medicinal fungus *Cordyceps militaris*. Plant. Pathol. J. 5:315-321.

Xiao JH, Chen DX, Liu JW, Liu ZL, Wan WH, Fang N, Xiao Y, Qi Y, Liang ZQ (2004). Optimization of submerged culture requirements for the production of mycelial growth and exo-polysaccharide by *Cordyceps jiangxiensis* JXPJ0109. J. Appl. Microbiol. 96:1105-1116.

Xiao YY, Chen C, Dong JF, Li CR, Fan MZ (2011). Morphologica observation of ascospores of *Ophiocordyceps sinensis* and its anamorph in growth process. J. Anhui. Agric. Univ. 38:587-591.

Xie F, Zhu ZX, Wei KL, Zhang SX, Zhang N, Tian W (2010). Effects of different culture media on the hyphal growth of *Cordyceps sinensis* strains. Chin. J. Microecol. 22:534-536.

Xu CP, Kim SW, Hwang HJ, Yun JW (2002). Application of statistically based experimental designs for the optimization of exopolysaccharide production by *Cordyceps militaris* NG3. Biotechnol. Appl. Biochem. 36:127-131.

Yang DR (1997). The exploitation and use state and develop potentialities of resourceful insects in Yunnan Province. Econ. Trib. 33:62-69.

Yin XW, Li ZH, Liu SC, Dong DX, Li RX (2009). Molecular identification of *Cordyceps sinensis* isolations and determination of main active constituents. Edible Fungi 5:20-22.

Zhang YJ, Liu XZ, Wang M (2008). Cloning, expression, and characterization of two novel cuticle-degrading serine proteases from the entomopathogenic fungus *Cordyceps sinensis*. Res. Microbiol. 159:462-469.

Zhao ZX, Cui XL, Cong YG (2006). Phylogeny analysis of 5.8S rDNA and ITS region of *Cordyceps sinensis* anamorph in Tibet. J. Yunnan. Univ. 28:83-87.

Zheng ZL, Huang CH, Cao L, Xie CH, Han RC (2011). *Agrobacterium tumefaciens*-mediated transformation as a tool for insertional mutagenesis in medicinal fungus *Cordyceps militaris*. Fungal. Biol. 115:265-274.

Zhu YP, Pan JR, Qiu JZ, Guan X (2008). Optimization of nutritional requirements for mycelial growth and sporulation of entomogenous fungus *Aschersonia aleyrodis* webber. Brazilian. J. Microbiol. 39:770-775.

20

Antagonistic confrontation of *Trichoderma* spp against fruit rot pathogens on Sapodilla (*Manilkara zapota* L.)

U. N. Bhale[1]*, P. M. Wagh[2] and J. N. Rajkonda[3]

[1]Research Laboratory, Department of Botany, Arts, Science and Commerce College, Naldurg, Tq. Tuljapur, Osmanabad District, 413602 (M.S.) India.
[2]Department of Biology, S. S. and L. S. Patkar College of Arts and Science, Goregaon (W), Mumbai- 400063 (M. S.) India.
[3]Department of Botany, Yeshwantrao Chavan College, Tuljapur, Osmanabad District, 413601 (M. S.) India.

Antagonistic potentials of five *Trichoderma* species that is *Trichoderma viride, Trichoderma harzianum, Trichoderma koningii, Trichoderma pseudokoningii* and *Trichoderma virens* were tested against fruit rots pathogens of sapodilla (*Manilkara zapota* L.) under laboratory conditions. Dual culture experiment of tested pathogens and *Trichoderma* spp revealed that, the percent inhibition of *T. koningii* (57.70%) and *T. harzianum* (54.40%) proved to be more than 50% antagonistic over control in case of *A. niger*. Similarly, in case of *R. solani, T. koningii* (67.07%) showed eloquent antagonistic activity as compared to others. In *G. candidum, T. pseudokoiningii* (75.07%) and *T. viride* (74.40%) have highly inhibited the radial growth over control followed by others. In case of *R. solani*, only *T. koningii* overgrew beyond 60% (R$_3$ scale). The results of this study identify *T. koningii* and *T. pseudokoningii* as promising biological control agents for further testing against post harvest disease in fruits.

Key words: *Manilkara zapota*, dual culture, fruit rot pathogen*s, Aspergillus niger, Rhizoctonia solani, Geotrichum candidum, Trichoderma viride, T. harzianum, T. koningii, T. pseudokoningii, T. virens.*

INTRODUCTION

Sapodilla (*Manilkara zapota* L.) is one of the edible fruits cultivated all over India. In India it ranks fifth position in production and consumption next to mango, banana, citrus and grapes. It is also commercially important because it is a source of chicle, the principle ingredient in chewing gum. It is a rich source of sugar, protein, phenol, carotenoids, amino acids, pectin, vitamin C and mineral like Phosphorus, Calcium, Iron and Magnesium (Moore and Stearn, 2007). At least 50% of total production of fruits and vegetables in the country is lost due to wastage and value destruction, and the cost of this wastage is estimated to be Rs. 23,000 crorers each year. As per the specifications of National Institute of Nutrition (NIN), at

least 300 g of fruits and vegetables are to be consumed by an individual per day for balanced diet (Roy, 2001). Sapodilla fruits are highly sensitive due to soft texture, therefore exogenous agents especially fungi, that affect physiology, morphology and biochemistry of fruits and thus ultimately causes severe loss to the fruit seller (Arya, 2011). Chemical control of pathogens provides certain degree of control but at the same time have adverse effects on environmental pollution (Charaya, 1993; Sankaram, 1999; Sokhi et al., 2000; Miller, 2004). In recent years the need to develop biologically ecofriendly disease control measures as an alternative to chemicals has become a priority of scientists worldwide. Therefore, it is important to find a practical, economic and non-toxic method to prevent fungal deterioration of stored food. Biological control of phytopathogens is an eco-friendly and cost effective approach. Hence, it should become an

*Corresponding author. E-mail: unbhale2007@rediffmail.com.

important component of plant disease management practices. Fungal antagonist that is, *Trichoderma* was evaluated as potential bio-control agent against number of fungal phytopathogens. Species of the genus *Trichoderma* are well documented fungal biocontrol agents (Papavizas, 1985; .Elad and Kapat, 1999; Howell, 2002). The antagonistic action of *Trichoderma* species against phytopathogenic fungi might be due to either by the secretion of extracellular hydrolytic enzymes (Chet, 1987; Di Pietro et al., 1993; Schirmbock et al., 1994) or by the production of antibiotics (Dennis and Webster, 1971a; Dennis and Webster, 1971b; Claydon et al., 1987; Howell, 1998). The effectiveness of biocontrol with *Trichoderma* spp. has also been shown by other investigators against *Penicillium digitatum* on citrus fruit (Borras and Aguilar, 1990), *B. cinerea* on grape berries (Elad, 1994), *Monilinia fructigena* on stone fruit (Hong et al., 1998), *B. cinerea*, *M. fructigena* and *P. expansum* on apple (Falconi and Mendgen, 1994), and *B. cinerea* and *P. expansum* on yams (*Dioscorea* spp.) (Okigbo and Ikediugwu, 2000).

The present investigation was made to evaluate *Trichoderma* spp against fruit rot pathogens such as *Aspergillus niger*, *Rhizoctonia solani* and *Geotrichum candidum* of sapodilla under laboratory conditions.

MATERIALS AND METHODS

The experiments of this work were carried out in the period 2008 to 2011.

Isolation and identification of test pathogen

Fruits showing symptoms of fungal infection were collected and symptomatology of the disease was studied under natural and *in vitro* conditions. Isolation of the pathogen was done from each of the distinct soft rots type of symptoms observed on fruits. Infected fruit parts (1 to 2 mm) were cut into small pieces by sterilized blade then surface sterilized with mercuric chloride (0.1%) for 1 min. The pieces were then washed thrice with sterilized distilled water and dried by sterilized blotting paper. These pieces were placed on Petri dishes (90-mm diameter) containing 20 mL potato dextrose agar (Peeled potato –200 g, Dextrose –20 g, Agar– 20 g and distilled water – 1000 ml, pH – 6.5) (PDA; Sd fine-CHEM Limited Mumbai, India) medium and incubated at 28 ± 2°C. The fungi namely, *Aspergillus niger*, *Rhizoctonia solani* and *Geotrichum candidum* were isolated and identified with the aid of standard literature available (Ellis, 1971; Barnett, 1960). The pathogenicity test of fungi was performed by the method of Thompson (1996).

Isolation of *Trichoderma* spp

Rhizospheric soils of irrigated and non irrigated plants were collected from different parts of Marathwada region of Maharashtra, India. From the rhizosphere soil samples, *Trichoderma* spp were isolated by using PDA and *Trichoderma* selective medium (TSM) by dilution plate technique (Johnson, 1957). The isolated species were identified up to species level based on colony characters, growth, structure of mycelium, conidiophores, phialides and conidia

(Kubicek and Harman, 2002). All *Trichoderma* spp were purified by hyphal tip technique (Tuite, 1996). The isolated *Trichoderma* spp were maintained throughout the study by periodical transfers on PDA and TSM slants under aseptic conditions to keep the culture fresh and viable.

Dual culture experiment

Antagonistic efficacy of *Trichoderma* spp namely, *T. viride*, *T. harzianum*, *T. koningii*, *T. pseudokoningii* and *T. virens* were tested against the isolated pathogenic fungi by dual culture experiment (Morton and Stroube, 1955). *Trichoderma* spp and test fungi were inoculated 6 cm apart. Three replicates were maintained for each treatment and incubated at 28 ± 2°C for 7 days. Monoculture plates of both served as control. Seven days after incubation (DAI), radial growth of test fungi and *Trichoderma* spp were measured. Color y diameter of test fungi in dual culture plate was observed and compared with control. Percentage of radial growth inhibition (%RGI) was calculated by using the formula: $100 \times [C - T / C]$, Where C = growth in control and T = growth in treatment (Vincent, 1947).

The degree of antagonism between each of the *Trichoderma* species and test pathogens in dual culture was scored on scale of R1 - R5 that is, R1=*Trichoderma* completely overgrew pathogens (100% over growth); R2= *Trichoderma* overgrew at least two-third pathogens (75% over growth); R3= *Trichoderma* colonizes on one half of the pathogens (50% over growth); R4= *Trichoderma* and the pathogens contact point after inoculation and R5= Pathogens overgrow bioagent - *Trichoderma* (Bell et al., 1982).

Statistical analysis

Data describing *in vitro* antagonisms were statistically analysed using the main factor was the *A. niger*, *R. solani* and *G. candidum* isolates of fruit rots pathogen and the sub-factors were the *Trichoderma* species. Arcsine transformation of biological control (*Trichoderma* species) percentage was calculated by using the following formula:

$$Y = \text{arcsine } \sqrt{p} = \text{Sin}^{-1}\sqrt{p}$$

Where, p is the percentage of inhibition and Y is the result of transformation.

Statistical analysis of the experiments was performed using the Handbook of Biological Statistics (Mungikar, 1997; McDonald, 2008).

RESULTS AND DISCUSSION

Isolation and identification of test pathogens

Infected soft rots fruits showed light brown coloured patch in the centre surrounded by white or creamish boundary and at severity complete rottening of fruit took place. These fruits with symptoms were collected from different locations of Thane District of Maharashtra. The fruit rot pathogens such as *Aspergillus niger* V. Tieghem, *Rhizoctonia solani* Kuhn and *Geotrichum candidum* Link ex Fries, were isolated following protocols (Ellis, 1971).

Isolation of *Trichoderma* spp

Five species of *Trichoderma*: *T. viride* Pers. ex. Gray, *T. harzianum* Rifai, *T. koningii* Oudemans *T. pseudokoningii* Rifai and *T. virens* J. Miller, Giddenand Foster A.A., were isolated from irrigated and non-irrigated rhizospheric soils of Marathwada region of Maharashtra. Isolates were deposited at Department of Botany, Arts, Science and Commerce College, Naldurg.

Taxonomical and morphological characters

Trichoderma viride Pers. ex.

Colony grows rapidly, white to greyish or rarely yellowish, surface smooth becomes hairy, typical coconut odour is emitted in old culture. Mycelium hyaline smooth, branched and septate. Chlamydospores intercalary, globose, rarely ellipsoidal, 10 to 15 µm in diameter. Conidiophores arise in compact or loose tuft, main branches produced several side branches. Phialides are in false whorls beneath each terminal phialides, usually more than 2 to 3 phialides, 8 to 15 × 2 to 3 µm in size, curved, pin shaped, narrower at the base. Conidia are Globose or short ovoid broadly ellipsoidal with minute roughing at their wall, 3.5 to 4.5 µm in size, accumulated at the tip of each phialides, pale green, smooth.

Trichoderma harzianum Rifai

Colony growing rapidly, white green, bright green to dull green. Mycelium is septate, colourless, smooth, 1.5 to 2.5 µm. Chlamydospores are mostly globose, smooth, 6 to 12 µm in diameter. Conidiophores are loose, tuft, main branch produced numerous side branches specially in lower portion. Phialides arise in false verticillate up to five in numbers, short, skittle shaped, narrow at the base, and attenuate abruptly sharp, pointed neck, 25 to 75 × 3 to 4 µm. Conidia are acuminated at the tip of the phialides, subglobose short, obovoid, often broad truncate base, smooth, pale green, much darker in mass, 2.8 to 3.2 × 2.5 to 2.8 µm.

Trichoderma koningii Oudemans

Colony fast growing, greenish white, dull to dark green.Mycelium is hyaline, highly ramified, 2 to 5 µm in size. Chlamydospores is formed in submerged hyphae, globose, ellipsoidal to barrel shaped, up to 12 µm in diameter. Conidiophores are branched, compact or in loose tuft, main branch produced several side branches, in group of 2 to 3 at wide angles. Phialides are pin shaped, narrower at the base, attenuate towards apex, 7.5 to 12 × 2.5 to 3.5 µm. Conidia are elliptical, oblong, truncate base and rounded apex, pale green appear much darker in mass, 3 to 5 × 1 to 2 µm in size.

Trichoderma pseudokoningii Rifai

Colony grows rapidly with very poor aerial growth. Mycelium is septate, smooth, colourless, 1 to 5 µm in size. Chlamydospores are infrequently in medium, globose, smooth, hyaline, 7 to 10 µm in diameter. Conidiophores are loosely tuft, may appear hairy at maturity, somewhat powdery with numerous long branches. Conidiophores branches irregularly formed, single or in opposite pairs or group of three. Phialides are in false whorls, opposite pair in group of four in apical portion, pin shaped, narrower at the base than middle, attenuated distinctly, abovate or spindle shaped, 5.5 to 8 × 2.5 to 3.5 µm in size. Conidia are short, sub cylindrical, almost oblong, ellipsoidal usually rounded, distally attenuate below, short, truncate base, green mass, 3.5 to 2 × 2 µm.

Trichoderma virens J. Miller, Gidden and Foster A. A.

Colony grows rapidly, floccose, white to grayish colouration. Mycelium is whitish in colour, turning grey at maturity, irregularly branched. Chlamydospores are mostly globose to subglobose, smooth, 7 to 12 µm in diameter. Conidiophores are conidiophores sub hyaline, 30 to 300 µm long, 2.5 to 4.5 µm in diameter, towards base frequently unbranched for about half of the length, towards the apex, branching irregular. Phialides are ampulliform to lageniform, 4.5 to 10 × 2.8 to 5.5 µm, swelling in the middle, mostly arising in closely verticils of 2 to 5 or terminal branches. Conidia are broadly ellipsoidal to obovoid, 3.5 to 6.0 × 2.8 to 4.1 µm, dark green.

Dual culture experiment

Trichoderma spp effectively inhibited the mycelial growth of the fruit rot pathogens. Table 1 illustrates that, in case of *A. niger*, *T. koningii* (57.70%) and *T. harzianum* (54.40%) spp were found to be more than 50% antagonistic over control. In *R. solani*, *T. koningii* (67.7%) showed highest mycelial growth inhibition, but others showed below 50% antagonism. Similarly, *T. pseudokoningii* (75.07%) and *T. viride* (74.40%) showed significant results followed by others in *G. candidum*. Among the three fruit rot pathogens, only *G. candidum* showed better inhibition by *Trichoderma* spp. (Figures 1 and 2).

According to modified Bell's scale, *T. harzianum and T. koningii* did not progress beyond 60% (R3 scale) but remaining species failed to overgrow *A. niger.* In case of

Table 1. Evaluation of *Trichoderma* spp against fruit rots pathogens of sapodilla.

| *Trichoderma* spp | Test Pathogens | | | | | |
	Radial growth of *A. niger* (mm)	% Inhibition	Radial growth of *R. solani* (mm)	% Inhibition	Radial growth of *G. candidum* (mm)	% Inhibition
T. viride	50	44.40 (50.57)	50	44.40 (51.43)	23	74.40 (88.12)
T. harzianum	41	54.40 (62.58)	55	38.80 (42.19)	35	61.10 (72.02)
T. koningii	38	57.70 (67.31)	29	67.07 (80.00)	30	66.07 (79.18)
T. pseudokoiningii	52	42.20 (48.06)	50	44.40 (51.00)	22	75.55 (88.90)
T. virens	50	44.40 (50.98)	50	44.40 (50.57)	28	68.9 (80.44)
Control	89.22		90		89	
SEm± CD(p=0.05)	7.53 19.35		8.11 20.83		10.42 26.76	

Radial growth and percent inhibition values are means of three replicates. Figures in parentheses are arcsine transformed values of % inhibition. ± = Standard Error

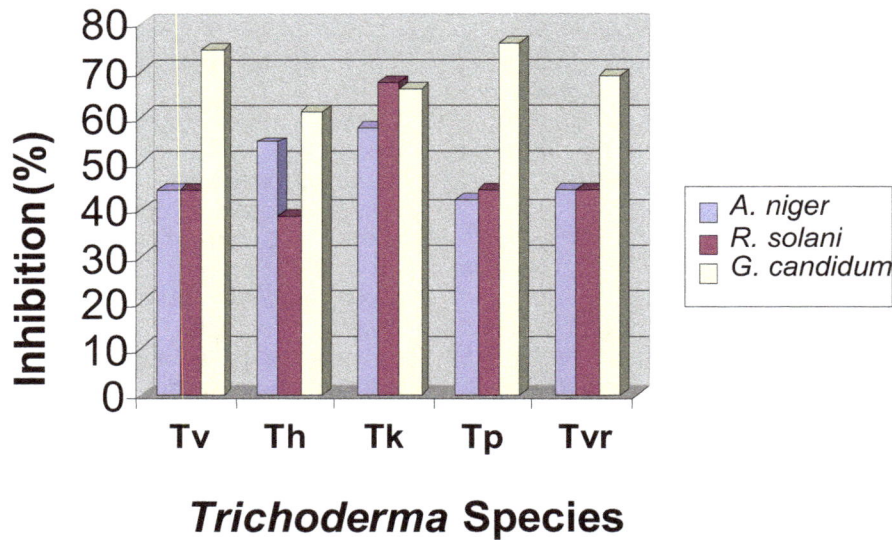

Figure 1. Evaluation of *Trichoderma* spp against fruit rots pathogens of sapodilla.

R. solani, only *T. koningii* overgrew beyond 60% (R_3 scale). In *G. candidum*, *T. pseudokoningii* and *T. viride* overgrew at least two third of pathogen (R_2 scale) but others were beyond 60% (R_3 scale) (Table 2).

Dual culture of pathogens and *Trichoderma* spp revealed that *T. viride* (Tv-2) (71.41%) highly inhibited the mycelia growth over control (Faheem et al., 2010). *T. viride* (86.2%) inhibited maximum growth of test fungus inciting collar rot of groundnut followed by *T. harzianum* (80.4%) (Harsukh et al., 2011). Seventeen *Trichoderma* strains were screened against *R. solani in vitro*, all strains including *T. harzianum*, *T. viride* and *T. aureoviride*, inhibited the growth of *R. solani* (Shalini, 2007). The antagonistic activity of the genus *Trichoderma* to *F. solani* and *R. solani* has been widely demonstrated (Lewis, 1998). The species of *Trichoderma* significantly inhibited the mycelial growth of plant pathogenic fungi (Rajkonda et al., 2011). Efficacy of *Trichoderma* species were reported against *Fusarium oxysporum* sp. *carthami* causing wilt of safflower and isolates no. 29 and 33 were found to minimize the growth of the pathogen as compared to others (Waghmare and Kurundkar, 2011).

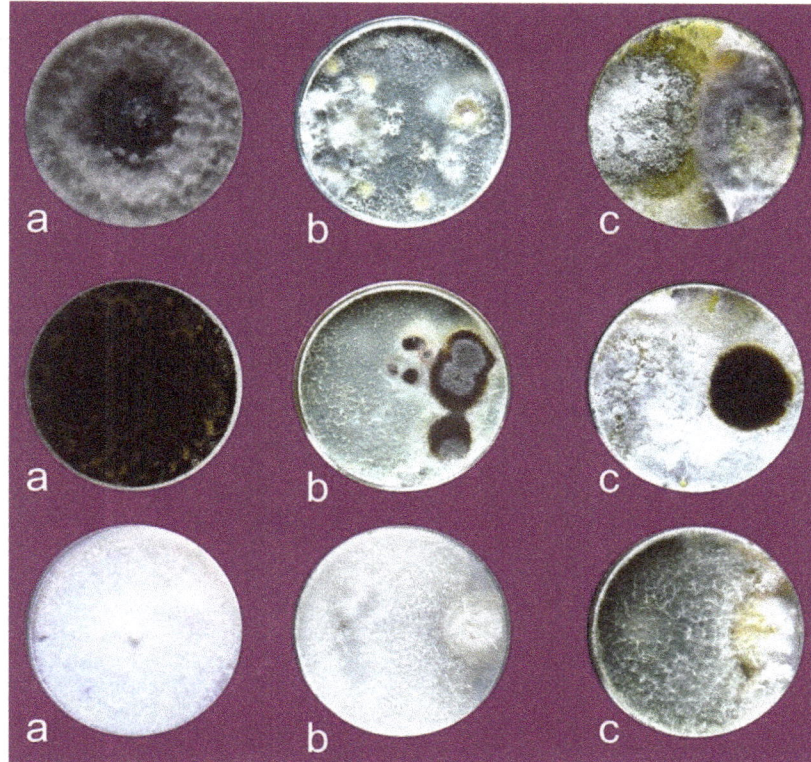

Figure 2. Antagonistic effects of *Trichoderma* spp on fruit rot pathogens of sapodilla. Row 1: *Rhizoctonia solani* (a-control, b-*T. koningii*, c-*T. harzianum*); Row 2: *Aspergillus niger* (a- control, b- *T.koningii*, c- *T.viride*); Row 3: *Geotrichum candidum* (a- control, b- *T.virens*, c- *T.koningii*).

T. harzianum was isolated from rambutan orchards in Sri Lanka and proved its antogonistic effect against *Botryodiplodia theobromae* (Sivakumar et al., 2000). *T. pseudokoningii* and *T. harzianum* have good antagonistic potentials against *C. destructivum* of cowpea (Akinbode and Ikotun, 2011). The results indicated that the treatment with the invert emulsion formulation of *T. harzianum* protected fruit from infection by the primary postharvest pathogens (*Rhizopus stolonifer*, *Botrytis cinerea*, and *Penicillium expansum)* of the fruits (grape, pear, apple, strawberry, and kiwifruit) tested for up to 2 months and reduced the diameters of decay lesion up to 86% and is a promising treatment to prolong the postharvest shelf-life of fresh fruit (Batta, 2007). Haran et al. (1996) reported dual culture experiments in which *T. harzianum* was overgrown by *R. solani* but hardly overgrown by *S. rolfsii* under the same conditions.

Conclusion

Trichoderma grows rapidly on a culture medium which should be beneficial during the confrontation. Our results concluded that the tested *Trichoderma* spp reduced the growth of all the tested three pathogens. *Trichoderma* spp showed significantly reduced the mycelial growth in *G. candidum*. We found an inhibition of mycelial growth of the pathogen tested. If there is direct contact between the two fungi, *Trichoderma* spp invaded colonies of fungal isolates sporulated there even after six days of confrontation. Therefore it can be incorporated for integrated disease management of fruit rot pathogens. Future research in this area should include *in vivo* studies on the effectiveness of the *Trichoderma* species as biocontrol agents. This could be done by *Tricoderma* into the soil of sapodilla plantations or by dipping the fruits into a suspension of *Trichoderma* after harvest. The effect of *Trichoderma* against other microorganisms especially against those that are beneficial to crops should also be investigated. Our future strategy will be to treat the soil by *Trichoderma* spp during plantation and after harvesting, sapodilla fruits can be dipped into a suspension of *Trichoderma* can be suggested.

ACKNOWLEDGEMENTS

Authors are thankful to the Director, Agharkar Research

Table 2. Evaluation of Trichoderma spp against fruit rots pathogens of sapodilla by dual culture,using Bell's scale*(R).

Trichoderma spp	Test pathogen		
	A. niger	R. solani	G. candidum
T. viride	R_4	R_4	R_2
T. harzianum	R_3	R_4	R_3
T. koningii	R_3	R_3	R_3
T. pseudokoiningii	R_4	R_4	R_2
T. virens	R_4	R_4	R_3

*Degree of antagonism. R1=Trichoderma completely overgrew pathogens (100% over growth); R2=Trichoderma overgrew at least two-third pathogens (75% over growth); R3=Trichoderma colonizes on one half of thepathogens (50% over growth); R4=Trichoderma and the pathogens contact point after inoculation; R5= Pathogens overgrow bioagent - Trichoderma.

Institute (ARI) Pune for the identification of Trichderma spp. and also thankful to UGC, New Delhi for financial assistance of major research project. The authors also thankfully acknowledged Principal Dr. S. D. Peshwe for providing laboratory facilities.

REFERENCES

Akinbode OA, T Ikotun (2011).Potentials of two Trichoderma species as antagonistic agents against Colletotrichum destructivum of cowpea. Afr. J. Microbiol. Res. 5(5):551-554.

Arya A (2011). Tropical Fruits- Diseases and Pests, KalyaniPublishers, New Delhi.

Barnett HL (1960). Illustrated genera of imperfect fungi. Burgress Publishing Company II eds, West Virginia.

Batta Yacoub A (2007). Control of postharvest diseases of fruit with an invert emulsion formulation of Trichoderma harzianum Rifai. Postharvest Biol. Technol. 43:143-150.

Bell DK,Wells HD, Markham CR (1982). In vitro antagonism of Trichoderma species against six fungal plant pathogens. Phytopathology 72(4):379-382.

Borras D, Aguilar RV (1990). Biological control of Penicillium digitatum on postharvest citrus fruit. Int. J. Food Microbiol. 11:179-184.

Charaya MU (1993). From DDT to Microbial pesticides. In: Pesticides pollution,(Eds.Kudesia, V.P. and Charaya, M.U.), Pragati Prakashan, Meerut, India pp. 207-220.

Chet I (1987). Trichoderma - application, mode of action and potential as a biocontrol agent of soil borne plant pathogenic fungi. In: Chet, I. (Ed.), Innovative Approaches to Plant Disease Control, John Wiley and Sons, New York, pp. 137-160.

Claydon N, Allan M, Hanson JR., Avent AG (1987). Antifungal alkyl pyrones of Trichoderma harzianum. Trans. Br. Mycol. Soc. 88:503-513.

Dennis C, Webster J (1971a). Antagonistic properties of species groups of Trichoderma. I. Production of non-volatile antibiotics. Trans. Br. Mycol. Soc. 57:25-39.

Dennis C, Webster J (1971b). Antagonistic properties of species groups of Trichoderma. II. Production of volatile antibiotics. Trans. Br. Mycol. Soc. 57:41-48.

Di Pietro A, Lorito M, Hayes C, Broadway K, Harman GE (1993). Endochitinase from Gliocladium virens. Isolation, characterization, synergistic antifungal activity in combination with gliotoxin. Phytopathology 83:308-313.

Elad Y (1994). Biological control of grape gray mold by Trichoderma harzianum. Crop Prot. 13:35-38.

Elad Y, Kapat A (1999). The role of Trichoderma harzianum protease in the biocontrol of Botrytis cinerea. Eur. J. Plant Pathol. 105:177-189.

Ellis MB (1971). Dematiaceous Hypomycetes. Commonwea th Mycological. Institute, Kew, Surrey, England.

Faheem Amin, Razdan VK, Mohiddin FA, Bhat KA, Saba Band y (2010). Potential of Trichoderma species as biocontrol agents of s il borne fungal propagules. J. Phytol. 2(10):38-41.

Falconi J, Mendgen K (1994). Epiphytic fungi on apple leaves and their value for control of the postharvest pathogens: Botrytis cinerea, Monilinia fructigena and Penicillium expansum. J. Plant Dis. Prot. 101:38-47.

Haran S, Schickler H, Oppenheim A, Chet I (1996). Different al expression of Trichoderma harzianum chitinase durin mycoparasitism. Phytopathology. 86:980-985.

Harsukh G, Kalu Rakholiya, Dinesh Vakharia (2011). Bioeff cacy f Trichoderma isolates against Aspergillus niger Van Tieghem incitin collar rot in Groundnut (Arachis hypogaea L.). J. Plant Prot. Res. 51(3):240-247.

Hong CX, Michailides TJ, Holtz BA (1998). Effects of woundin inoculums density, and biological control agents on postharvest brown rot of stone fruits. Plant Dis. 82:1210-1216.

Howell CR (2002). Cotton seedling pre-emergence damping-off incitec by Rhizopus oryzae and Pythium spp. and its biological control wi h Trichoderma spp. Phytopathology 92:177-180.

Howell CR (1998). The role of antibiosis in biocontrol. In: Harman GE, Kubicek CP(eds) Trichoderma & Gliocladium, vol. 2.Taylor & Francis, Padstow, pp. 173-184.

Johnson LA (1957). Effect of antibiotics on the number of bacteria ard fungi isolated and fungi isolated from soil by dilution plate method. Phytopathology 47:21-22.

Kubicek CP, Harman GE (2002). Trichoderma and Gliocladium (vol. [. Basic biology, taxonomy and genetics. pp. 14-24.

Lewis JA, Larkin RP, Rogers DL (1998). A formulation of Trichoderma and Gliocladium to reduce damping-off caused by Rhizoctonia sola i and saprophytic growth of the pathogen in soil less mix. Plant Dis. 82:501-506.

McDonald JH (2008). Handbook of Biological statistics. Sparky House Publishing, Baltimore, Maryland. pp. 160-164.

Miller GT (2004). Sustaining the Earth, 6th edition. Thompson Learning. Inc. Pacific Grove, California. Chapter 9, pp. 211-216.

Moore HE, Stearn (2007). Post harvest physiology and technology o Sapote mamey fruit. Post-harvest Biology and Technology.45:28 -297.

Morton DJ, Stroube WH (1955). Antagonistic and stimulating effects o soil microorganisms upon sclerotium. Phytopathology 45:417-420.

Mungikar AM (1997). An Introduction to Biometry. Saraswati Printin Press, Aurangabad, pp. 57-63.

Okigbo RN, Ikediugwu FE (2000). Studies on biological control c postharvest rot in yams (Dioscorea spp.) using Trichoderma viride. Phytopathol. 148:351-355.

Papavizas GC (1985). Trichoderma and Gliocladium: biology, ecolog and potential for biocontrol. Ann. Rev. Phytopathol. 23:23-54.

Rajkonda JN, Sawant VS, Ambuse MG, Bhale UN (2011). Inimical potential of Trichoderma species against pathogenic fungi. Plant Sc . Feed. 1(1):10-13.

Roy SK (2001). Strategic post harvest Management of fruits and vegetables. Indian Hortic. 45(4):4-7.

Sankaram A (1999).Integrated pest management: Looking ba k and forward. Curr. Sci. 77:26-32.

Schirmbock M, Lorito M, Wang YL, Hayes CK, Arisan-Atac I, Scala F, Harman GE, Kubicek CP (1994). Parallel formation and synergism of hydrolytic enzymes and peptaibolantibiotics, molecular mechanisms involved in the antagonistic action of Trichoderma harzianum against phytopathogenic fungi. Appl. Environ. Microbiol. 60:4364-4370.

Shalini S, Kotasthane AS (2007). Parasitism of Rhizoctonia solani by strains of Trichoderma spp. EJEAF Chem. 6:2272-2281.

Sivakumar D, Wilson Wijeratnam RS, Wijesundera RLC, Abeyesekere M (2000). Antagonistic effect of Trichoderma harzianum or Postharvest Pathogens of Rambutan (Nephelium lappaceum) Phytoparasitica 28(3):240-247.

Sokhi SS, Singh PP, Grewal RK (2000). Environmental impact of

Fungicides in agroecosystem. In Pesticides and Environment, (Eds. Dhaliwal,G.S. and Balwinder Singh),Commonwealth Publishers, New Delhi,India. pp. 254-278.

Thompson AK (1996). Post -harvest technology of fruits and vegetables. Blackwell Science Ltd.London.

Tuite J (1996). Plant Pathological Methods. Fungi and Bacteria Burgess Pub. Co. Minneapolis, Minn. USA. 293 pp.

Vincent JM (1947). Distortion of fungal hyphae in the presence of certain inhibitors. Nature 150:850.

Waghmare SJ, Kurundkar BP(2011). Efficacy of local isolates of *Trichoderma* spp against *Fusarium oxysporum sp. carthami* causing wilt of safflower. Adv. Plant Sci. 24(1):37-38.

21

Biochemical characterisation of two novel laccases from *Magnaporthe grisea*

Kaminee Ranka and Bharat B. Chattoo*

Centre for Genome Research, Department of Microbiology and Biotechnology Centre, Faculty of Science, the Maharaja Sayajirao University of Baroda, Vadodara -390002, India.

Laccases are widely distributed oxido-reductases that catalyse the biological oxidation-reduction of polyphenols with a concomitant reduction of molecular oxygen to water. Genome analysis of *Magnaporthe grisea* using bioinformatic approach showed the presence of multiple laccases, which encode proteins with three domains of multicopper oxidase. The transcript levels of all *M. grisea* laccases were analysed by quantitative RT-PCR, in order to study their expression patterns in normal and nitrogen starved conditions. The highest relative expression was observed for MGG_08127 (*Mg*Lac1) in normal conditions. The highest induction was observed for MGG_02876 (*Mg*Lac2) in nitrogen starvation. Since total fungal protein extracts would contain multiple laccases, heterologous gene expression, purification and further enzyme characterisation was carried out to analyse the function of these two laccases from *M. grisea*. Thus, we identified a novel multifunctional laccase, *Mg*Lac2, in *M. grisea* which showed lignin-like dye decolourising activity, 1, 8-dihydroxynapthalene (DHN) polymerisation ability and also ferroxidase activity. Its optimum pH and maximum thermostability were at 4 to 4.5 and 30°C, respectively. *Mg*Lac1 also showed dye decolourization activity, its optimum pH and maximum thermostability were at 4 to 5 and 30°C, respectively. We found that the laccases expressed in normal conditions and in conditions which mimic pathogenicity are different biochemically.

Key words: *Magnaporthe grisea*, laccase, dye decolourization, 1, 8-dihydroxynapthalene (DHN) polymerisation, ferroxidase activity, glutathione-S-transferase (GST).

INTRODUCTION

Laccases are copper-containing enzymes catalyzing the monoelectronic oxidation of a number of phenolic compounds and aromatic amines at the expense of molecular oxygen. These enzymes are widely distributed among plants and fungi (Hatakka, 1994). The proposed physiological roles of these enzymes, biosynthetic in plants or biodegradative in fungi are not yet clear. Plant laccases participate in the radical-based mechanisms of lignin polymer formation (Sterjiades et al., 1992; Liu et al., 1994; Boudet, 2000; Ranocha et al., 2002; Hoopes and Dean, 2004), whereas in fungi, laccases probably have roles in morphogenesis, fungal plant pathogen-host interaction, stress responses and lignin degradation (Thurston, 1994). Laccase occurrence has also been reported in prokaryotes (Claus, 2003). Among prokaryotes, enzymes with laccase-like activity have been described participating in pigmentation, electron transport, sporulation, metal oxidation, and UV and H_2O_2 resistance among others (Claus, 2003; Givaudan et al., 1993; Faure et al., 1994; Faure et al., 1996; Alexandre et al., 1999; Francis and Tebo, 2001; Hullo et al., 2001; Kim et al., 2001; Enguita et al., 2003).

Laccases belong to the multi-copper oxidase family

---

*Corresponding author. E-mail: bharat.chattoo@bcmsu.ac.in.

Abbreviations: *Mg*Lac1, *Magnaporthe grisea* laccase 1; ***Mg*Lac2,** *Magnaporthe grisea* laccase 2; **GST,** glutathione-S-transferase; **DHN,** 1, 8-dihydroxynapthalene; **ABTS,** 2, 2'-azinobis-bis-3-ethylbenzthiazolinesulphonate.

containing at least four copper ions essential for enzyme activity, coordinated in three different redox sites: Type 1, site where substrate oxidation takes place; and Types 2 and 3, trinuclear cluster where reduction of molecular oxygen and release of water take place (Claus, 2004). Laccases are suitable for use in biotechnology; they may be used as a pulp-bleaching agent in the paper industry, in dye decolourization in textile industries, for bioremediation and drug analysis, among other uses (Mayer and Staples, 2002; Jordaan and Leukes, 2003; Ryan et al., 2003). Laccases have the advantage that they do not need H_2O_2 for substrate oxidation like peroxidases and have a broader substrate spectrum compared to tyrosinases (Saito et al., 2003).

Genome analysis of *Magnaporthe grisea* using bioinformatic approach revealed the presence of twelve putative laccases having three multicopper oxidase domains. *M. grisea* is the causal agent of rice blast disease, which is one of the most serious diseases on cultivated rice (Ou, 1985). It is capable of destroying enough rice to feed 60 million people every year (Zeigler et al., 1994). Due to the experimental tractability and socioeconomic impact of rice blast, the fungus has served as an important model organism in the studies aimed at understanding the biology of fungal plant pathogens (Talbot, 2003; Valent, 1990). Several studies have suggested that lack of nutrients is one of the signals that control expression of pathogenicity genes in various fungal pathogens of plants (Snoeijers et al., 2000) and humans (Lengeler et al., 2000). Starvation stress has also been implicated as a key influence on fungal gene expression during growth of *M. grisea* within the host plants, that is, starvation stress mimics pathogenicity in *M. grisea* (Talbot et al., 1997).

In this report, we describe heterologous expression, purification and characterisation of two laccases, MGG_08127 and MGG_02876 from *M. grisea*. MGG_08127 and MGG_02876 were named as *Mg*Lac1 and *Mg*Lac2, respectively, for our own convenience. In the present study, for the first time, we demonstrate that *Mg*Lac2 of *M. grisea* oxidises a lignin-like dye, polymerises dihydroxynapthalene and also has ferroxidase activity.

MATERIALS AND METHODS

Fungal strain, media and culture conditions

M. grisea isolate B157, belonging to the international race IC9 was previously isolated in our laboratory from infected rice leaves (Kachroo et al., 1994). The fungus was grown and maintained on YEG medium (glucose, 1 g; yeast extract, 0.2 g; H_2O to 100 ml) or oatmeal agar (Hi-Media, Mumbai, India) at 28°C. The composition of complete media (CM), minimal media (MM) and MM-N (used for nitrogen starvation) were as reported earlier (Talbot et al., 1997). Mycelia used for RNA extraction and total protein extraction were obtained by growing the mycelia in liquid medium for 3 days at

28°C.

Total RNA extraction and cDNA synthesis

Fungal biomass grown in liquid media was frozen in liquid nitrogen. Total RNA was isolated using TRIZOL reagent (Invitrogen Life Technologies, California, USA). The quality of isolated RNA was checked by electrophoresis on formaldehyde gels and quantified by UV spectrophotometry. 5 µg of total RNA was used to synthesise the first strand cDNA using MuMLV reverse transcriptase (Fermentas GmBH, St. Leon-Rot, Germany) and random hexamer in 20 µl reaction system.

Quantitative real-time PCR (qRT-PCR)

Quantitative real-time PCR (qRT-PCR) was performed by monitoring the increase in fluorescence of the SYBR Green dye on LightCycler system (Roche Applied Science, Mannheim, Germany), according to the manufacturer's instructions. Each qRT-PCR quantification was carried out in triplicate using primers for each individual gene (Table 1). Thermal cycling conditions consisted of 2 min at 95°C followed by 40 cycles of 10 s at 95°C, 10 s at 54°C, and 10 s at 72°C. The data were normalized against *tubulin* gene.

GenBank ID, national center for biotechnical information (NCBI)

GenBank ID of *Mg*Lac1 is MGG_08127 and that of *Mg*Lac2 is MGG_02876.

Heterologous expression of *Mg*Lac1 and *Mg*Lac2 in *Saccharomyces cerevisiae*

*Mg*Lac1 was amplified by PCR using the following forward primer: 5'ATGAATCTTCGGGACACCATCT 3' and reverse primer 5' TTATCTCCTCAAACCAGACTCCA 3'. *Mg*Lac2 was amplified by PCR using the following forward primer: 5' GGGGATCCCGATGGGTATCATGCAGGGGATG 3' and reverse primer 5' GCAAGCTTGGTTAAACACCGCTGTCGATCTG 3'. The amplified PCR products were gel purified and sequenced. The vectors pEG(KT)*Mg*Lac1 and *p*EG(KT)*Mg*Lac2 were constructed by cloning the coding region of *Mg*Lac1 and *Mg*Lac2 translationally in frame with glutathione-S-Transferase (GST) gene in pEG(KT) (Mitchell et al., 1993). *S. cerevisiae* strain, s288C, was transformed using the recombinant plasmids by 'one step transformation' method (Chen et al., 1992).

Enzyme purification and characterisation

Enzyme purification

The GST*Mg*Lac1 and GST*Mg*Lac2 fusion proteins were purified from total protein extract of the transformed *S. cerevisiae*. Two days old cultures of *S. cerevisiae* cells (OD$_{600}$ ~1) harbouring the expression constructs was inoculated into 100 ml 1X YNB (yeast nitrogen base) containing 2% ethanol. The cultures were incubated at 28°C by shaking for 30 h, until the OD$_{600}$ was about 0.6 and expression was induced by the addition of 4% galactose, followed by further incubation at 28°C for 5 h. Cells were harvested by centrifugation (4°C, 10000 g, for 15 min) and pellets were resuspended in 1 ml of ice-cold PBS. Cell lysis was achieved by

Table 1. List of primers of 12 multicopper oxidases and tubulin used for the qRT-PCR.

Gene id	Primer sequence	
MGG_08127.5	Forward	5' CCTGCCAGCGCGAATTACG 3'
	Reverse	5' CGACCTCCACTGCCTTTGGG 3'
MGG_02876.5	Forward	5' AAGACGGTGTGCCTGGTGTGA 3'
	Reverse	5' AGAAGACCATTGGGCCAACG 3'
MGG_13464.5	Forward	5' ACCACTCTCACTTCTCCGGG 3'
	Reverse	5' GGATCAGCTCGTACTGGATGCG 3'
MGG_0579.5	Forward	5' CGGGCTCGACCGTGACTTA 3'
	Reverse	5' TTCCCAGGTCCTCGTCGTAGTT 3'
MGG_09139.5	Forward	5' ATGTATGGCGGCATCGTCATCA 3'
	Reverse	5' GGCCGTTGGACAGAATCGGAG 3'
MGG_11608.5	Forward	5' ACGTGACCAACAACATGCAGAC 3'
	Reverse	5' GGGTGGCGCGGAACTTGTA 3'
MGG_00551.5	Forward	5' CAGATGGTGTTGTCCGCGAT 3'
	Reverse	5' TTCTGCTGAATACCGTGCCAGTG 3'
MGG_07771.5	Forward	5' AGCGGGACAGCGTTCAAAT 3'
	Reverse	5' ATAGGGCCGTACAACCCATCG 3'
MGG_07220.5	Forward	5' CGGCTCCAACGAGATGGAT 3'
	Reverse	5' CGATGATAAGAGGACCACGCAG 3'
MGG_02156.5	Forward	5' GTCCCATGAAGGAGCTGATCGC 3'
	Reverse	5' ACGTTGACCATCCTGACCAGGTA 3'
MGG_14307.5	Forward	5' ACTATGAGATCCGACCCGACATT 3'
	Reverse	5' CCAGGCCAACAACAACGTCC 3'
MGG_09102.5	Forward	5' GCATCGACGAGCACGAGTTC 3'
	Reverse	5' GTCACCCTCACGGCGTAGT 3'
β Tubulin	TubL1	5' GAGTCCAACATGAACGATCT 3'
	TubR1	5' GTACTCCTCTTCCTCCTCGT 3'

sonication. Lysed cells were centrifuged (13000 g, for 10 min) at 4°C to remove insoluble material and cleared supernatants were used to purify fusion protein. The GST fusion proteins were purified by affinity chromatography using a glutathione-sepharose matrix under mild conditions, using the GST purification module as per manufacturer's instructions (GE Healthcare, Buckinghamshire, UK). Protein concentrations were estimated by Bradford method (Bradford 1976). Around 200 to 300 ng of purified proteins were loaded on 12% SDS–PAGE.

Thrombin protease cleavage

Removal of the GST moiety from the proteins of interest was accomplished through a thrombin protease cleavage site located between the GST moiety and the recombinant proteins. GST from the fusion protein was removed using Thrombin Cleavage Capture Kit (Novagen, Merck KGaA, Darmstadt, Germany) as per the manufacturer's instructions. Cleaved GST was easily removed by a second round of chromatography on the glutathione column. The purified proteins were checked on 12% SDS–PAGE.

Raising polyclonal antibodies

The purified protein fraction was used to raise polyclonal antibodies in rabbit. 5 µg of purified MgLac1 and MgLac2 in 0.5 ml of 1X PBS were emulsified with an equal volume of Freund's complete adjuvant and used immediately for subcutaneous injection in rabbit. Two booster doses of antigen- adjuvant mixture were given after every 4 weeks. 15 ml of blood were drawn 7 days after the second booster dose. The antibody titre was estimated by indirect-enzyme linked immuno sorbent analysis (ELISA) using anti-rabbit IgG (Bangalore Genie, India) as the secondary antibody.

Enzyme activity

Kinetic measurements were carried out at 30°C, with initial velocity measurements performed in 3 ml glass cuvettes with 1 cm path lengths. Reactions were initiated by the addition of purified proteins. The velocities of protein catalyzed reactions were measured at 420 nm for 2,2'-azinobis-bis-(3-ethylbenzthiazolinesulphonate) (ABTS), 530 nm for syringaldazine 477 nm for 2,6-dimethoxyphenol (DMP), 390 nm for catechol, 390 nm for pyrogallol, 320 nm for phloroglucinol, 390 nm for hydroquinone, 468 nm for guaiacol, and 275 nm for L-tyrosine. Km and Vmax for the afore-mentioned were determined using the Michaelis-Menten's equation. Graph Pad Prism Software (La Jolla, CA, USA) was used for the calculations.

Enzyme inhibition

Enzyme inhibition studies were performed by pre-incubation of the enzyme with the inhibitor for a certain period of time (2 to 10 min) to ensure that the inhibition is complete. The subsequent kinetic measurements were carried out by the addition of proper substrate and then monitoring the substrate's decline or the formation of the product. Four potential inhibitors (sodium azide, cysteine, EDTA (ethylenediaminetetraacetic acid) and Cl (chloride)) were evaluated to test the inhibition of both laccases.

pH optimum

Three substrates (ABTS, catechol, and pyrogallol) were used to determine the effect of pH on enzyme activity of the purified protein. The pH optima were determined over the range of pH 3 to 9 and a 0.1 M Britton- Robinson buffer (Xu, 1996) was used to check the enzyme activity.

Temperature effect and thermostability

The effect of temperature on laccase activity was determined spectrophotometrically following the laccase-catalysed oxidation of ABTS for 2 min at temperatures ranging from 25 to 90°C, at 5°C intervals. Britton-Robinson buffer (Xu, 1996) was used for all the reactions. Thermostability of both the laccases was determined by following the oxidation of ABTS after pre-incubation of laccase for 1 h at different temperatures (20 to 60°C). The reactions were initiated by the addition of substrate.

Protein extraction and western blotting

Total intracellular protein was extracted from wild type *M. grisea* mycelium grown in CM for 24 h. *M. grisea* biomass was crushed into fine powder in liquid nitrogen and resuspended in 1X PBS containing 1 mM PMSF. The extract was clarified by centrifuging at 13000 g for 15 min at 4°C. For secretory proteins, culture filtrate was concentrated with an Amicon Stirred Cell protein concentrator with a 3 kDa cutoff. For western blot analysis 15 µg of intracellular and secretory proteins were electrophoresed on a 12% SDS–PAGE and transferred to a nitrocellulose membrane (Hybond ECL, GE Healtcare, Buckinghamshire, England). The membrane was probed with polyclonal antibodies raised in rabbit followed by a 1:1000 dilution of goat anti-rabbit alkaline phosphatase labelled secondary antibody (Bangalore Genei, Bangalore) and detected with nitroblue tetrazolium (NBT) and 5-bromo-4-chloro-3-indolyl phosphate (BCIP).

Dye decolourisation assay

Dye decolourising activity was estimated by diluting 20 ml of a 0.2% (w/v) Remazol brilliant blue dye solution (in H_2O) to 4 ml citrate-phosphate buffer, pH 4.5. Laccases (400 U) were added and decolourisation was assayed after 0, 10, 20, 30, 45, 60, 75 and 90 min at 23°C. Decolourisation was monitored by diluting 0.5 ml of the dye solutions to 2 ml with a 10 mM sodium azide solution in 10 mM citrate-phosphate buffer, pH 4.5. At the selected time intervals, the ratio of absorbance (A_{591}/A_{500}) was measured for the azide-dye solution (Lozovaya et al., 2006). Decolourisation assays were performed in triplicate, and absorbances were compared with that of a control, that is, identical solution without the enzyme (Edens et al., 1999).

Polymerisation of dihroxynapthalene (DHN)

DHN polymerisation was estimated using chromotropic acid

disodium salt dehydrate (1,8- dihroxynapthalene-3,6-disulfonic acid disodium salt; 4,5- dihroxynapthalene-2,7-disulfonic acid disodium salt) (Sigma Chemical, St. Louis, MO, USA), by the addition of 15 U of laccase to a 2 ml solution of 1 mM DHN in 5 mM citrate-phosphate buffer, pH 4.5, in 50% ethanol. The polymerisation solution was incubated at 23°C and monitored on nanodrop spectrophotometer (Nanodrop Technologies, USA) by scanning the solution from 320 to 520 nm at 0 to 120 min. An identical control solution without the enzyme that auto-oxidized was also scanned (Edens et al., 1999).

Ferroxidase activity

Ferroxidase activity was determined using ferrous sulfate as the electron donor and 3-(2-pyridyl)- 5,6-bis(4-phenylsulfonic acid)-1,2,4-triazine (ferrozine) as a specific chelator to bind ferrous iron remaining at the end of the reaction. Reactions were carried out in disposable cuvettes containing 100 µM ferrous sulfate in 100 mM sodium acetate buffer, and pH 5.0. The reactions were quenched by the addition of ferrozine to a final concentration of 1.5 mM, and the rate of Fe^{2+} oxidation was calculated from the decreased absorbance at 560 nm using a molar absorptivity of $e_{560}= 25,400/M$ /cm for the Fe^{2+}-ferrozine complex (Hoopes and Dean, 2004).

RESULTS

Expression profiling of laccases of *M. grisea*

High laccase activity was reported in liquid cultures of *M. grisea* as early as 24 h (Iyer and Chattoo, 2003). Therefore, quantitative real time PCR (qRT-PCR) was carried out to monitor the mRNA expression of all the twelve multicopper oxidase genes. Nitrogen starvation-stress induces expression of a large number of genes expressed during growth of the fungus in plant tissue, particularly during disease symptom outbreak (Talbot et al., 1997). We, therefore, checked the relative expression of 12 multicopper oxidases *of M. grisea* in normal condition versus nitrogen starved condition. qRT-PCR was carried out using 12 laccase specific primers (Table 1). The calibrator used was *TUB* (β-tubulin), where the primers were designed to amplify 78 bp fragments.

Relative expression of MGG_08127 was highest in the normal hyphal condition (Figure 1A). Most of the multicopper oxidases showed 2 to 4 fold induction under nitrogen starvation condition, except *MgLac2* (MGG_02876) which showed 15-fold induction (Figure 1B). Therefore these two genes were selected for further analysis and were named as *MgLac1* (MGG_08127) and *MgLac2* (MGG_02876), respectively.

Enzyme purification

Bioinformatics analysis of *Mg*Lac1 and *Mg*Lac2 protein revealed them to be laccases. To characterise their enzymatic properties, we cloned the full-length coding

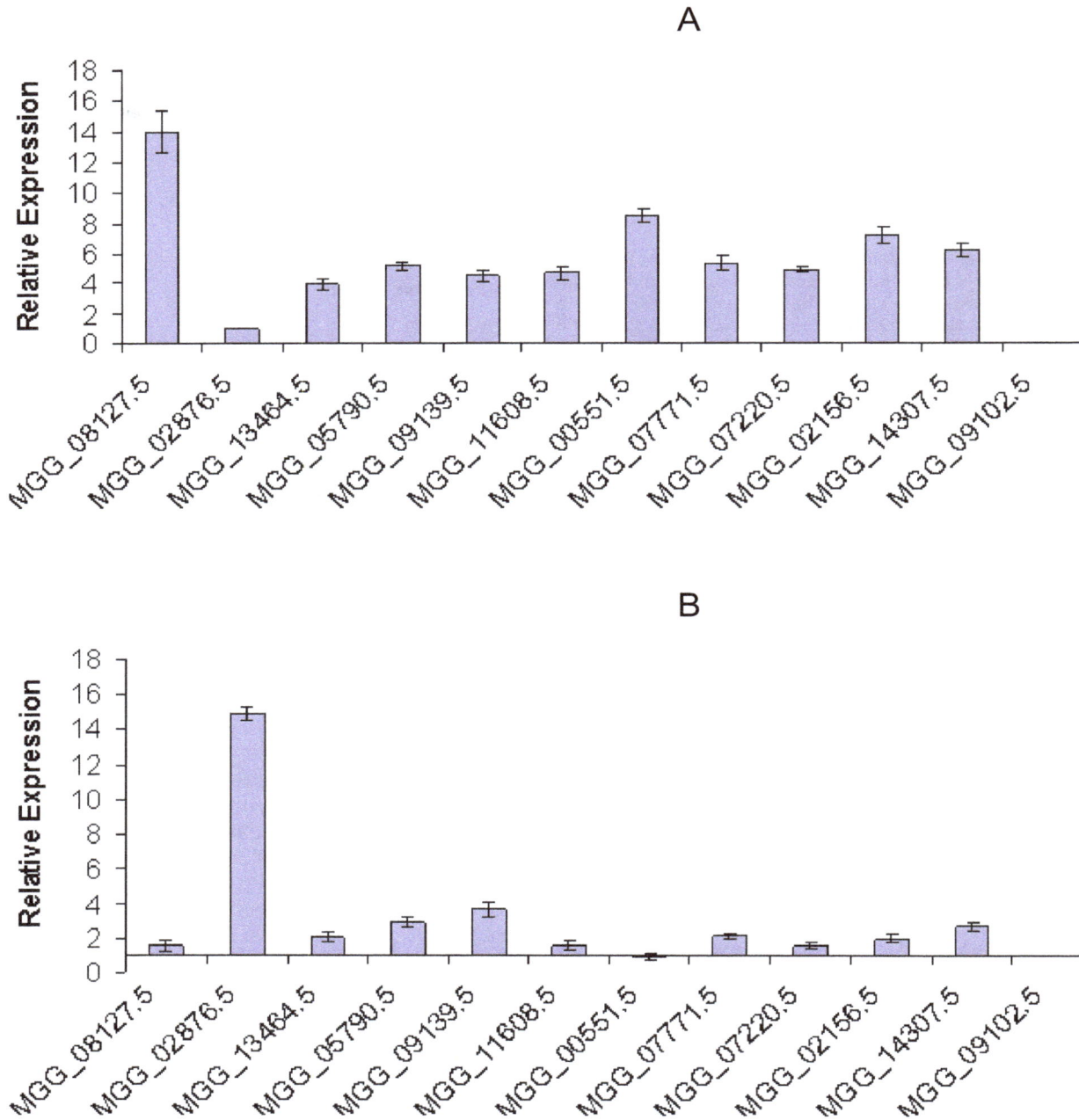

Figure 1. (A) Relative expression of 12 multicopper oxidases in *M. grisea*. The relative expression level was compared with the lowest expressing multicopper oxidase. MGG_08127 (*MgLac1*) was the highest expressing multicopper oxidase. The values given are the average of triplicates. Bars indicate mean ± SD. (B) Response of the 12 multicopper oxidases to nitrogen starvation. *M. grisea* was grown in CM for 48 h and then in nitrogen starvation media for 24 h. The transcript levels were analysed by qRT-PCR. The relative induction levels were compared with the basal levels observed in the normal condition. 15-fold induction was observed for MGG_02876 (*MgLac2*). The values given are the average of triplicates. Bars indicate mean ± SD.

sequence of *Mg*Lac1 and *Mg*Lac2. The expression plasmid constructs were generated containing *Mg*Lac1 and *Mg*Lac2 translationally in frame with glutathione-

S-Transferase (GST) gene in pEG(KT) vector. The constructs pEG(KT) *MgLac1* and pEG(KT) *MgLac2* were expressed in *S. cerevisiae* (s288C) and induced by 4%

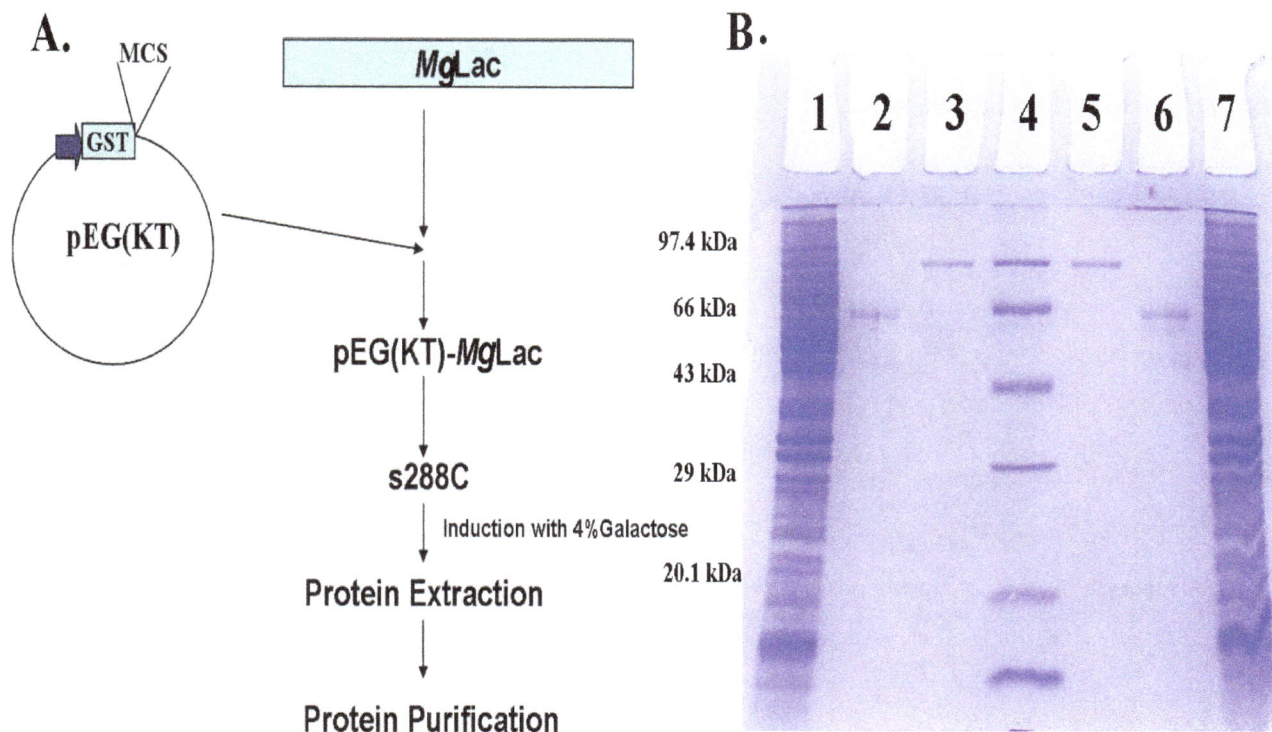

Figure 2. Heterologous expression and purification of MgLac1 and MgLac2. (A) Steps involved in the construction of expression vectors pEG(KT)MgLac1 and pEG(KT)MgLac2 which were cloned in pEG(KT) vector and expressed in S. cerevisiae strain, s288C. (B) The fusion proteins were purified by affinity chromatography using glutathione-sepharose as affinity matrix and checked on 12% SDS–PAGE. Lane 1. Crude protein extract of s288C transformed with pEG(KT)MgLac1 plasmid; Lane 2. Purified MgLac1; Lane 3. Purified GST-MgLac1; Lane 4. Molecular weight marker; Lane 5. Purified GST-MgLac2; Lane 6. Purified MgLac2; Lane 7. Crude protein extract of s288C transformed with pEG(KT)MgLac2 plasmid.

galactose (Figure 2A). MgLac1 and MgLac2 were expressed from the pEG(KT) vector as a fusion protein with GST at its N-terminus; we will refer to these fusion proteins as GST-MgLac1 and GST-MgLac2. The fusion proteins were purified using glutathione sepharose affinity matrix. The GST tag was removed from fusion proteins as described in materials and methods. Both MgLac1 and MgLac2 resolved as a single fragment of 66 kDa on SDS-PAGE when stained with Coomassie Blue (Figure 2B). The purified MgLac1 and MgLac2 were subsequently used for in-vitro analysis of enzyme properties.

Localisation of MgLac1 and MgLac2

MgLac1 and MgLac2 were predicted to be extracellular using program Wolf PSORT (Horton et al., 2007). To validate extracellular localisation of MgLac1 and MgLac2 western blot analysis was carried out using antibodies raised in rabbit. The polyclonal antibodies titre was found to be 1600.

Our analysis revealed a clear band of 66 kDa in culture filtrate of M. grisea which corresponds to the relative molecular weight (Mr) of both MgLac1 and MgLac2. Such band was not observed in case of intracellular proteins of M. grisea confirming that the MgLac1 and MgLac2; both are secretory proteins (Figure 3).

MgLac1 and MgLac2 are laccases

Enzyme activity

The purified MgLac1 and MgLac2 were further tested for their laccase activity using common laccase substrates such as ABTS, syringaldazine and DMP. MgLac1 K_m values were observed to be 0.1035 ± 0.02 mM (ABTS) and 1.113 ± 0.28 mM (Syringaldazine). The V_{max} values were observed to be 0.0205 ± 0.0007 s^{-1}(ABTS) and 0.0158 ± 0.0042 s^{-1}(Syringaldazine). MgLac2 K_m values were observed to be 0.1301 ± 0.02 mM (ABTS) and 1.243 ± 0.32 mM (Syringaldazine). The V_{max} values were observed to be 0.0168 ± 0.0004 s^{-1}(ABTS) and 0.0143 ± 0.0048 s^{-1}(Syringaldazine). However, both MgLac1 and

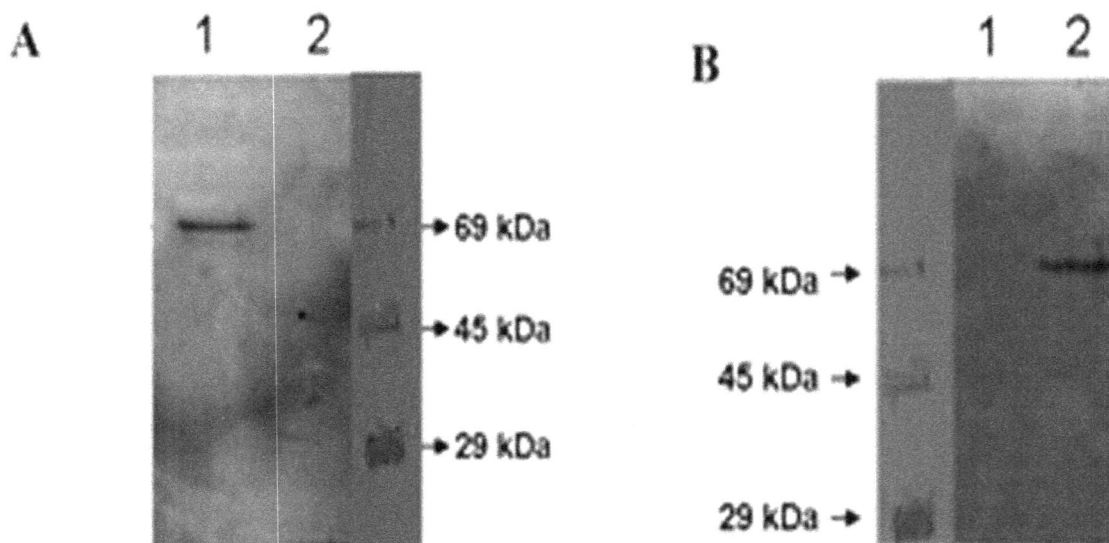

Figure 3. Western blot analysis. (A) Western blot analysis of *Mg*Lac1 of *M. grisea*. Lane 1. culture filtrate; Lane 2. intracellular proteins. 15 µg of protein was loaded in each well. (B) Western blot analysis of *Mg*Lac2 of *M. grisea*. Lane 1. intracellular proteins; Lane 2. culture filtrate. 15 µg of protein was loaded in each well.

*Mg*Lac2 were unable to oxidise a common laccase substrate, DMP. Significant differences were found in the affinity constants (K_M) and substrate preferences of both laccases towards other phenolic substrates. *Mg*Lac1 oxidised catechol, pyrogallol and phloroglucinol with K_m values 1.047 ± 0.16, 1.301 ± 0.16 and 0.389 ± 0.05 mM, respectively. The Vmax values for *Mg*Lac1 were observed to be 0.0267 ± 0.0008 s^{-1} (Catechol), 0.0168 ± 0.0004 s^{-1} (Pyrogallol) and 0.0115 ± 0.0006 s^{-1} (Phloroglucinol). *Mg*Lac2 oxidised catechol, pyrogallol and hydroquinone with K_m values 7.432 ± 1.06, 2.509 ± 0.29 and 2.088 ± 0.39 mM, respectively. The V_{max} values for *Mg*Lac2 were observed to be 0.0110 ± 0.0016 s^{-1} (Catechol), 0.0187 ± 0.0007 s^{-1} (Pyrogallol) and 0.0097 ± 0.0005 sec^{-1} (Hydroquinone). Both enzymes did not oxidise tyrosine, thus, proving that they are not tyrosinases.

Enzyme inhibition

To characterise the laccases further, general inhibitors which form stable copper complexes, were tested using ABTS as substrate. It was observed that *Mg*Lac1 was more susceptible to common laccase inhibitor sodium azide with IC_{50} of 0.63 ± 0.03 mM whereas other inhibitors, cysteine (IC_{50}, 0.73 ± 0.01 mM) and EDTA (IC_{50}, 2.41 ± 0.21 mM) showed inhibition to a lesser extent. In comparision with *Mg*Lac1, *Mg*Lac2 was more susceptible to inhibition by cysteine with IC_{50} of 0.28 ± 0.07 mM than sodium azide (IC_{50}, 0.37 ± 0.02 mM) and EDTA (IC_{50}, 3.55 ± 0.27 mM).

pH optimum

The enzyme activity of *Mg*Lac1 and *Mg*Lac2 were measured in pH range of 3 to 8. The optimum pH for *Mg*Lac1 activity was at pH 4 to 5 (Figure 4A). The optimum pH for *Mg*Lac2 activity was at pH 4 to 4.5 (Figure 4B). The enzyme activity of both the laccases decreased abruptly at pH higher than 5, with no activity after pH 6. The pH optimums determined for both laccases were acidic, which is representative of typical laccases.

Temperature effects and thermostability

Both proteins showed enzymatic activity over a remarkably wide range of temperatures. Significant enzyme activity was observed over the temperature range of 25 to 80°C. Maximum enzyme activity of *Mg*Lac1 was observed at 80°C and maximum thermostability at 30°C, when stored for 60 min (Figure 5A). However, maximum enzyme activity of *Mg*Lac2 was observed at 70°C with maximum thermostability at 30°C, when stored for 60 min (Figure 5B).

MgLac1 and *MgLac2* showing dye decolourisation activity

The purified *Mg*Lac1 and *Mg*Lac2 were also used for determining decolourisation activity for Remazol brilliant blue dye. To control for the presence of possible

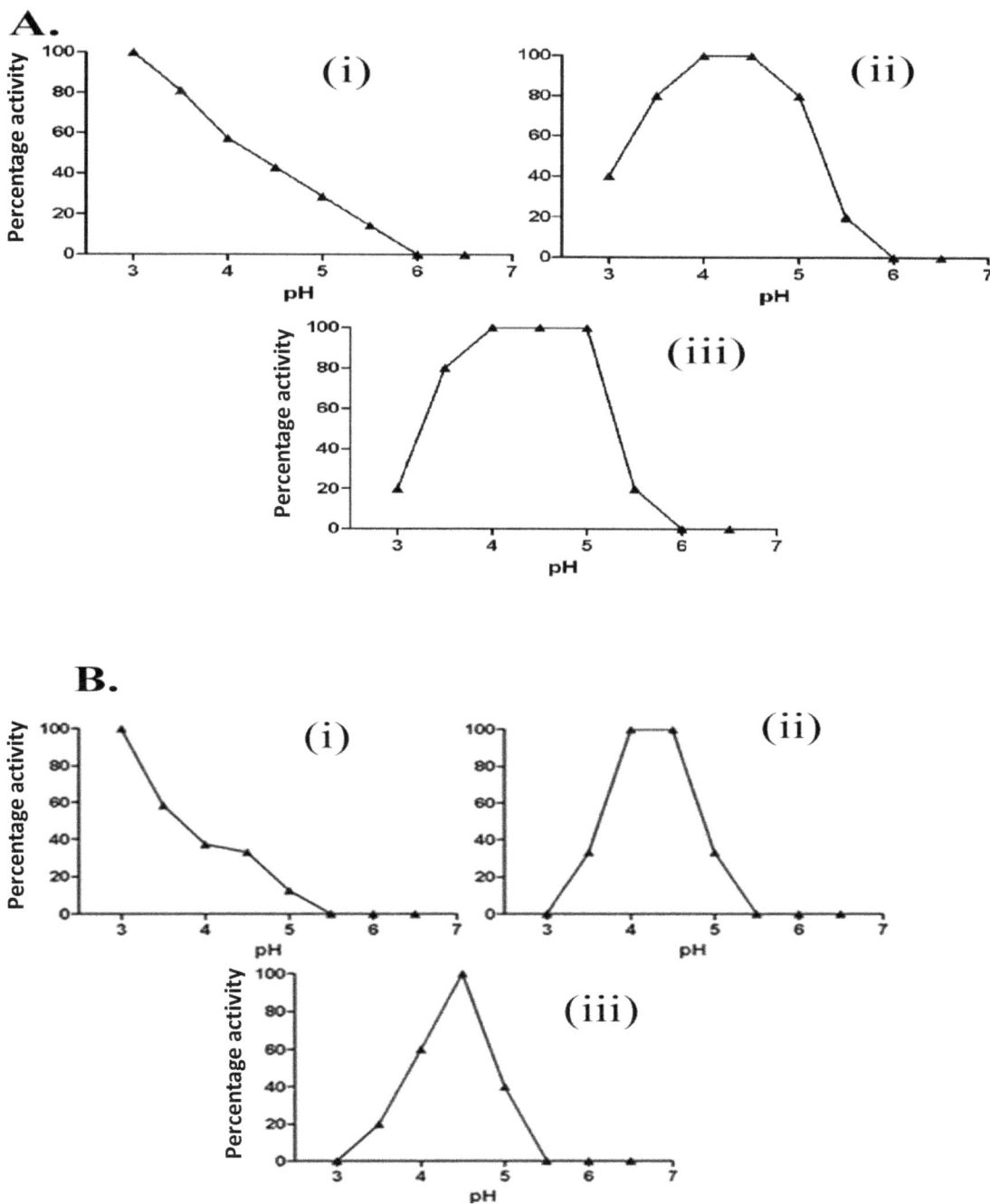

Figure 4. Effect of pH on activity of *Mg*Lac1 and MgLac2. (A) Effect of pH on activity of *MgLac1* with different substrates. (i) ABTS; (ii) Catechol; (iii) Pyrogallol. The pH optimum was determined over a range of pH 3 to 9. (B) Effect of pH on activity of *MgLac2* with different substrates. (i) ABTS;(ii) Catechol; (iii) Pyrogallol. The pH optimum was determined over a range of pH 3 to 9.

contaminating enzyme activity not detectable by SDS-PAGE, a fraction of empty vector transformed *S. cerevisiae* strain, s288C purified in an identical fashion was used as a negative control. Under our experimental conditions *Mg*Lac1 and *Mg*Lac2 were able to decolourise Remazol Brilliant Blue dye (Figure 6A). We utilised this lignin-like dye decolourisation assay to assess a possible role of these laccases in lignin degradation. Laccases are

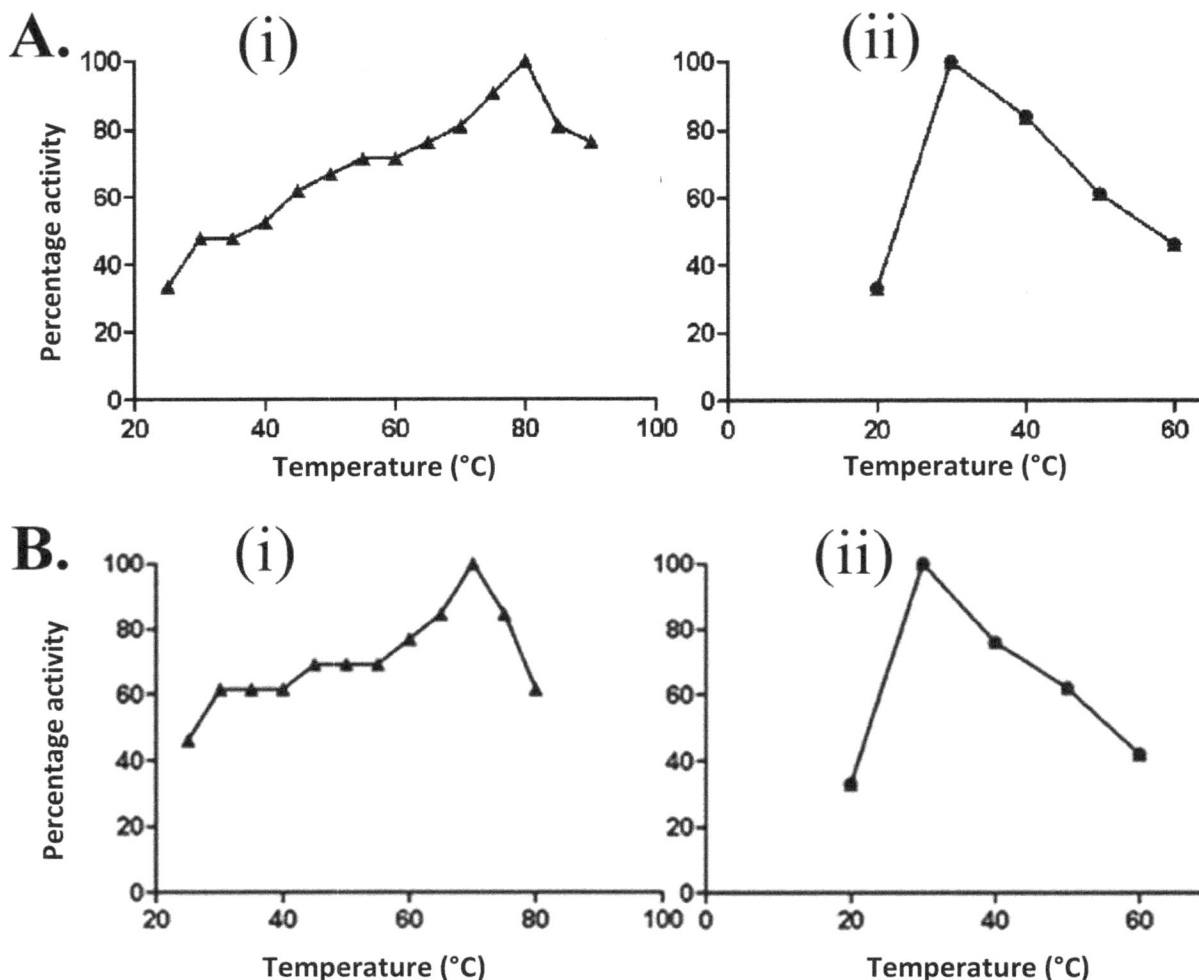

Figure 5. Effect of temperature on activities of *Mg*Lac1 and *Mg*Lac2. (A) (i) Temperature of *Mg*Lac1 maximum activity was determined at temperatures ranging from 25 to 90°C. (ii) Thermostability of *Mg*Lac1 was determined at temperatures ranging from 20 to 70°C. (B) (i) Temperature of *Mg*Lac2 maximum activity was determined at temperatures ranging from 25 to 90°C. (ii) Thermostability of *Mg*Lac2 was determined at temperatures ranging from 20 to 60°C.

known to be involved in lignin degradation in fungi (Crestini and Argyropoulos, 1998; Goodell et al, 1998; Ardon et al, 1998).

*Mg*Lac2 shows DHN polymerised potential

*Mg*Lac2 polymerised DHN to a higher molecular mass melanin. An increase of 0.1 optical density units was observed with maximum absorbance at 348 nm (Figure 6B). In contrast, *Mg*Lac1 did not show any DHN polymerisation. As a negative control, identical fraction from empty vector transformed *S. cerevisiae* strain, s288C was used. However, no enzyme activity was observed

in the empty vector transformed *S. cerevisiae* fraction. Such DHN polymerisation ability has also been observed in a secretory laccase of *Gaeumannomycesgraminis* var. *tritici* (Edens et al., 1999).

*Mg*Lac2 also shows ferroxidase activity

In addition to the earlier properties of *Mg*Lac2, we found that this laccase possessed iron oxidase activity as well. Laccase-catalyzed Fe(II) oxidation was measured by the addition of purified laccase to a reaction mixture containing ferrous sulfate in metal-free buffer. Similar fraction from empty vector transformed *S. cerevisiae*

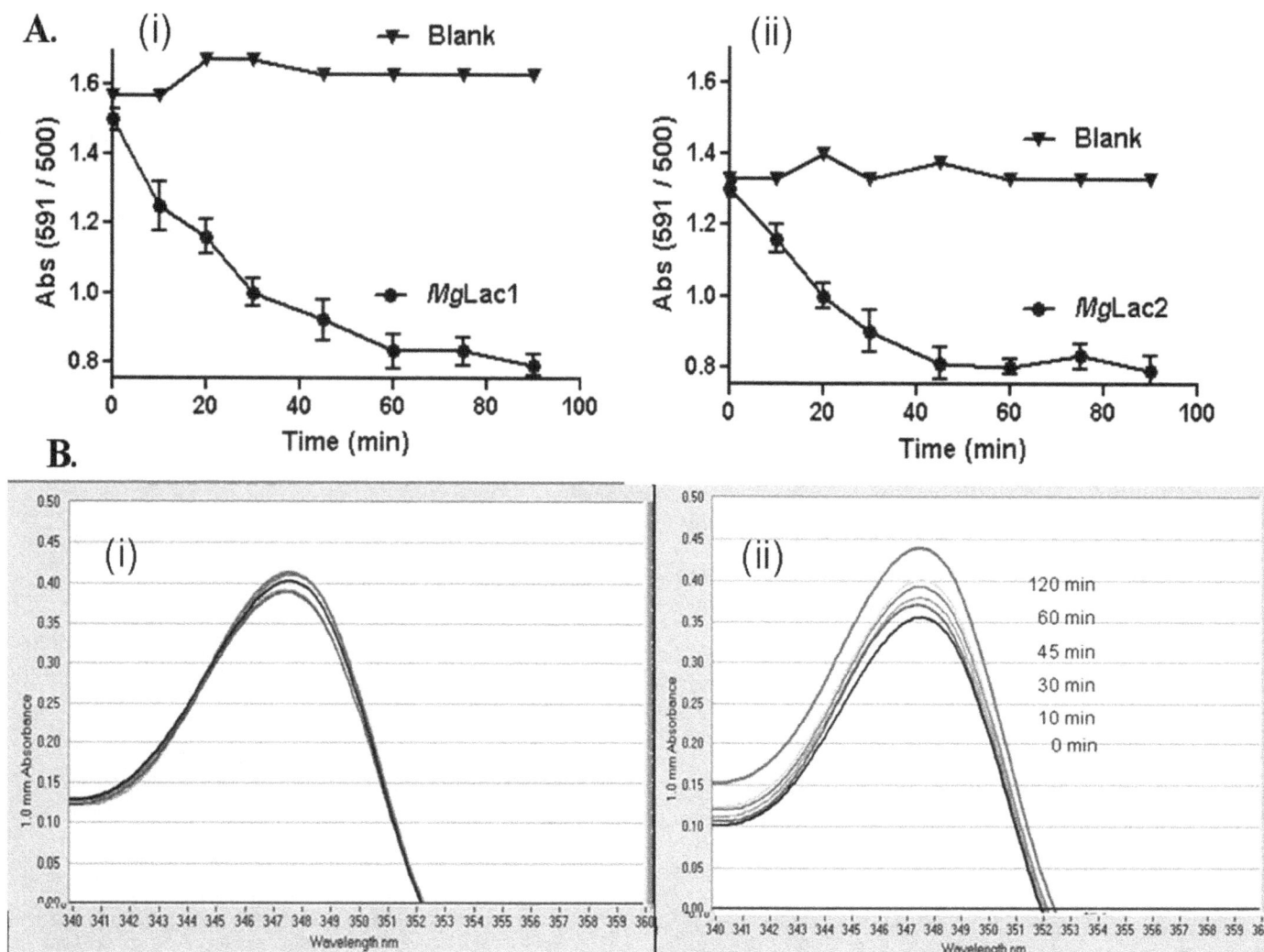

Figure 6. Dye decolourisation and DHN polymerisation. (A) Dye decolourisation activity of *Mg*Lac1 (i) and *Mg*Lac2 (ii) was estimated using Remazol brilliant blue R. Purified laccases were added, and decolourisation was monitored spectrophotometrically after indicated time points at 23°C. (B) DHN polymerisation of *Mg*Lac2 was estimated using Chromotropic acid disodium salts dehydrate. Purified laccase was added, and polymerisation was monitored spectrophotometrically by scanning the solution from 320 to 520 nm at 0 to 120 min. (i) Control (ii) *Mg*Lac2.

served as a negative control and did not show any enzyme activity. The specific ferroxidase activity was $1.574 \times 10^{-6} \pm 0.26$ U/mg protein. Such ferroxidase activity was also observed in laccase of *Cryptococcus neoformans* (Liu et al., 1999).

DISCUSSION

Laccases are ubiquitous enzymes in fungi and higher plants and have been shown to play important roles in developmental cycle of various fungi (Thurston, 1994). High laccase activity was detected in liquid cultures of *M.*

grisea as early as 24 h (Iyer and Chattoo, 2003). Therefore, we hypothesised that the high expression of laccase during very early stages may be important for infection. *MgLac2* was induced by 15-fold in nitrogen starvation in *M. grisea*.

The molecular masses of *Mg*Lac1 and *Mg*Lac2 were found to be in the range of 60 and 80 kDa, which are typical for laccases (Thurston, 1994). *Mg*Lac1 and *Mg*Lac2 exhibited very low K_m values for ABTS indicating very high binding affinity towards this substrate and oxidised syringaldazine efficiently, but there was no activity towards L-tyrosine which proved that both of them were not tyrosinases. Thus, *Mg*Lac1 and *Mg*Lac2 were classified as a true laccases based on their substrate specificity. Laccases are known to act on *p*-diphenol as well as *o*-diphenol, but they mostly have a higher affinity and activity toward the *p*-diphenols (Xu, 1996). Under current investigation *Mg*Lac2 showed affinity towards para-substituted phenol, that is, hydroquinone, but this was not the case with the *Mg*Lac1, since no laccase activity was observed towards hydroquinone. *Mg*Lac2 had higher affinity (according to K_m values) toward pyrogallol than the compounds having two adjacent hydroxyl groups such as catechol. To characterise the enzyme further, several general inhibitors, especially compounds able to form stable copper complexes were tested using ABTS as substrate. These inhibitors are not laccase specific and their application for phenoloxidase originates from results obtained with other metalloenzymes (Johannes and Majcherczyk, 2000). Sodium azide (NaN_3), that complex with the coppers in the active site, was observed to inhibit *Mg*Lac1 efficiently. Sulfhydryl organic compound, cysteine, which has a reducing effect on the copper-containing active site of laccases, was found to be the most efficient inhibitor of *Mg*Lac2.

The pH optimums obtained for *Mg*Lac1 and *Mg*Lac2 were acidic range and thus represented typical laccases. Laccases tend to react differently to varying pH with different substrates. The reasons for dependency of laccases to act on different substrates have been previously elucidated (Xu, 1996). The dependence of laccase on pH usually renders a bell-shaped profile as with catechol and pyrogallol. With ABTS there is rather a monotonic decline than a bell-shaped profile as reported (Xu, 1996).

Laccases are reported to be involved in lignin degradation in white-rot basidiomycetes *Trametes versicolors* (Crestini and Argyropoulos, 1998), *Lentinus edode* (Goodell et al, 1998) and *Pleurotus ostreatus* (Ardon et al, 1998). In phytopathogen, *Gaeumannomyces graminis* var. *tritici,* laccase is known to be involved in melanisation (Edens et al., 1999). Ferroxidase activity was also observed in *Cryptococcus neoformans* laccase (Liu et al., 1999). In our experimental conditions *Mg*Lac1 and *Mg*Lac2 both were able to decolourise Remazol Brilliant Blue suggesting a possible role in lignin degradation.

In addition to the decolourisation activity, *Mg*Lac2 also showed DHN polymerisation potential and ferroxidase activity. This is the first report of a single fungal laccase showing three above mentioned activities. Role of laccase in iron metabolism was reported in *C. neoformans*. It was speculated that laccase oxidation of ferrous iron may be important during infection in addition to its role in melanin production. The fact that *C. neoformans* laccase is expressed only during very early stages of infection would be consistent with its role in iron acquisition and in protection of fungal cells from hydroxyl radical attack from host cells during the initial adaptation to the host environment (Jung et al., 2006).

In this study we present biochemical evidence that laccases expressed in normal conditions and n conditions which mimic pathogenicity are different, that is, their substrate preferences and inhibition. Laccase which is highly induced in nitrogen starvation that is, *Mg*Lac2, produced by *M. grisea*, may be involved in lignin degradation and/or melanin synthesis and/or iron oxidation. May be multicopper oxidase-dependent iron uptake systems are present in this fungus and this laccase-iron oxidase activity might protect the fungus in concert with other laccase products (melanin) by maintaining Fe in an oxidized form, thereby decreasing production of antifungal hydroxyl radicals. Future studies will focus on the role(s) of other laccases of *M. grisea* in melanization, delignification, and/or protection from antimicrobial compounds.

ACKNOWLEDGEMENT

We thank Department of Biotechnology (DBT), Ministry of Science and Technology, Government of India for providing the facilities.

REFERENCES

Alexandre G, Bally R, Taylor BL, Zhulin IB (1993). Loss of cytochrome oxidase activity and acquisition of resistance to quinone analogs in a laccase-positive variant of *Azospirillum lipoferum*. J. Bacteriol., 181: 6730-6738.

Ardon O, Kerem Z, Hadar Y (1998). Enhancement of lignin degradation and laccase activity in *Pleurotus ostreatus* by cotton stalk extract Can. J. Microbiol., 44: 676-680.

Bradford MM (1976). A rapid and sensitive method for the quantitation of microgram quantities of protein utilizing the principle of protein-dye binding. Anal Biochem., 72:248-254.

Boudet AM (2000). Lignins and lignification: selected issues. Plant Physiol Biochem., 38: 81-96.

Chen DC, Yang BC, Kuo TT (1992). One-step transformation of yeast in stationary phase. Curr. Genet., 21: 83-84.

Claus H (2003). Laccases and their occurrence in prokaryotes. Arch Microbiol. 179:145-150.

Claus H (2004). Laccases: structure, reactions, distribution. Micron, 35: 93-96.

Crestini C, Argyropoulos DS (1998). The early oxidative biodegradation steps of residual kraft lignin models with laccase. Bioorg. Med. Chem., 6: 2161-2169.

Edens WA, Goins TQ, Dooley D, Henson JM (1999). Purification and characterization of a secreted laccase of *Gaeumannomyces graminis* var. tritici. App.l Environ. Microbiol., 65: 3071-3074.

Enguita FJ, Martins LO, Henriques AO, Carrondo MA (2003). Crystal structure of a bacterial endospore coat component: a laccase with enhanced thermostability properties. J. Biol. Chem., 278: 19416-19425.

Faure D, Bouillant ML, Bally R (1994). Isolation of *Azospirillum lipoferum* 4T Tn5 mutants affected in melanization and laccase activity. Appl. Environ. Microbiol., 60: 3413-3415.

Faure D, Bouillant ML, Jacoud C, Bally R (1996). Phenolic derivates related to lignin metabolism as substrate for *Azospirillum* laccase activity. Phytochemistry. 42: 357-360.

Francis CA, Tebo BM (2001). CumA multicopper oxidase genes from diverse Mn(II)-oxidazing and non-Mn(II)-oxidazing *Pseudomonas* strains. Appl. Environ. Microbiol., 67: 4272-4278.

Givaudan A, Effose A, Faure D, Potier P, Bouillant ML, Bally R (1993). Polyphenol oxidase in *Azospirillum lipoferum* isolated from rice rhizosphere: evidence of laccase activity in non-motile strains of *Azospirillum lipoferum*. FEMS Microbiol. Lett., 108: 205-210.

Goodell B, Yamamoto K, Jellison J, Nakamura M, Fujii T, Takabe K (1998). Laccase immunolabelling and microanalytical analysis of wood degraded by *Lentinus edodes*. Holzforschung, 52:345-350.

Hatakka A (1994). Lignin-modifying enzymes from select white-rot fungi: production and role in lignin degradation. FEMS Microbiol. Rev., 13: 125-135.

Hoopes JT, Dean JFD (2004). Ferroxidase activity in a laccase like multicopper oxidase from *Liriodendron tulipifera*. Plant Physiol. Biochem., 42: 27-33.

Horton P, Park KJ, Obayashi T, Fujita N, Harada H, Adams-Collier CJ, Nakai K (2007). WoLF PSORT: protein localization predictor. Nucleic Acids Res., 35:585-587.

Hullo MF, Moszer I, Danchin A, Martin-Verstraete I (2001). CotA of *Bacillus subtilis* is a copper-dependent laccase. J. Bacteriol., 183: 5426-5430.

Iyer G, Chattoo BB (2003). Purification and characterization of laccase from the rice blast fungus, *Magnaporthe grisea*. FEMS Microbiol. Lett., 227:121-126.

Johannes C, Majcherczyk A (2000). Laccase activity tests and laccase inhibitors. J. Biotechnol., 78: 193-199.

Jordaan J, Leukes WD (2003). Isolation of a thermostable laccase with DMAB and MBTH oxidative coupling activity from a mesophilic white rot fungus. Enzyme Microb. Technol., 33: 212-219.

Jung WH, Sham A, White R, Kronstad JW (2006). Iron regulation of the major virulence factors in the AIDS-associated pathogen *Cryptococcus neoformans*. PLoS Biol., 4: e410.

Kachroo P, Leong SA, Chattoo BB (1994). Pot2, an inverted repeat transposon from the rice blast fungus *Magnaporthe grisea*. Mol. Gen. Genet., 245: 339-348.

Kim C, Lorenz WW, Hoopes J, Dean JF (2001). Oxidation of phenolate siderophores by the multicopper oxidase encoded by the *Escherichia coli yacK* gene. J. Bacteriol., 183: 4866-4875.

Lengeler KB, Davidson RC, D'Souza C, Harashima T, Shen WC, Wang P, Pan X, Waugh M, Heitman J (2000). Signal transduction cascades regulating fungal development and virulence. Microbiol. Mol. Biol. Rev., 64: 746-785.

Liu L, Dean JFD, Friedman WE, Eriksson KEL (1994). Laccase like phenoloxidase is correlated with lignin biosynthesis in *Zinnia elegans* stem tissue. Plant J., 6: 213-224.

Liu L, Tewari RP, Williamson PR (1999). Laccase protects *Crypyococcus neoformans* from antifungal activity of alveolar marophages. Infect. Immun., 67: 6034-6039.

Lozovaya VV, Lygin AV, Zernova OV, Li S, Widholm JM and Hartman GL (2006). Lignin degradation by *Fusarium solani* f. sp. *glycines*. Plant Dis., 90: 77-82.

Mayer AM, Staples RC (2002) Laccase: new function for an old enzyme. Phytochemistry, 60: 551–565.

Mitchell DA, Marshall TK, Deschenes R J (1993). Vectors for the inducible overexpression of glutathione S-transferase fusion proteins in yeast. Yeast, 9: 715-722.

Ou SH (1985). *Rice Diseases*, second ed. Commonwealth Mycological Institute, Kew, United Kingdom, pp. 109-201.

Ranocha P, Chabannes M, Chamayou S, Danoun S, Jauneau A, Boudet AM, Goffner D (2002). Laccase down-regulation causes alterations in phenolic metabolism and cell wall structure in poplar. Plant Physiol., 129: 1-11.

Ryan S, Schnitzhofer W, Tzanov T, Cavaco-Paulo A, Gübitz GM (2003). An acid-stable laccase from *Sclerotium rolfsii* with potential for wool dye decolourization. Enzyme Microb. Technol., 33: 766-774.

Saito T, Hong P, Kato K, Okazaki M, Inagaki H, Maeda S (2003) Purification and characterization of an extracellular laccase of a fungus (family Chaetomiaceae) isolated from soil. Enzyme Microb. Technol., 33: 520-526.

Snoeijers SS, Perez-Garcia A, Joosten MH, De Wit PJ (2000). The effect of nitrogen on disease development and gene expression in bacterial and fungal pathogens. Eur. J. Plant Pathol., 106: 493-506.

Sterjiades R, Dean JF, Eriksson KE (1992). Laccase from Sycamore Maple (Acer pseudoplatanus) Polymerizes Monolignols. Plant Physiol. 99: 1162-1168.

Talbot NJ (2003). On the trail of a cereal killer: Exploring the biology of *Magnaporthe grisea*. Annu. Rev. Microbiol., 57: 177-202.

Talbot NJ, McCafferty HRK, Ma M, Moore K, Hamer JE (1997). Nitrogen starvation of the rice blast fungus *Magnaporthe grisea* may act as an environmental cue for disease symptom expression. Physiol. Mol. Plant Pathol., 50: 179-195.

Thurston CF (1994). The structure and function of laccases. Microbiology, 140: 19-26.

Valent B (1990). Rice blast as a model system for plant pathology. Phytopathology, 80: 33-36.

Xu F (1996). Oxidation of phenols, anilines, and benzenethiols by fungal laccases: correlation between activity and redox potentials as well as halide inhibition. Biochemistry, 35: 7608-7614.

Zeigler RS, Leong SA, Teeng PS (1994). Rice Blast Disease. Wallingford: CAB Int., pp. 381-408.

22

Antifungal effects of the essential oil from *Thymus vulgaris* L. and comparison with synthetic thymol on *Aspergillus niger*

M. Moghtader

Department of Biodiversity, International Center for Science, High Technology and Environmental Sciences, Kerman, Iran. E-mail: moghtader18@yahoo.com.

The antifungal effects of the essential oil from *Thymus vulgaris* L. and comparison with synthetic thymol on *Aspergillus niger* growth was studied. The chemical composition of the essential oil of *T. vulgaris*, the aerial parts of this plant which is grown in a village in Kerman Province at full flowering stage in June 2012 were collected. The sample was cleaned, dried in the shade and hydro distillation method was performed for the extraction of essential oil. The main oil content from the plant of *T. vulgaris* was 2.25% (v/w). Essential oil was analyzed by capillary gas chromatography (GC) using flame ionization (FID) and capillary gas chromatography coupled mass spectrometry (GC/MS) for detection. Thirty-two compounds were identified in the essential oil of *T. vulgaris* that concluded 99.56% of the total oil. The major components were Thymol (32.67%), *P*-cymene (16.68%), γ-terpinene (12.65%) and Carvacrol (8.32%). The study of antifungal effects of the oil sample was tested against strain of *A. niger* (PTCC=5223) fungi by disc diffusion method via average inhibition zone. The results show that essential oil from *Thymus* plant at 1, 1/2 and 1/4 oil dilutions exhibits strong antifungal activity than Streptomycin sulphate (72% SP) and gentamycin (8 mg/ml) antibiotics on *A. niger* and that exhibited on strong synthetic thymol was at 10% dilution. The high percentage antifungal activities of *Thymus* oil are related with thymol is a natural monoterpene phenol as the main compound.

Key words: *Thymus vulgaris* L., *Aspergillus niger*, essential oil, antifungal activity, fungal growth, thymol.

INTRODUCTION

Aspergillus is a genus consisting of several hundred mold species found in various climates worldwide. *A. niger* is a fungus and one of the most common species of the genus *Aspergillus*. It causes a disease called black mold on certain fruits and vegetables such as grapes, onions, and peanuts, and is a common contaminant of food. It is ubiquitous in soil and is commonly reported from indoor environments, where its black colonies can be confused with those of *Stachybotrys* species which is also called black mould (Samson et al., 2001). Some strains of *A. niger* have been reported to produce potent mycotoxins called ochratoxins (Abarca et al., 1994) but other sources disagree, claiming this report is based upon misidentification of the fungal species. Recent evidence suggests that some *A. niger* strains produce ochratoxin A (Schuster et al., 2002). *A. niger* is one of the most common causes of otomycosis (fungal ear infections), which can cause pain, temporary hearing loss, and in severe cases, damage to the ear canal and tympanic membrane.

The genus Thymus (thyme) contains about 350 species of aromatic perennial herbaceous plants and sub shrubs 40 cm tall in the family Lamiaceae, native to temperate regions in Europe, North Africa and Asia. Stems tend to be narrow or even wiry; leaves are evergreen in most species, arranged in opposite pairs, oval, entire, and small, 4 to 20 mm long, and usually aromatic. Thyme flowers are in dense terminal heads, with an uneven calyx, with the upper lip three-lobed, yellow, white or purple. Several members of the genus are cultivated as culinary herbs or ornamentals, when they are also called thyme after its best-known species, *T. vulgaris* or Thyme Green. *T. vulgaris* L. or common thyme is a low growing herbaceous plant, sometimes becoming somewhat woody. It is an evergreen shrub growing to 0.2 m (0 ft 8 in) by 0.3 m (1 ft). It is hardy to zone 7 and is not frost tender. It is in leaf 12 January it is in flower from June to August. The flowers are hermaphrodite (have both male and female organs) and are pollinated by Bees, flies, Lepidoptera. It is noted for attracting wildlife. Herbalists of the middle Ages regarded thyme as a stimulant and antispasmodic and recommended sleeping on thyme and inhaling it as a remedy for melancholy and epilepsy.

In 1725, a German apothecary discovered that the plant's essential oil contains a powerful disinfectant called thymol that is effective against bacteria and fungi. Thymol also acts as an expectorant, loosening phlegm in the respiratory tract so it can be coughed up. Later herbalists listed thyme for these uses and as remedy for numerous other complaints, including diarrhea and fever. They prescribed the oil externally as an antiseptic for fungal infections such as athlete's foot. Thymol (also known as 2-isopropyl-5-methylphenol, IPMP) is a natural monoterpene phenol derivative of cymene, $C_{10}H_{14}O$, isomeric with carvacrol, found in oil of thyme, and extracted from *T. vulgaris* (common thyme) and various other kinds of plants as a white crystalline substance of a pleasant aromatic odor and strong antiseptic properties. Thymol also provides the distinctive, strong flavor of the culinary herb thyme, also produced from *T. vulgaris*. Thymol is part of a naturally occurring class of compounds known as biocides, with strong antimicrobial attributes when used alone or with other biocides such as carvacrol. thymol has been shown to be an effective fungicide (Ahmad et al., 2010). Thymol has microbial activity because of its phenolic structure. There is evidence supporting the belief that thymol, when applied two to three time's daily, can eliminate certain kinds of fungal infections that affect fingernails and toenails in humans. Regular application to the affected nail over periods of about three months has been shown to eliminate the affliction by effectively preventing further progress by simply cutting the nail as one normally would, all infected material is eventually eliminated.

The antifungal nature of thymol is caused by thymol's ability to alter the hyphal morphology and cause hyphal aggregates, resulting in reduced hyphal diameters and lyses of hyphal wall (Numpaque et al., 2011). Additionally, thymol is lipophilic, enabling it to interact with the cell membrane of fungus cells, altering cell membrane permeability by permitting the loss of macromolecules (Segvic et al., 2007). Oil of thyme, the essential oil of common thyme (*T. vulgaris*), contains 20 to 54% thymol. Thyme essential oil also contains a range of additional compounds, such as *p*-cymene, myrcene, borneol and linalool. Thymol, an antiseptic, is the main active ingredient in various commercially produced mouthwashes such as Listerine (Pierce, 1999). Before the advent of modern antibiotics, oil of thyme was used to medicate bandages. Thymol has also been shown to be effective against various fungi that commonly infect toenails (Ramsewak et al., 2003). Thymol can also be found as the active ingredient in some all-natural, alcohol-free hand sanitizers. A tea made by infusing the herb in water can be used for coughs and bronchitis. One study by Leeds Metropolitan University found that thyme may be beneficial in treating acne. This study evaluated and identified the chemical compounds of *T. vulgaris* mainly. Also antifungal activity of *T. vulgaris* has been compared with synthetic thymol and Streptomycin sulphate (72% SP), gentamicin (8 mg/ml) antibiotics standard on culture of *A. niger* (PTCC=5223) fungi.

MATERIALS AND METHODS

Plant material collection and isolation of their essential oil

The aerial parts of *T. vulgaris* L. were obtained from this plant grown in a village in Kerman Province, Iran at full flowering stage in June 2012. This plant identify by the botany herbarium in university of Kerman. The sample was cleaned in shade condition to prevent volatility of the plant material constituents and to keep the natural color of the sample fixed. Then they were air-dried and were powdered using a milling machine and kept in a cool dry place until ready for extraction of the essential oil. Afterwards, essential oil was taken from 150 g of the powdered sample in hydro distillation method with the help of Clevenger set for three hours. Following the sample oils were dried with anhydrous sodium sulfate and kept in sterile sample tubes in refrigerator. The oil yield from aerial parts of *T. vulgaris* plant was calculated.

Analysis of essential oil

Gas chromatography

GC analysis was performed using a model HP-439 gas chromatograph equipped with column CP Sil. 5 CB in 25 m length, internal diameter of 0.25 mm and film thickness 0.39 μm. Oven temperature was from 60 to 220°C at a rate of 7°C slope per minute. Injector temperature was 280°C and detector (FID) temperature was 270°C and carrier gas was helium.

Gas chromatography / mass spectrometry

In order to analyze and identify the combinations forming the essential oil, the chromatograph gas set attached to a mass spectrometry, Model Hewlett Packard-5973 was used. The conditions of analysis and specifications of the GC/MC set were as follows: Capillary column HP 5MS in 60 m length, internal diameter of 0.25 mm and layer thickness of 0.25 μM, thermal program of oven (3 min) in 60°C, then 60 to 220°C with a 6°C slope per minute, then 3 min in 220°C, the temperature of place of injection 280°C, gas conveying helium, the speed of gas move 1.0 milliliter per minute, the ratio of fission 1 to 43, the rate of injection 0.1 μL, temperature of the reservoir of ionization 230°C, ionization mode EI, Ionization energy 70eV. The series of normal Alkans C8-C17 were also injected to the set under the same condition with that of

Table 1. Combinations identified in the essential oil of *Thymus vulgaris* L.

Compound name	Restrictive index (RI)	Percentage (%)
Tricyclen	925	0.68
α-thujene	928	1.35
α-pinene	935	2.24
Camphene	957	0.72
Sabinene	978	0.56
β-pinene	984	1.07
Myrcene	992	0.34
α-phellandrene	1021	0.18
α-terpinene	1025	1.45
P-cymene	1029	16.68
Limonene	1035	1.47
1,8-cineole	1038	1.43
γ-terpinene	1067	12.65
Terpinolene	1094	0.73
Linalool	1126	1.36
Camphore	1138	0.69
Borneol	1154	2.85
Terpinene-4-ol	1192	0.34
α-terpineol	1203	1.56
Methyl thymol	1218	0.37
Thymoquinone	1224	0.43
Methyl carvacrol	1228	0.62
Geraniol	1246	2.32
Thymol	1278	32.67
Carvacrol	1287	8.32
Thymyl acetate	1362	0.22
β-caryophyllene	1425	2.39
α-humulene	1466	0.2
Germacrene D	1482	1.46
γ-cadinene	1521	0.67
Spathulenol	1569	0.18
Caryophyllene oxide	1579	1.36
Total		99.56

essential oil injection to calculate restrictive index (RI) of components of essential oil. The Restrictive Index of components of the sample was calculated by using a computerized program. Finally, the components of essential oil was identified by comparing the mass spectrums obtained with the existing standard mass spectrums at electronic library of Wiley 2000 existing in Absolution software of GC/Ms set and calculation of standard restrictive index in accordance with C8-C17 Alkans and comparing them with the existing standard figurers in references (Adams, 2001).

Antifungal assay

The solvent showing no antifungal activity from DMSO was selected as a diluting medium for the oil. Undiluted oil was taken as dilution 1, 1/2, 1/4, 1/8 and 1/16 dilutions of the oil were made DMSO. For antifungal activity 50 µL of each dilution was used. The antifungal activity of the essential oil was evaluated by disc diffusion method using Mueller Hinton Agar (Baron and Finegold, 1995) and determination of inhibition zones at different oil dilutions against *A. niger* (PTCC=5223). The fungal strains under experiment were obtained from the Center for Fungi and Bacteria of Iranian Scientific and Industrial Researches Organization. The antifungal property of the oil was tested by agar well diffusion method using Sabouraud Dextrose Agar (SDA). Standard reference antibiotics were used in order to control the sensitivity of the tested fungi (Streptomycin sulphate 72% SP and gentamicin 8 mg/ml). The incubation conditions used was 48 to 72 h at 24°C for fungi. All the experiments were carried out in triplicate and averages were calculated for the inhibition zone diameters.

RESULTS

The study of the analysis of *T. vulgaris* L. essential oil under investigation showed that the output of essential oil is 2.25% (v/w). The identified combination in essential oil, restrictive index (RI), and quantitative percentage of the compounds. Thirty-two compounds being identified in the essential oil of this plant with 99.56%, the combinations of Thymol (32.67%), P-cymene (16.68%), γ-terpinene (12.65%) and Carvacrol (8.32%) with 70.32% constitute the highest percentage of essential oil (Table 1).

Table 2. The zone diameter of inhibition of antibiotics, thyme oil and synthetic thymol on *Aspergillus niger* (mm).

Antibiotics		Dilutions of thyme oil					Synthetic thymol	
Streptomycin sulphate (72% SP)	Gentamycin (8 mg/ml)	1	1/2	1/4	1/8	1/16	1 %	10%
20	18	32	27	22	18	12	16	26

The indexes of restrictive have been calculated by injecting the mixture of normal hydrocarbons (C8-C17) to HP-5MS column. The results of studying the antifungal impacts of the *T. vulgaris* essential oil shows that the oil of this plant has an inhibitory effect in 1, 1/2, 1/4, 1/8 and 1/16 dilution with average diameter growth of 32, 27, 22, 18 and 12 mm, respectively. the results with standard antibiotics Streptomycin sulphate (72% SP) with a diameter of 20 mm and gentamicin (8mg/ml) with a diameter of 18 mm had inhibitory effects. Thymol synthetic in 1% dilution had moderate inhibitory (16 mm) effect on *A. niger* growth but at 10 % dilution had a strong inhibitory (26 mm) of fungi growth. The results showed essential oil from *Thymus* plant at 1, 1/2 and 1/4 oil dilutions exhibited strong antifungal activity than Streptomycin sulphate (72% SP) and gentamycin (8 mg/ml) antibiotics on *A. niger* and exhibited strong of thymol was at 10% dilution. The high percentage antifungal activities of fumaria oil are related with thymol is a natural monoterpene phenol as the main compound (Table 2).

DISCUSSION

Identify chemical composition and antifungal effect of *T. vulgaris* L. in this paper compared with other researcher. In comparison our results with other researchers the essential oil of *Thymus vulgaris* L. has been studied in Iran and world. Essential oil of the aerial parts of *Z. multiflora, T. vulgaris* and *T. kotschyanus* were collected from growing fields in their natural habitats in Iran surveyed the effectiveness, Minimum inhibitory concentration (MIC) and Minimum fungicide concentration (MFC) of three medicinal plant essential oils of *Zataria multiflora, T. vulgaris* and *T. kotschyanus* on the mycelial growth of four pathogenic fungi including *Pythium aphanidermatum, Rhizoctonia solani* (AG4), *Fusarium graminearum* and *Sclerotinia sclerotiorum*. Essential oils were very effective on the four studied plant pathogenic fungi with growth inhibition average of 100% at 200µl/l concentration. In this research MIC and MFC of the essential oils were variable depending to species of fungi. *P. aphanidermatum* and *S. sclerotiorum* were the most sensitive and most resistant to the studied essential oils with average growth inhibition 89.54 and 75.35%, respectively (Amini et al., 2012). The composition and antimicrobial activity of the essential oil of *T. daenensis* Wild, an endemic species from Iran, was studied. Thirty compounds, accounting for 97.5% of the total oil, were identified. The main constituents were thymol (29.8%) and carvacrol (13.6%), *p*-cymene (11.3), borneol (6.8%) and 1, 8-cineole (5.89%).

The antimicrobial activity of essential oil of *T. daenensis* was tested against two Gram-negative and Gram-positive bacteria. The results of the bioassays showed the interesting antimicrobial activity, in Gram-positive bacteria, *S. aureus*, was the most sensitive to the oil, and as well the oil exhibited a remarkable antifungal activity against all the tested fungi. (Teimouri, 2012). In a research *Thymus vulgari* and *Citrus aurantifolia* were found to inhibit *A. parasiticus* (Razzaghi-Abyaneh et al., 2009). Essential oils *T. eriocalyx* could be safely used as preservatives (Rasooli et al., 2006). It is concluded that essential oils of *Thymus* and *Mentha* species possess great antifungal potential and could be used as natural preservatives and fungicides In related report the potential antifungal effects of *T. vulgaris* L., *T. tosevii* L., *Mentha spicata* L., and *Mentha piperita* L. (Labiatae) essential oils thymol (48.9%) and *p*-cymene (19.0%) were the main components of *T. vulgaris,* while carvacrol (12.8%), α-terpinyl acetate (12.3%), *cis*-myrtanol (11.2%) and thymol (10.4%) were dominant in *T. tosevii. Thymus* species showed very strong antifungal activities. In *M. piperita* oil menthol (37.4%), menthyl acetate (17.4%) and menthone (12.7%) were the main components, whereas those of *M. spicata* oil were carvone (69.5%) and menthone (21.9%). *Mentha* sp. showed strong antifungal activities, however lower than *Thymus* sp. (Soković et al., 2009). The thyme oil absolutely inhibited the mycelial growth of *A. flavus* at 0.7 µL ml^{-1} (Kumar et al., 2008).

Thymol is one of main compound of *T.spicata, Satureja thymbra, Salvia fruticosa, Laurus nobilis, Mentha pulegium, Inula viscosa, Pimpinella anisum, Eucalyptus camaldulensis,* and *Origanum minitiflorum* plants growing wild in southern Turkey. (Muller-Riebau et al., 1995). The relatively high antifungal activity of thymol and carvacrol against *C. acutatum* and *B. theobromae* and the low levels of microbial transformation indicate that both compounds could be an alternative to traditional chemical fungicides for control of pre- and post harvest phytopathogenic fungi on fruits or vegetables (Numpaque et al., 2011). In a study thymol exhibited approximately three-times stronger inhibition than essential oil of thyme (Segvic et al., 2007). Essential oils of thyme, cinnamon, anise and spearmint have more effect on fungal development and subsequent mycotoxin production in wheat grains. The extent of inhibition of fungal growth and mycotoxin production was dependent on the concentration of essential oils used (Soliman and Badeaa, 2002). Essential oil of *T. vulgaris* thymol chemotype potentiates the antifungal action of amphotericin B suggesting a possible utilization of this essential oil in addition to antifungal drugs for the treatment of mycoses (Giordani et al., 2005). The essential oils of *T. vulgaris* and *T. zygis* showed similar antifungal activity, which was greater than *T. mastichina* (Pina-Vaz et al., 2004). The composition of the essential oil of *T. pulegioides* showed high contents of carvacrol and thymol. *T. pulegioides* essential oil exhibited a significant activity against clinically relevant fungi (Pinto et al., 2006). Thymol and carvacrol that are the two main constituents of *T. glandulosus* and *Origanum compactum* exhibited the strongest antifungal activity (Chebli et al., 2003).

The essential oil of *T. revolutus* C. exhibited a significant antibacterial and antifungal activity (Karaman et al., 2001). *P-cymene*, linalool, terpinen-4-ol and thymol were main content in essential oils from two clonal types of *T. vulgaris* (Bhaskara Reddy et al., 1998). The most effective oil was that of thyme, with a fungicidal activity attributable to thymol, found in a concentration of 50.06% in the oil tested (Zambonelli et al., 1996). The fungicidal activity of the oils was correlated with their thymol content (Zambonelli et al., 2004). *A. flavus* was more sensitive to thyme essential oil than *A. niger*. Clove essential oil was a stronger inhibitor against *A. niger* than against *A. flavus* (Viuda-Maartos et al., 2007). The chemical composition of *Cinnamomum zeylanicum* (bark), *Cinnamomum zeylanicum* (leaf), *Cinnamomum cassia*, *Syzygium aromaticum* and *Cymbopogon citratus* had most active essential oils (Pawar and Thaker, 2006). *A. niger* was shown to be more difficult to destroy of essential oils *Thymus broussonettii*, *T. zygis* and *T. satureioides* than the most resistant bacterium *E. coli* (Tantaoui-Elaraki et al., 1993). The essential oils extracted from thyme (*T. vulgaris* L.), basil (*Ocimum basilicum* L.), coriander (*Coriandrum sativum* L.), rosemary (*Rosmarinus officinalis* L.), sage (*Salvia officinalis* L.), fennel (*Foeniculum vulgare* L.), spearmint (*Mentha spicata* L.) and carraway (*Carum carvi* L.) were investigated for their antimicrobial activity against eleven different bacterial and three fungal strains *A. niger* ATCC 16404, *Penicillium* sp. CICC 251 and two *E. coli* and *Salmonella enterica* serovar Enteritidis clinical isolates. The majority of the tested essential oils exhibited considerable inhibitory capacity against all the organisms tested, as supported by growth inhibition zone diameters, MICs and MBC's. Thyme, coriander and basil oils proved the best antibacterial activity, while thyme and spearmint oils better inhibited the fungal species (Lixandru et al., 2010).

Conclusion

The thyme essential oil recommended for large scale application based its strong antifungal as well as on *A. niger* efficacy. In this study we find out that regard to the antifungal effects of *T. vulgaris* essential oil under investigation as compared with synthetic thymol on *A.* *niger* and observed exhibited strong of synthetic thymol. The high percentage antifungal activities of *T.*oil are related with thymol is a natural monoterpene phenol as the main compound. This essential oil can be used as a combination with antifungal effects and natural origin. The effectiveness oil concentration of the oil depends on the target pathogen and effects of natural compounds on fungus.

ACKNOWLEDGEMENT

This study was conducted under the financial support of the International Center for Science, High Technology and Environmental Sciences in Kerman, Iran. Hereby, the executive and colleagues of the research project expresses their thanks for the support extended to them.

REFERENCES

Abarca M, Bragulat, M, Castellá, G, Cabañes F (1994). Ochratoxin A production by strains of *Aspergillus niger* var. *niger*. Appl. Environ Microbiol. 60 (7):2650-2652.

Adams RP (2001). Identification of essential oil components by gas chromatography mass spectroscopy. Illinois Allured Publication Corporation.

Ahmad A, Khan A, Yousuf S, Khan LA, Manzoor N (2010). Proton translocating ATPase mediated fungicidal activity of eugenol and thymol. Fitoterapia 81(8):1157-1162.

Amini M, Safaie N, Salmani MJ, Shams-Bakhsh M (2012). Antifungal activity of three medicinal plant essential oils against some phytopathogenic fungi. Trakia J. Sci. 10(1):1-8.

Baron EJ, Finegold SM (1995). Bailey and Scott's Diagnostic Microbiology, 8th ed. Mosby, St. Louis, MO, USA, pp. 171-193.

Bhaskara Reddy MV, Angers P, Gosselin A, Arul J (1998). Characterization and use of essential oil from *Thymus vulgaris* against *Botrytis cinerea* and *Rhizopus stolonifer* in strawberry fruits. Phytochemistry 47(8):1515-1520.

Chebli Bouchra CH, Achouri M, Hassani LMI, Hmamouchi M (2003). Chemical composition and antifungal activity of essential cils of seven Moroccan Labiatae against *Botrytis cinerea* Pers: Fr. J. Ethnopharmacol. 89(1):165-169.

Giordani R, Regli P, Kaloustian J, Mikaïl C, Abou L, Portugal H (2005). Antifungal effect of various essential oils against *Candida albicans* potentiation of antifungal action of amphotericin B by essential oil from *Thymus vulgaris*. Phytother. Res. 18(12):990-995.

Karaman S, Digrak M, Ravid U, Ilcim A (2001). Antibacterial and antifungal activity of the essential oils of *Thymus revolutus* Celak from Turkey. J. Ethnopharmacol. 76(2):183-186.

Kumar A, Shukla R, Singh P, Shekhar Prasad CH, Kishore Dubey N (2008). Assessment of *Thymus vulgaris* L. essential oil as a safe botanical preservative against post harvest fungal infestation of food commodities. Innov. Food Sci. Emerg. Technol. 9(4):575-580.

Lixandru BE, Drăcea NO, Dragomirescu CC, Drăgulescu EC, Coldea IL, Anton L, Dobre E, Rovinaru C, Codiță I (2010). Antimicrobial activity of plant essential oils against bacterial and fungal species involved in food poisoning and/or food decay. Roum Arch. Microbiol. Immunol. 6(4):224-230.

Mueller-Riebau F, Berger B, Yegen O (1995). Chemical composition and fungitoxic properties to phytopathogenic fungi of essential oils of selected aromatic plants growing wild in Turkey. J. Agric. Food Chem. 43(8):2262-2266.

Numpaque MA, Oviedo LA, Gil JH, Garcia CM, Durango DL (2011). Thymol and carvacrol: biotransformation and antifungal activity against the plant pathogenic fungi *Colletotrichum acutatum* and *Botryodiplodia theobromae*. Trop. Plant Pathol. 36:3-13.

Pawar VC, Thaker VS (2006). *In vitro* efficacy of 75 essential oils against *Aspergillus niger*. Mycoses 49(4):316-323.

Pierce A (1999). American Pharmaceutical Association Practical Guide to Natural Medicines. New York: Stonesong Press. pp. 338-340.

Pina-Vaz C, Rodrigues AG, Pinto E, Costa-de-Oliveira S, Tavares C, Salgueiro L, Cavaleiro C, Gonçalves MJ, Martinez-de-Oliveira J (2004). Antifungal activity of *Thymus* oils and their major compounds. J. Eur. Acad. Dermatol. Venereol. 18(1):73-78.

Pinto E, Pina-Vaz C, Salgueiro L, Gonçalves MJ, Costa-de-Oliveira S, Cavaleiro C, Palmeira A, Acácio Rodrigues A, Martinez-de-Oliveira J (2006). Antifungal activity of the essential oil of *Thymus pulegioides* on *Candida*, *Aspergillus* and dermatophyte species. J. Med. Microbiol. 55(10):1367-1373.

Ramsewak RS, Nair MG, Stommel M, Selanders L (2003). *In vitro* antagonistic activity of monoterpenes and their mixtures against 'toe nail fungus' pathogens. Phytother. Res. 17(4):376-379.

Rasooli I, Rezaei MB, Allameh A (2006). Growth inhibition and morphological alterations of *Aspergillus niger* by essential oils from *Thymus eriocalyx* and *Thymus x-porlock*. Food Control 17(5):359-364.

Razzaghi-Abyaneh M, Shams-Ghahfarokhi M, Rezaee MB, Jaimand K, Alinezhad S, Saberi R, Yoshinari T (2009). Chemical composition and antiaflatoxigenic activity of *Carum carvi* L., *Thymus vulgaris* L. and *Citrus aurantifolia* essential oils. Food Control. 20(11):1018-1024.

Samson RA, Houbraken J, Summerbell RC, Flannigan B, Miller JD (2001). Common and important species of fungi and actinomycetes in indoor environments. In: Micro-ogransims in Home and Indoor Work Environments. New York: Taylor & Francis. pp. 287-292.

Schuster E, Dunn-Coleman N, Frisvad JC, Van Dijck PW (2002). On the safety of *Aspergillus niger* review. Appl. Microbiol. Biotechnol. 59(4-5):426-35.

Segvic Klaric M, Kosalec I, Mastelic J, Pieckova E, Pepeljnak S (2007). Antifungal activity of thyme (*Thymus vulgaris* L.) essential oil and thymol against moulds from damp dwellings. Lett. Appl. Microbiol. 44(1):36-42.

Soković MD, Vukojević J, Marin PD, Brkić DD, Vajs V, Leo JL, Griensven DV (2009). Chemical composition of essential oils of *Thymus* and *Mentha* species and their antifungal activities. J. Mol. 14:238-249.

Soliman KM, Badeaa R (2002). Effect of oil extracted from some medicinal plants on different mycotoxigenic fungi. Food Chem. Toxicol. 40(11):1669-1675.

Tantaoui-Elaraki A, Lattaoui N, Errifi A, Benjilali B (1993). Composition and antimicrobial activity of the essential oils of *Thymus broussonettii, T. zygis* and *T. satureioides*. J. Essent Oil. Res. 5(1):45-53.

Teimouri M (2012). Antimicrobial activity and essential oil composition of *Thymus daenensis* Celak from Iran. Journal of Med. Plants Res. 6(4):631-635.

Viuda-Maartos M, Ruiz-Navajas Y, Fernandez-Lopez J, Perez-Alvarez JA (2007). Antifungal activities of thyme, clove and oregano essential oils. J. Food Safety 27(1):91-101.

Zambonelli A, D'Aulerio AZ, Bianchi A, Albasini A (1996). Effects of essential oils on phytopathogenic fungi *In vitro*. J. Phytopathol. 144(9-10):491-494.

Zambonelli A, D'Aulerio AZ, Severi A, Benvenuti S, Maggi L, Bianchi A (2004). Chemical composition and fungicidal activity of commercial essential oils of *Thymus vulgaris* L. J. Essent. Oil Res. 16(1):69-74.

23

In vitro antifungal effects of *Fumaria vaillantii* Loisel. essential oil on *Aspergillus flavus*

M. Moghtader

Department of Biodiversity, Institute of Science and High Technology and Environmental Sciences, Graduate University of Advanced Technology, Kerman, Iran. E-mail:moghtader18@yahoo.com.

In order to identify the chemical composition of essential oil of *Fumaria vaillantii*, the leaves with young branches of this plant which grows in a village in Kerman Province at full flowering stage in May 2012 were collected. The sample was cleaned and then dried in the shade, and essential oil hydrodistillation method was performed. The main oil content from the plant of *F. vaillantii* was 0.25% (v/w) and that essential oil was analyzed by capillary gas chromatography (GC) using flame ionization (FID) and capillary gas chromatography coupled mass spectrometry (GC/MS) for detection. Eighteen compounds were identified in the essential oil of *F. vaillantii* that included 99.62% of the total oil. The major components were Parfumidine (18.94%), Fumaricine (16.30%), Thymol (12.45%) and Fumaritine (10.78%). The study of the antifungal effects of the oil sample was carried against strain of *Aspergillus flavus* (PTCC=5004) by disc diffusion method via average inhibition zone. The results show that the essential oil from fumaria plant at 1 and 1/2 oil dilutions exhibited strong antifungal activity than gentamycin antibiotic on *A. flavus* and synthetic thymol exhibited good inhibition at 10% dilution. Large percentage antifungal activities of fumaria oil are related with thymol as a natural monoterpene phenol is the main compound.

Key words: *Fumaria vaillantii* Loisel. essential oil, *Aspergillus flavus*, aflatoxin, thymol.

INTRODUCTION

Aspergillus is a genus of moulds reproduced only asexually. Some *Aspergillus* species function as plant and/or animal pathogens (Bennett, 2010; Geiser, 2009). More than 60 *Aspergillus* species are medically relevant pathogens (Thom, 1926). The most common pathogenic species are *Aspergillus fumigatus* and *Aspergillus flavus*. *A. flavus* produces aflatoxin which is both a toxin and a carcinogen, and which can contaminate foods such as nuts. The most common that cause allergic disease are *A. fumigatus* and *Aspergillus clavatus*. Other species are important as agricultural pathogens. *Aspergillus* spp. cause disease on many grain crops, especially maize, and synthesize mycotoxins including aflatoxin. Thymol (2-isopropyl-5-methylphenol) is a natural monoterpene phenol extracted from *Thymus vulgaris* (common thyme) and various other kinds of plants as a white crystalline substance of a pleasant aromatic odor and strong antiseptic properties. Thymol is part of a naturally occurring class of compounds known as biocides, with strong antimicrobial attributes when used alone or with other biocides such as carvacrol (Ahmad et al., 2010). The antifungal nature of thymol is caused by thyme's ability to alter the hyphal morphology and cause hyphal aggregates, resulting in reduced hyphal diameters and lyses of hyphal wall (Numpaque et al., 2011). Additionally, thymol is lipophilic, enabling it to interact with the cell membrane of fungus cells, altering cell membrane permeability permitting the loss of macromolecules (Segvic et al., 2007). This study evaluated and identified the chemical compounds of *F. vaillantii*. Also, antifungal activity of *F. vaillantii* has been compared with synthetic thymol and standard gentamicin antibiotic on culture of A. flavus.

MATERIALS AND METHODS

Plant material collection and isolation of their essential oil

The leaves and young branches of F.vaillantii were collected at Kerman Province (Iran) at full flowering stage in May 2012. The samples were air-dried and powdered using a milling machine and kept in a cool dry place until ready for extraction of the essential oil. Afterwards, essential oil was taken from 150 g of the powdered sample in hydro distillation method with the help of Clevenger set for 3 h. The sample oils were dried with anhydrous sodium sulfate and kept in sterile sample tubes in a refrigerator.

Analysis of essential oil

Gas chromatography

GC analysis was performed using a HP-439 gas chromatograph equipped with a CP-Sil 5CB capillary column (25 m × 0.25 mm id, 33 µm film thickness). Oven temperature was from 60 to 220°C at 7°C min. Injector temperature 280°C detector (FID) temperature 270°C and carrier gas was helium (ml/min).

Gas chromatography/mass mass spectrometry

GC-MS analyses were carried out using a Hewlett Packard-5973 apparatus which was equipped with a MS reference library and a HP 5MS cross linked fused-silica capillary column (60 m × 0.25 mm i.d., 0.25 µm phase thickness). The oven temperature program was 60°C for 3 min, rising to 220°C at 6°C/min, 220°C at 20°C/min, 220°C (3 min); the injector temperature was 280°C; the carrier gas was helium at 1 mL7min; the injection mode was split with a split ratio of 1:43; the sample volume injected was 0.1 µL; the interface temperature 230°C. Identification and quantification of oil components, the components of the oil were identified by comparison of their linear retention indices (LRIs) on the CP-Sil 5CB column (determined in relation to a homologous series of n-alkanes, C8-C17) with those of pure standards or as reported in the literature (Adams, 2001). The percentages of each component were reported as raw percentages without standardization.

Antifungal assay

Antifungal activity of the essential oil was assayed using the agar disc diffusion method using Mueller Hinton Agar (Baron and Finegold, 1995) and the measure of inhibition zones at different oil dilutions against A. flavus (PTCC=5004) from Center for Fungi and Bacteria of Iranian Scientific and Industrial Researches Organization was done. A sample of 50 µL of a suspension of the tested microorganism was spread onto the surface of Mueller-Hinton agar plates. Filter paper discs (5 mm in diameter) were placed on the surface of inoculated plates, and then they were soaked with 50 µL of essential oil dilutions in DMSO (1, 1/2, 1/4, 1/8 and 1/16). Gentamicin (8 mg/ml) was used as positive control. After incubation at 24°C for 48 and 72 h, the diameters of the inhibition zones were measured in millimeters. Each test was carried out in triplicate.

RESULTS AND DISCUSSION

The F. vaillantii essential oil yield was 0.25% (v/w). 18 compounds were identified in the essential oil of this plant with 99.62%; the combinations of Parfumidine (18.94%); Fumaricine (16.30%), Thymol (12.45%) and Fumaritine (10.78%) with 58.47% constitute the highest percentage of essential oil (Table 1).

The results of studying the antifungal impacts of the F. vaillantii essential oil shows that the oil of this plant has an inhibitory effect in 1, 1/2, 1/4, 1/8 and 1/16 dilution with average diameter growth of 28, 22, 15, 12 and 8 mm respectively. The results of standard antibiotic gentamicin (8mg/ml) with a diameter of 19 mm had inhibitory effect. Synthetic thymol in 1% dilution had no inhibitory effect on A. flavus growth but at 10% dilution had a good inhibition (18 mm) of the fungi growth (Table 2). The results show that the essential oil from fumaria plant at 1 and 1/2 oil dilutions exhibited strong antifungal activity than gentamycin antibiotic on A. flavus and thymol exhibited good inhibition at 10% dilution. The large percentage antifungal activities of fumaria oil is related to thymol a natural monoterpene phenol which is the main compound.

The essential oil of F.vaillantii plant has been studied less in Iran and in the world. In a report, Eugenol (1) and thymol (2) exhibited excellent fungicidal activity against pathogenic yeasts, including isolates resistant to azoles. The rapid irreversible action of compound-1 and compound 2 on fungal cells suggested a membrane-located target for their action (Ahmad et al., 2010). Fumaria officinalis is approved for the indication of colicky pain affecting the gallbladder and biliary system, together with the gastrointestinal tract (Hentschel et al., 1995). In a report, the essential oil was classified into three groups. The first group, composed of citron, lavender and tea tree oils, stopped the apical growth in a loading dose of 63 µg ml[-1] air, but allowed the regrowth of the hyphae after removal of the vapor, indicating fungi static action. The second group, consisting of perilla and lemon-grass oils, stopped the apical growth in a loading dose of 6.3 µg ml[-1] air, and did not allow the regrowth after gaseous contact at 63 µg ml[-1] air, indicative of fungicidal action. The third group, consisting of cinnamon bark and thyme oils, retarded the growth in a dose of 6.3 µg ml[-1] air, stopped it in a dose of 63 µg ml[-1] air, and incompletely suppressed regrowth of the hyphae (Inouye et al., 2000). In a research, minimum inhibitory concentration (MIC) of thymol on A. flavus was highly effective at doses as low as 250 ppm (Mahmoud, 1994). In a paper, total quinolizidine alkaloid contents were 426 mg/100 g (F. capreolata) and 521 mg/100 g (F. bastardi). The isoquinoline alkaloids, stylopine, protopine, fumaritine, fumaricine, fumarophycine, fumariline and fumarofine were determined. In the first species, an ester of phtalic acid was identified, and in the second species a peak seems to be a benzophenanthridine was identified, probably dehydro derivative and three other peaks were identified as phtalidisoquinoline; one of them seems to be dihydrofumariline. The chemotaxonomic significance of

Table 1. Compounds identified in the essential oil of *Fumaria vaillantii* Loisel.

Compound Name	Restrictive Index (RI)	Percentage (%)
Cryptopine	1093	0.29
Corydamine	1135	0.82
Fumarophycine	1176	2.96
α-terpineol	1185	1.89
Hydrastine	1206	5.06
Unknown	1252	0.16
Unknown	1287	0.22
Thymol	1287	12.45
Carvacrol	1306	8.53
Parfumine	1342	1.68
Stylopine	1474	5.73
Sinactine	1488	2.32
Fumariline	1492	3.42
Protopine	1502	4.29
Fumaritine	1516	10.78
Parfumidine	1533	18.94
Fumaricine	1575	16.3
Adlumine	1589	0.63
Bicuculline	1607	0.3
Dihydrofumariline	1622	3.23
Total		99.62

The indexes of restrictive have been calculated by injecting the mixture of normal hydrocarbons (C8-C17) to HP-5MS column.

Table 2. The zone diameter (mm) of inhibition of antibiotic, fumaria oil and synthetic thymol on *aspergillus flavus* (mm)

Antibiotic	Dilutions of fumaria oil					Synthetic thymol (%)	
Gentamycin (8 mg/ml)	1	1/2	1/4	1/8	1/16	1	10
19	28	22	15	12	8	0	18

the results is discussed (Maiz-Benabdesselam et al., 2007). In a report, thymol and carvacrol inhibited the radial growth of *Colletotrichum acutatum* and *Botryodiplodia theobromae* completely and this effect remained for 240 h. Furthermore, thymol and carvacrol were metabolized by the plant pathogenic fungi in low proportion to several compounds, including thymoquinone, thymohydroquinone, thymyl and carvacryl acetate, thymyl and carvacryl methyl ether. The transformations affect the structural requirements of thymol and carvacrol related to their antimicrobial activity and mode of action. The relatively high antifungal activity of thymol and carvacrol against *C. acutatum* and *B. theobromae* and the low levels of microbial transformation indicate that both compounds could be an alternative to traditional chemical fungicides for the control of pre- and postharvest phytopathogenic fungi on fruits or vegetables (Numpaque et al., 2011). The

essential oil of the aerial parts of *Rosmarinus officinalis* collected from Konya, Turkey was analyzed by gas chromatography and gas chromatography mass spectrometry. The oil yield of dried plant (volume/dry weight) obtained by hydro distillation was 1.9%. 20 compounds representing 99.93% of the oils were identified. The main constituents of the oils were *p*-cymene (44.02%), linalool (20.5%), γ-terpinene (16.62%), thymol (1.81%), β-pinene (3.61%), α-pinene (2.83%) and eucalyptol (2.64%). The oil consisted of monoterpenic hydrocarbons, oxygenated monoterpenes and sesquiterpene hydrocarbons. Also, the inhibition effect of rosemary oil was investigated against *Alternaria alternata*, *Botrytis cinerea* and *Fusarium oxysporum*. The experiment was carried out *in vitro* using disc diffusion to investigate the antifungal action of the oil. Oil tested on potato dextrose agar plates exhibited an inhibitory effect. The extent of inhibition of fungal growth varied depending on the levels of essential oil used in experiment (Ozcan et al., 2008). In a research, the 50% ethanolic extract of Fumaria indica was investigated for its anti-inflammatory and antinociceptive potential in animal models. The extract (400 mg kg^{-1}) exhibited maximum anti-inflammatory effects of 42.2 and 42.1% after 3 h with carrageenean and histamine, respectively. The same dose of extract showed 38.9% reduction in granuloma mass in a chronic condition. A significant anti-nociceptive activity was evidenced in mice; 6.6 to 67.7% (p < 0.01) protection in mechanical, 33.9 to 125.1% (p < 0.05) protection in thermal induced pain and 22.2 to 73.9% (p < 0.05) protection in acetic acid-induced writhing (Rao et al., 2007). In a study, minimum inhibitory concentrations (MIC) of both thymol and essential oil (Thymus vulgaris L.) were below 20 µg ml^{-1}, except for Mucor spp. (µg ml^{-1}). Thymol exhibited approximately three-times stronger inhibition than essential oil of thyme. The vaporous phase of the thyme essential oil (82 µg l^{-1}) in glass chambers strongly suppressed the sporulation of moulds during 60 days of exposure. The thyme essential oil possesses a wide range spectrum of fungicidal activity. The vaporous phase of the oil exhibited long-lasting suppressive activity on moulds from damp dwellings. Essential oil of thyme and thymol could be used for disinfection of mouldy walls in the dwellings in low concentration (Segvic et al., 2007). In a report, essential oils of 12 medicinal plants were tested for inhibitory activity against A. flavus, A. parasiticus, A. ochraceus and Fusarium moniliforme. The oils of thyme and cinnamon (< or = 500 ppm), marigold (< or = 2000 ppm), spearmint, basil, and quyssum (3000 ppm) completely inhibited all the test fungi. Caraway was inhibitory at 2000 ppm against A. flavus, A. parasiticus and 3000 ppm against A. ochraceaus and F. moniliforme. A. flavus, A. ochraceus, A. parasiticus and F. moniliforme were completely inhibited by anise at< or = 500 ppm. However, chamomile and hazanbul at all concentrations were partially effective against the test toxigenic fungi. The results indicate that the test toxigenic fungi are sensitive to the 12 essential oils, and particularly sensitive to thyme and cinnamon. The results also show that the essential oils of thyme, cinnamon, anise and spearmint have more effect on fungal development and subsequent mycotoxin production in wheat grains. The extent of inhibition of fungal growth and mycotoxin production was dependent on the concentration of essential oils used (Soliman et al., 2002). In a research, the isoquinoline alkaloids protopine, cryptopine, sinactine, stylopine, bicuculline, adlumine, parfumine, fumariline, fumarophycine, fumaritine, dihydrofumariline, parfumidine and dihydrosanguinarine were determined in Fumaria agraria, F. bastardii, F. capreolata, F. sepium, F. densiflora, F. faurei, F. officinalis subsp. officinalis, F. parviflora, F. petteri subsp. calcarata and F. macrosepala (Suau et al., 2002). Essential oil of white wood (Melaleuca cajeputi) gave the highest inhibition followed by the essential oils of cinnamon (Cinnamomum cassia)

and lavender (Lavandula officinalis), respectively. Furthermore, the inhibitory effects of these three essential oils at different concentrations were examined. It was found that the essential oil of white wood at 1.5625% (v/v) and of cinnamon and lavender at 50% (v/v) were the optimum concentrations for fungal growth inhibition. The essential oil of white wood at 25% (v/v) completely inhibited the growth of A. flavus IMI 242684 on PDA for 28 days (Thanaboripat et al., 2007). In this study, we find out that regarding the antifungal effects of F.vaillantii essential oil under investigation as compared with synthetic thymol and gentamycin antibiotic, this essential oil can be used as a combination with antifungal effects of natural origin.

ACKNOWLEDGEMENT

This study was conducted under the financial support of the Department of Biodiversity, Institute of Science and High Technology and Environmental Sciences, Graduate University of Advanced Technology in Kerman. Hereby, the executive and colleagues of the research project expresses their thanks for the support extended to them.

REFERENCES

Adams RP (2001). Identification of essential oil components by gas chromatography mass spectroscopy. Illinois Allured Publication Corporation.

Ahmad A, Khan A, Yousuf S, Khan LA, Manzoor N (2010). Proton translocation ATPase mediated fungicidal activity of eugenol and thymol. Fitoterapia 81(8):1157-1162.

Baron EJ, Finegold SM (1995). Bailey and Scott's Diagnostic Microbiology, 8th ed. Mosby, St. Louis, MO, USA, pp. 171-193.

Bennett JW (2010). An Overview of the Genus Aspergillus. Aspergillus: Molecular Biology and Genomics. Caister Academic Press. ISBN 978-1-904455-53-0.

Hentschel C, Dressler S, Hahn EG (1995). Fumaria officinalis and clinical applications. Forts. Cher. Med. 113(19):291-292.

Inouye S, Tsuruoka T, Watanabe M, Takeo K, Akao M, Nishiyama Y, Yamaguchi H, (2000). Inhibitory effect of essential oils on apical growth of Aspergillus fumigatus by vapour contact. Pub. Med. Gov. 43(1-2):17-23.

Mahmoud ALE (1994). Antifungal action and antiaflatoxigenic properties of some essential oil constituents. Lett. Appl. Microbiol. 19(2):110-113.

Maiz-Benabdesselam F, Chibane M (2007). Determination of isoquinoline alkaloids contents in two Algerian species of Fumaria. Afr. J. Biotechnol. 6(21):2487-2492.

Numpaque MA, Oviedo LA, Gil JH, Garcia CM, Durango DL (2011). Thymol and carvacrol: biotransformation and antifungal activity against the plant pathogenic fungi Colletotrichum acutatum and Botryodiplodia theobromae. Trop. Plant Pathol. 36:3-13.

Ozcan MM, Chalchat JC (2008). Chemical composition and antifungal activity of rosemary (Rosmarinus officinalis L.) oil from Turkey. Int. J. Food Sci. Nutr. 59(7-8):691-698.

Rao ChV, Verma AR, Gupta PK, Vijayakumar M (2007). Anti-inflammatory and anti-nociceptive activities of Fumaria indica whole plant extract in experimental animals. Acta Pharm. 57(4):491-498.

Segvic KM, Kosalec I, Mastelic J, Pieckova E, Pepeljnak S (2007). Antifungal activity of thyme (Thymus vulgaris L.) essential oil and thymol against moulds from damp dwellings.Lett. Appl. Microbiol. 44(1):36-42.

Soliman KM, Badeaa R (2002). Effect of oil extracted from some medicinal plants on different mycotoxigenic fungi. Food Chem. Toxicol. 40(11):1669-1675.

Suau R, Cabezudo B (2002). Direct determination of alkaloid contents in *Fumaria* species by GC-MS. Phytochem. Anal. 13(6):363-367.

Thanaboripat D, Suvathi Y, Srilohasin P, Sripakdee S, Patthanawanitchai O, Charoensettasilp S (2007). Inhibitory effect of essential oils on the growth of *Aspergillus flatus*. KMITL Sci. Tech. J. 7:1.

Thom C, Church M (1926). The Aspergilli. Baltimore: The Williams & Wilkins Company.

24

Comparative study on solubilization of tri-calcium phosphate (TCP) by phosphate solubilizing fungi (PSF) isolated from Nsukka pepper plant rhizosphere and root free soil

Onyia, Chiadikobi Ejikeme[1] and Anyanwu, Chukwudi Uzoma[2]

[1]South-East Zonal Biotechnology Centre, University of Nigeria, Nsukka, Enugu State, Nigeria.
[2]Department of Microbiology, University of Nigeria, Nsukka, Enugu State, Nigeria.

Phosphate solubilizing fungi (PSF) possessing the ability to solubilize insoluble phosphate was isolated from the rhizospheric soil of Nsukka pepper plant. Twelve (12) fungal strains were isolated from both root-free soil (non-rhizosphere) and rhizosphere of Nsukka pepper plant. Of the 12 fungi isolated, 6 strains were from Nsukka pepper plant rhizosphere whereas the other six were from root-free soil. PF isolates were designated to be isolates from the rhizosphere of Nsukka pepper whereas NF isolates were designated to be isolates from non-rhizosphere soil. The former isolates were able to solubilize insoluble phosphate, both in PVK agar and NBRIP broth, whereas the latter isolates were unable to solubilize insoluble phosphate on Pikovskaya (PVK) agar medium but solubilize slightly on NBRIP broth. Fungal phosphate solubilization was analyzed by determining the quantity solubilized using NBRIP broth and calculation of the phosphate solubilization efficiency (E) on PVK agar was estimated using E = solubilization diameter/growth diameter x 100. The highest phosphate solubilization efficiency was demonstrated by PF7 isolate as 240.0% followed by PF2 isolate as 137.5% with 4.12 and 4.06 mg/ml, respectively, as the NBRIP broth analysis, whereas the non-rhizosphere isolates showed no solubilization effect on PVK agar medium but slight solubilization on NBRIP.

Key words: Phosphate solubilizing fungi, Pikovskaya, phosphate solubilization efficiency, rhizosphere.

INTRODUCTION

Phosphate fertilizers might increase phosphate availability initially, but will promote the formation of insoluble phosphate compounds making phosphorus unavailable to plants. Therefore, phosphate solubilizing microorganisms may be a choice for maintaining the steady supply of plant available phosphate (Xie, 2008). Root exudates are important for microbial attraction and fungal establishment on roots rhizosphere. Some of the metabolites known to excrete as the exudates from plants are the sugars and the amino acids which serve as required nutrients

for growth of microorganisms (Odunfa, 1979). From the root exudates of pepper plants, twelve amino acids and seven sugars were detected (Naqvi and Chauhan, 1980). The exudation of carbon-containing metabolites and amino acids from the roots into the soil
 matrix serves as substrates for the fungi, resulting in an increased microbial biomass which gives rise to organic acid production (Rogier et al., 2012). Root exudates serve as an important source of nutrients for microorganisms in the rhizosphere and induce an aggregation of microbes around the root region, this forms a mutual relationship between the plant and the microbe, whereby the plant provides nutrients to the microbes and the microbes solubilize phosphate for plant uptake (Shukla et al., 2011). The presence of total organic acid (TOA) posed the concept of phosphate solubilization by the process of acidification and chelation (Akintokun et al., 2007; Khan et al., 2009). Microorganisms are known to solubilize insoluble phosphate (for example, tri-calcium phosphate) through the production of organic acids (Jayandra et al., 1999) and these acids lower the pH of the medium and bring about the dissolution of bound forms of phosphate (Sharma et al., 2012). Phosphorus deficiency is a major constraint for crop production (Aadarsh, 2011) and many soils are deficient in readily available forms of phosphorus for plant uptake, therefore in order to circumvent phosphorus deficiency, phosphate solubilizing microorganism could play an important role in supplying phosphate to plants in a more environmentally-friendly and sustainable manner. Since the indiscriminate and excessive application of chemical fertilizers has led to high accumulation of insoluble phosphate, unhealthy nature and environmentally unfriendly nature of soil, we were desperate to find an alternative strategy by exploring the potentials in Nsukka pepper plants rhizosphere.

MATERIALS AND METHODS

Sample collection

Soil samples were collected from rhizosphere of pepper plants and non rhizosphere region, at University of Nigeria, Nsukka agriculture farm. The soil samples were taken within 1 and 10 cm radius by 5 to 10 cm depths for rhizosphere collection and 500 to 1000 cm by 5 to 10 cm depth away from the rhizosphere.

Isolation of indigenous rhizospheric and non rhizospheric fungi

From each soil sample, 10 g were transferred to 250 mL Erlenmeyer flask containing 90 ml of sterile distilled water. The flasks were shaken for about 20 minand allowed settling for few minutes. A 1 ml of the suspension were transferred to 9 ml of distilled water in test tube and serially diluted. The appropriate dilution was plated out using 0.1 ml aliquot on potato dextrose agar (PDA) medium. The soil samples were also sprinkled directly on some plates and incubated.

Inoculums size determination

A haemocytometer (Improved Neubauer counting chamber) and a

cover glass were thoroughly cleaned with moist cloth. The chamber was placed on a flat horizontal surface and the cover glass, slides into position using firm pressure. A 0.1ml cell suspension of a 5 day culture was transferred to the chamber using pasteur pipette held at 45°C. The filled chamber was placed on the microscope and the cells were counted using 40x objective lens. The area counted (A), depth of the counting chamber (D), average number of cell counted (N) and the dilution factor (DF) of cells per ml was calculated as Cell/mL =N 'x' DF 'x' 10^5/ A 'x' D.

Composition of PVK medium (Pikovskaya, 1948), supplemented with bromo-phenol-blue

Glucose 10.0 g, $Ca_3(PO_4)_2$ 5.0 g*, $(NH_4)_2SO_4$ 0.5 g, NaCl 0.2 g, $MgSO_4.7H_2O$ 0.1 g, KCl 0.2 g, yeast extract 0.5 g, $MnSO_4.H_2O$ 0.002 g, $FeSO_4.7H_2O$ 0.002 g, Agar 15.0 g, distilled H_2O 1000 ml bromo-phenol-blue (BPB) 0.025 g* and pH 7.0 were used.

Isolation of phosphate solubilizing fungi (PSF)

Each fungal isolate were aseptically transferred onto Pikovskaya (1948) medium (PVK) supplemented with bromo-phenol-blue (BPB) and tri-calcium phosphate (TCP) using point inoculation and incubated at 28°C for 7 days. The solubilizations of phosphate were observed as a zone of clearance with a diameter that was measured in millimeters. The phosphate solubilization ability of the fungi was analyzed by determining the phosphate solubilization efficiency (E) of each isolate [E = solubilization diameter / growth diameter x 100] (Nguyen et al., 1992) sited by Qurban (2012). After confirming the phosphate solubilizing ability on solid medium, the phosphate solubilization were also carried out using National Botanical Research Institute's Phosphate - Bromo Phenol Blue (NBRIP-BPB) broth (Pradhan and Sukla, 2005).

Composition of NBRIP-BPB liquid medium

NBRIP-BPB liquid medium consist of $(NH_4)_2SO_4$(0.1 g/l), $Ca_3(PO_4)_2$ (5.0 g/l), $MgSO_4.H_2O$ (0.25 g/l), $MgCl_2.6H_2O$ (5.0 g/l), KCl (0.2 g/l), BPB (0.025 g/l), glucose (10 g/l) (Nautiyal, 1999) inoculated with a 1% (v/v) inoculums pre-culture grown in the same medium. The phosphate solubilization activity of each of the isolates was determined by growing the isolates in NBRIP medium containing a pH indicator (bromophenol blue) for 12 days (taking reading at 4 days intervals) at 29°C. At the end of the incubation period, spectrophotometric readings were taken at OD_{600} and the final values were subtracted from the initial values (control) (Nautiyal, 1999).

Quantification of total organic acids produced by each isolate

The broth cultures of the PSF isolates were used for this assay. Titrable acidity was estimated by titrating one milliliter of the culture supernatant against 0.1 M NaOH in the presence of phenolphthalein as an indicator (Whitelaw, 2000). The titrable acidity was expressed as milliliter (ml) of 0.1 M NaOH consumed per 1.0 ml of culture filtrate.

Statistical analysis

All experiments were conducted in triplicates and mean of the values reported. Means were calculated using GENSTAT statistical package. Significant difference was set at P < 0.05 level.

RESULTS

Isolation and enumeration of phosphate solubilizing fungi from the rhizosphere and non rhizosphere soil

Out of the twelve (12) fungi strains isolated from both soil samples, only six (6) showed significant zone of phosphate solubilization on PVK-BPB agar medium. Clear halos were formed around the colonies after 5 to 7 days of incubation on solidified PVK medium supplemented with TCP, indicating phosphate-solubilizing ability of the fungal isolates. The phosphate solubilizing fungi strains are PF2, PF3, PF4, PF5, PF6 and PF7, with PF7 (Figure 3) showing the highest percentage of phosphate solubilization efficiency. The non-rhizospheric fungi (NF) isolates showed no significant zone of phosphate solubilization on PVK-BPB agar medium as shown in Table 1.

Effect of cell size on phosphate solubilization

Higher fungal populations were found in rhizospheric region of Nsukka pepper plant when compared with non-rhizosphere soil (Table 2) and may contribute to its solubilization effect. Quantitative assays conducted on efficiency of the phosphate solubilizing fungi were based on the lowering of pH, owing to production of organic acids into the surrounding medium. The highest amount of insoluble phosphate solubilized was found to be 4.12 mg/ml by isolate PF7 (cell-size 1.74×10^8), followed by 4.06 mg/ml solubilized by isolate PF2 (cell-size 1.34×10^8) and 4.04 mg/ml solubilized by isolate PF6 (cell-size 9.80×10^7), whereas the least amount of insoluble phosphate solubilized was 0.28 mg/ml by NF20 (cell-size 1.54×10^3).

Effect of solubilization with time

As shown in Figure 1, none of the isolates was able to solubilize insoluble phosphate on the day of inoculation (day zero) because no metabolites have been produced, but as the growth phase entered its exponential phase, they utilize the available nutrient thereby excreting metabolites (eg. organic acid) for insoluble phosphate solubilization. When compared with Figure 2, decrease in pH increased the solubilization, with the isolates highest solubilization of insoluble phosphate on the 12th day of incubation.

Reduction in pH, as a result of total organic acid produced by PSF

Increases in titrable acidity were observed to correlate with reduction in pH of the medium as shown in Figure 2. Results indicated that all the fungi isolates that produced organic acid also solubilizes insoluble phosphate and quantity solubilized is commensurate to organic acid produced. These pH reductions are due to secretion of organic acids excreted by PSF (Sharma, 2012).

DISCUSSION

The result obtained in Table 1 highlights the existence of phosphate solubilizing fungi in the rhizospheric soils of Nsukka pepper plants.

Baby et al. (2001) sited by Sharma et al. (2012) carried out an investigation on microbial dynamics in the rhizosphere of tea plants and reported that there were significant difference on the population level of PSB in different seedlings of tea. Higher PSB populations were found in rhizosphere other than non-rhizosphere soil (Qurban et al., 2012) and Lynch (1986) in his work also observed a high percentage of PSB population concentrates in the rhizosphere of plants. All these findings supported our work, which explained the high microbial level in the rhizospheric region of Nsukka pepper plants.

The production of halos around the colony of the organism is an indication of the presence of phosphate solubilizing organisms. Owing to confirmation on agar solubilization, NBRIP broth modification medium were used, where BPB produces a blue-colored dye that decolorizes to light yellow, due to a drop in pH of the medium. In the course of our work, NBRIP-BPB broth medium at pH 7, dropped to pH 2.42 by the action of PF7 isolate, which showed the maximum phosphate solubilization of 4.12 mg/ml while NF20 isolate dropped slightly to pH 6.25 showing the least phosphate solubilization of 0.28 mg/ml, although the NF-isolates did not give any detectable solubilization zone (halos) in plate PVK agar assay.

Laboratory study reviewed by Kucey et al. (1989) and sited by Ramachandran et al. (2003) have shown that the microbial solubilization of insoluble phosphate in liquid medium has been due to the excretion of diffusible organic acids as a result of which a decrease in pH were obtained. This is consistent with our work where, all the PF isolates, NF11 and NF14 were able to produce organic acids that solubilize insoluble phosphate. But contradictory results were produced when isolates NF17, NF18, NF19 and NF20 slightly solubilizes insoluble phosphate in NBRIP broth without organic acid production. The slight variation in some NF-isolates' phosphate solubilization without organic acid production, may be as a result of phosphatase activity that directly affects the decomposition of soil organic and inorganic insoluble phosphate (DaWei et al., 2011). The solubilization strength of the isolates might be attributed to microbial population size, pH and soil enzyme activity Ponmurugan and Gopi, 2006).

The results that were obtained in this study suggested the existence of phosphate solubilizing fungi in the rhizospheric region of Nsukka pepper plants, which might be organism to produce more organic acid, thereby decreasing the pH of the medium to an extent that causes solubilization of insoluble phosphate. Varsha et al. (2010) reported that the production of organic acids by some fungi brings about the drop in pH and dissociation of insoluble phosphate, termed phosphate solubilization Figure 2 is consistent with Pradhan and Sukla (2005) and

Table 1. Phosphate solubilization efficiency using PVK agar.

Isolate	Solubilization diameter [SD] (mm)	Growth diameter [GD] (mm)	Solubilization efficiency [SE]
Control	0.0	0	0.0
PF2	66.0	48.0	137.5
PF3	43.0	32.0	134.4
PF4	43.0	35.0	122.9
PF5	24.0	22.0	109.1
PF6	41.0	31.0	132.3
PF7	60.0	25.0	240.0
NF11	0.0	16.0	0.0
NF14	0.0	11.0	0.0
NFI7	0.0	13.0	0.0
NF18	0.0	06.0	0.0
NF19	0.0	15.0	0.0
NF20	0.0	14.0	0.0

Table 2. Comparative analysis of cell numbers to solubilization.

Isolate	Solubilization efficiency	Cell numbers (cell/ml/)	pH value	Quantity solubilized (mg/ml)
Control	0.0	0.0	7.00	0.0000
PF2	137.5	1.34×10^8	2.42	4.0587
PF3	134.4	6.41×10^7	2.57	3.6897
PF4	122.9	7.10×10^7	3.60	3.6933
PF5	109.1	3.50×10^7	4.17	3.1067
PF6	132.3	9.80×10^7	2.46	4.0433
PF7	240.0	1.74×10^8	2.42	4.1233
NF11	0.0	4.22×10^4	4.75	1.4433
NF14	0.0	2.63×10^4	4.24	2.8633
NF17	0.0	4.61×10^3	6.80	0.9690
NF18	0.0	6.20×10^3	6.02	0.8033
NF19	0.0	2.21×10^3	5.97	0.7633
NF20	0.0	1.54×10^3	6.25	0.2833

Varsha et al. (2010) where decrease in pH increases quantities of insoluble phosphate solubilized.

The result of the phosphate solubilization activity showed that PF-isolates from Nsukka pepper plant rhizosphere had higher solubilizing activity than the NF-isolates from the non-rhizosphere region. However, there is a positive correlation between phosphate solubilizing capacity and organic acid activity (Figure 2).

Based on the results of this study, Nsukka pepper plant exudates contributes greatly to the conglomeration of phosphate solubilizing fungi as shown in Table 2, and therefore can be suggested as plant booster for phosphate solubilization, making soluble phosphate available for plants uptake and improvement of agricultural sustainability.

ACKNOWLEDGEMENT

The authors are grateful to Director of South East Zonal Biotechnology Centre, University of Nigeria, Nsukka, Prof. C. E. A. Okezie for his support.

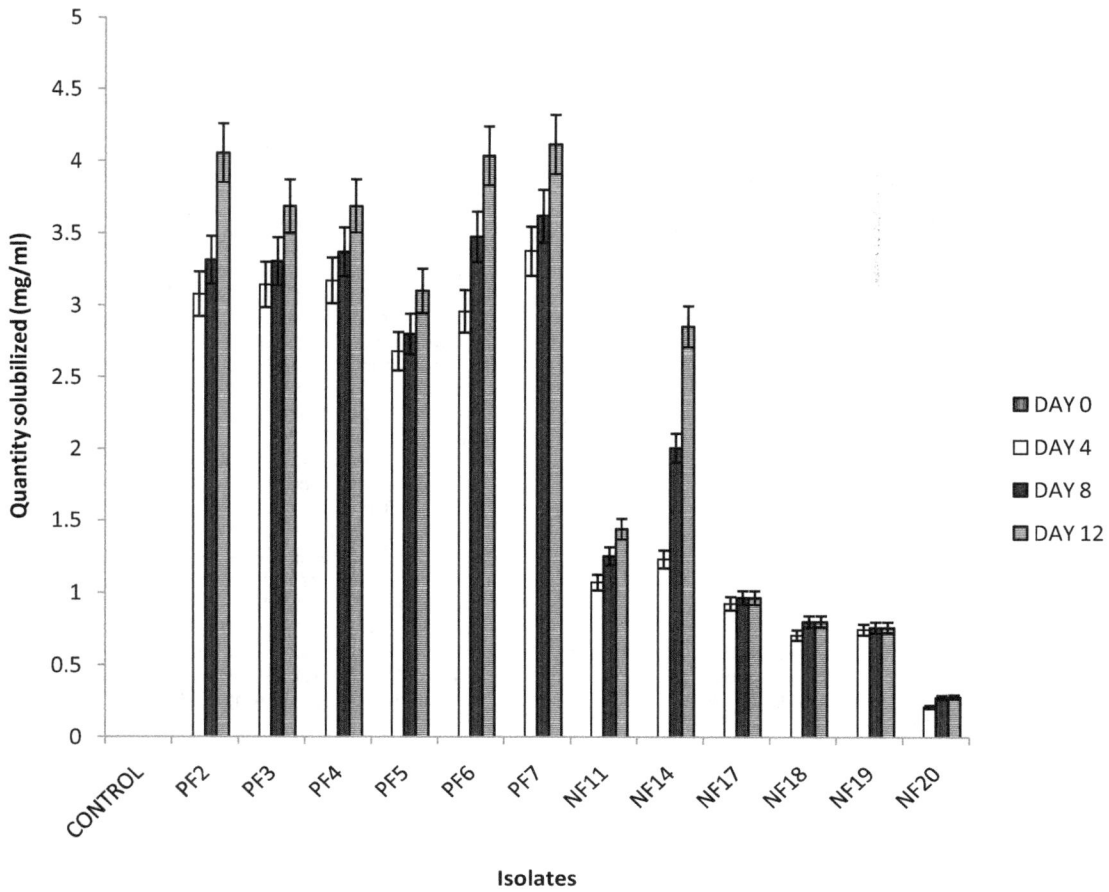

Figure 1. Quantity of phosphate solubilized (mg/ml) with time, using NBRIP-BPB broth.

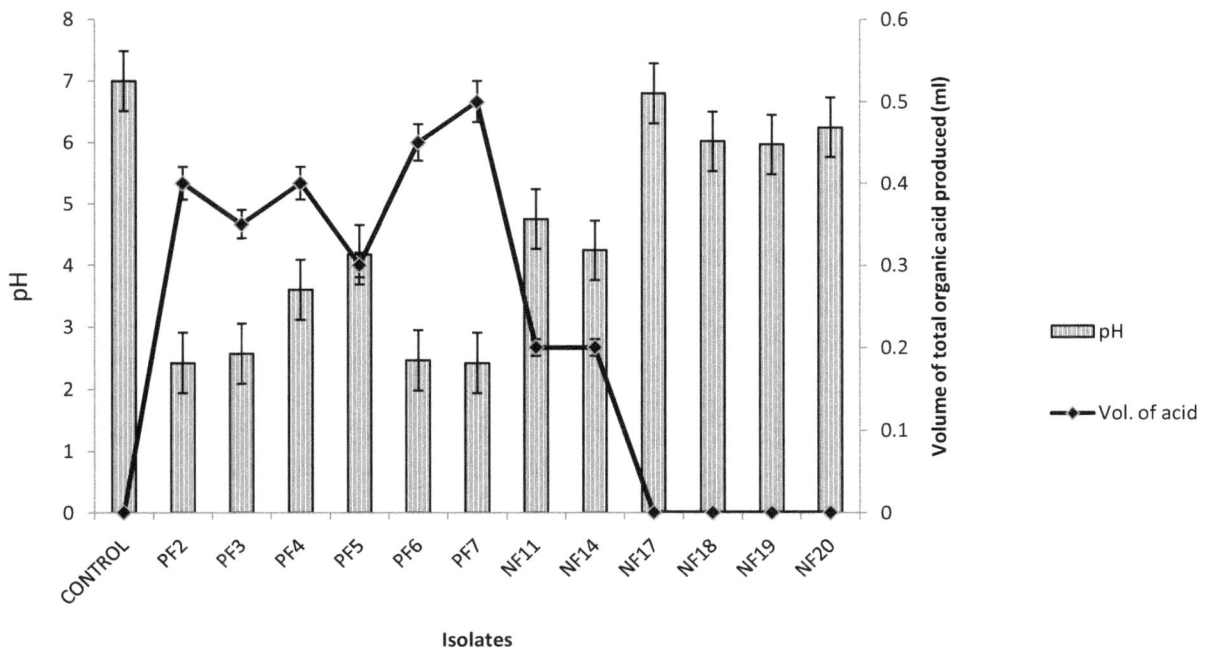

Figure 2. Comparative assay of total organic acids and pH of each isolate.

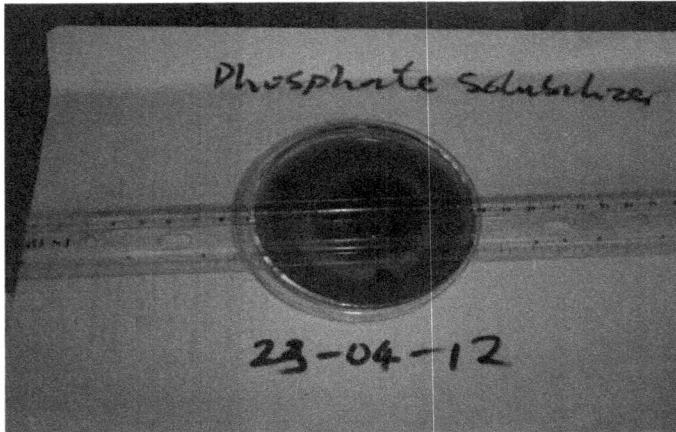

Figure 3. Formation of halo on PVK-BPB'S medium by the PF7 isolate.

REFERENCES

Aadarsh P, Deepa V, Murthy PB, Deecaramna M, Sridhar R, Dhandapani P (2011). Effect of halophilic phosphobacteria on *Avicennia officinalis* seedlings. Int. J. Soil Sci. 6 (2):134-141.

Akintokun AK, Akande PO, Popoola TOS, Babalola AO (2007). Solubilization of insoluble phosphate by organic acid-producing fungi isolated from Nigerian soil. Int .J. Soil Sci. 2:301-307.

DaWei MA, RenBin Z, Wei D, JianJun S, YaShu L, LiGuang S (2011). Alkaline phosphatase activity in ornithogenic soils in polar tundra. Adv. Polar Sci. 22 (2):92-100.

Jayandra KJ, Sanjay S, Nautiyal CS (1999). Occurrence of salt, pH, and temperature-tolerant, phosphate-solubilizing bacteria in alkaline soils. Curr. Microbiol. (39): 89-93.

Nautiyal CS (1999). An efficient microbiological growth medium for screening phosphate solubilizing microorganisms. FEMS Microbiol. Lett. 170:65-270.

Naqvi SMA, Chauhan SK (1980). Effect of root exudates on the spore germination of rhizosphere and rhizoplane mycoflora of chilli (Capsicum annuum L.). Plant Soil. 55: 397- 402.

Odunfa AVS (1979). Free amino acids in the seed and root exudates in relation to the nitrogen requirement of rhizosphere soil *Fusaria*. Plant and Soil, 52: 491-499.

Pikovskaya RI (1948). Mobilization of phosphorus in soil connection with vital capacity of source microbial species. Microbiologiya 17:362-370.

Ponmurugan P, Gopi C (2006). In vitro production of growth regulators and phosphatase activity by phosphate solubilizing bacteria. Afr .J. Biotechnol. 5 (4):348-350.

Pradhan N, Sukla LB (2005). Solubilization of inorganic phosphates by fungi isolated from agricultural soil. Afr .J. Biotechnol. 5 (10): 850-854.

Qurban AP, Radziah O, Zaharah AR, Sariah M, Mohd RI (2012). Isolation and characterization of phosphate-solubilizing bacteria from aerobic rice. Afr. J. Biotechnol. 11(11):2711-2719.

Rogier FD, Leendert CVL, Peter AHM (2012). Impact of root exudates and plant defense signaling on bacterial communities in the rhizosphere. Agron. Sustain. Dev. 32:227-243.

Ramachandran K, Srinivasan V, Hamza S, Anandaraj M (2003). Phosphate solubilizing bacteria isolated from the rhizosphere soil and its growth promotion on black pepper (*Piper nigrum L*) cuttings. Springer. pp. 325-331.

Sharma BC, Subba R, Saha A (2012). In-vitro solubilization of tricalcium phosphate and production of IAA by phosphate solubilizing bacteria isolated from tea rhizosphere of Darjeeling Himalaya. Plant Sci Feed. 2 (6):96-99.

Shukla KP, Sharma S, Singh NK, Singh V, Tiwari K, Singh S (2011). Nature and role of root exudates: efficacy in bioremediation. Afr .J Biotechnol. 10 (48):9717-9724.

Lynch JM, Whipps JM, (1986). The influence of the rhizosphere in crop productivity. Adv. Microbial. Ecol. 9:187-244.

Whitelaw MA (2000). Growth promotion of plants inoculated with phosphate solubilizing fungi. Edited by Donald L. Spark. Adv. Agron. 69:99-151.

Varsha N, Pratima D, Tithi S, Shalini R (2010). Isolation and characterization of fungal isolate for phosphate solubilization and plant growth promoting activity. J. Yeast Fungal Res. 1(1):009-014.

Xie J (2008). Screening for calcium phosphate solubilizing *Rhizobium leguminosarum*. M.Sc, dissertation, Department of Soil Science, University of Saskatchewan, Saskatoon, SK.

25

Production cellulase by different co-culture of *Aspergillus niger* and *Tricoderma viride* from waste paper

Ali A. Juwaied, Suhad Adnan and Ahmed Abdulamier Hussain Hussain Al-Amiery*

Department of Applied Science, Biochemical Division, University of Technology, Baghdad, Iraq.

In Iraq, there is attempts to transfer the various industrial carbon waste to veterinary proteins depend on microorganisms by using of chemical process. Five different co-culture combinations (1:1 ratio, 1 × 10^6 conidia)of *Aspergillus niger* and *Trichoderma viride*, mixing of *A. niger* and *T. viride*, in 24 and 48 h old monocultures of *Aspergillus* similar mixing of A. in 24 and 48 h old monoculture of *Trichoderma* and the monocultures of both were evaluated for their potential performance of cellulases production. The study indicates that the cellulases obtained from compatible mixed cultures simultaneous mixing of both fungi have more enzyme activity as compared to their pure cultures and other combinations. The fermentation experiments were performed in solid stat fermentation (SSF). Incubation time, carbon sources and initial pH of fermentation medium was optimized with simultaneous mixed culture. It was revealed that the newspaper at pH = 5 and 40°C was the best source of carbon for the enhanced production of cellulase in the compatible mixed culture experiments after 8 days of incubation with 5.70 U/ml. Based on the reported results, it may be concluded that industrial carbon waste can be a potential substrate for production of cellulase, incorporation of co-culturing *A. niger* and *T. viride*. The aim of this work is to produce of Cellulase from waste paper and reduce the pollution.

Key words: *Aspergillus niger*, *Tricoderma viride*, cellulase, culture.

INTRODUCTION

Plant cell walls are composed of various polysaccharides and lignin, forming a rigid and complex matrix recalcitrant to microbial degradation. This structure is strengthened by crosslinkages such as diferulic acid bridges between adjacent hemicelluloses chains (Oosterveld et al., 1997) or between lignin and hemicellulose (Lam et al., 1992), increasing its resistance to microbial invasion. The products of degradation constitute both nutrients for growth and regulators of the production of lignocellulolytic enzymes (de Vries et al., 1999; Peij et al., 1998). Among these microorganisms, filamentous fungi such as *Trichoderma* spp. and *Aspergillus* spp. are especially good secretors of lignocellulolytic enzymes (Archer and Peberdy, 1997). Grasses and cereals contain substantial amounts of cell wall-bound hydroxycinnamate esters linked to polysaccharides (Ishii, 1997). Particularly, ferulates play an important role in cross-linking cell wall polysaccharides (Ralph et al., 2004). The cotton strip assay (Latter and Howson, 1977; Howard and Howard, 1985) was put into practical use before it was properly evaluated and before the relationships, if any, between tensile strength change of cotton cloth and soil processes relevant to soil research programmes were examined. Perhaps the clearest expression of the reasons for using the assay is given by (Walton and Allsopp, 1977): (i) Cellulose is a major constituent of plant remains; (ii) the decomposition of dead plant remains is a major biological process, of great interest to many scientists studying soil processes; (iii) cellulose provides an important food source

*Corresponding author. E-mail: dr.ahmed1975@gmail.com.

for a wide variety of organisms and (iv) a method is needed to compare rates of breakdown of cellulose in different soils. Cotton is a natural substrate and degradation of any material must begin with bond breaking, which leads to changes in tensile strength. Walton and Allsopp considered that, as long as this technique is used for comparative assessments of biological activity in different soils, it will remain a powerful tool for field research. Cellulose is a homopolymer, consisting of glucose moieties joined in 0 - 1, 4 linkages, the number of glucose moieties in a molecule being the degree of polymerization (DP). Native cellulose from higher plants has a DP of about 14000 and this value appears to be remarkably constant. X-ray and infra-red data suggest that the basic structure of cellulose involves the anhydrocellobiose unit ($C_{12}H_{20}O_{10}$) rather than the anhydroglucose unit (C61-11005), so that the shortest cellulose-like molecule would be cellotetrose (DP = 4):

$(C_6H_{11}O_5)$, $(C_{12}H_{20}O_{10})$, $(C_6H_{11}O_5)$

Here, (C61-11105) represents a glucose moiety with a free secondary hydroxyl group on C-4 and (C61-11106) is a glucose molecule with a reducing group. Cellobiose, cellotriose, and cellotetrose are water-soluble, cellopentose is very sparingly soluble and cellulose molecules with DP greater than 6 are insoluble (Ljungdahl and Eriksson, 1985). Because cellulose is a polymer, we have to think rather carefully about how we define cellulose decomposition. In nature, cellulose is degraded by a range of aerobic and anaerobic fungi and bacteria, many of which occur in extreme conditions of temperature and pH. A list of fungi which have been examined in cellulose decomposition studies is given in (Ljungdahl and Eriksson, 1985). Many fungi can utilize oligosaccharides and polysaccharides as carbon sources, but, in general, these molecules are too large to be transported directly into the cell and must first be hydrolyzed to their subunits. Cellulose represents an important potential carbon and energy source for fungi, many of which produce a series of enzymes, collectively called cellulase (Sagar, 1988), which facilitate the degradation of cellulose to glucose units. Cellulose-degrading enzyme systems have been studied in detail in 2 fungi, *Sporotrichum pulverulentum,* and the conidial state of the white-rot fungus *(Phanerochaeate chrysosporium)* (Eriksson, 1981) and the mould *Trichoderma reesei* (Ryu and Mandels, 1980). They have similar hydrolytic enzyme systems with at least 3 components: (i) Endo-β-1,4-glucanases, which split randomly β-1,4-glucosidic linkages within the cellulose polymer. (ii) Exo-β-1,4-glucanase, which split off either cellobiose or glucose from the non-reducing end of the cellulose polymer. (iii) β-glucosidase, which hydrolyze cellobiose and water-soluble cellodextrins to glucose. Cellulase is a synergistic enzyme that is used to break up cellulose into glucose or other oligosaccharide compounds (Chellapandi and Jani, 2008; Acharya et al.,

2008). Cellulases have a wide range of applications. Potential applications are in food, animal feed, textile, fuel, chemical industries, paper and pulp industry, waste management, medical/ pharmaceutical industry, protoplast production, genetic engineering and pollution treatment (Tarek and Nagwa, 2007; Beguin and Anberl, 1993; Mandels, 1985).

MATERIALS AND METHODS

General

All chemical used were of reagent grade (supplied by either Merck or Fluka) and used as supplied.

Substrates

Waste paper was obtained from Alnajaf factory for producing paper in Alnajaf city in Iraq.

Microorganisms

A. niger procured from biotechnology division were maintained on Potato dextrose ager (PDA) slants ager and *T. viride* from biotechnology division was maintained on PDA slants. All cultures were subcultured every 4 weeks, incubated at 30°C for 7 days and subsequently stored at 4°C for inoculums preparations.

Conidial count

The conidial count was made on a Haemacytometer slide bridge (Sharma, 1989).

Culture media

A suspension containing *A. niger* and *T. viride* were used to initiate growth in 250 ml Erlenmeyer conical flask supplemented with $K_2H_2PO_4$ (0.15 g/L), KH_2PO_4 (0.20 g/L), $NaHPO_4$ (1.50 g/L), $NaHPO_4$ (2.00 g/L), $NaNO_3$ (3.80 g/L), H_3BO_3 (0.057×10^{-3} g/L), $MnSO_4.H_2O$ (5.5×10^{-6} g/L), $CuSO_4.7H_2O$ (2.5×10^{-6} g/L), $ZnSO_4.7H_2O$ (0.5×10^{-3} g/L), $Fe(SO_4)_3.6H_2O$ (4.5×10^{-6} g/L), $MgSO_4.7H_2O$ (5.5×10^{-6} g/L), $NH4NO3$ (0.60 g/L), $(NH_4)_6Mo_7O_{24}$ (0.025×10^{-3} g/L).

Fermentation procedure

Using for cellulase production in 250 ml Erlenmeyer conical flask (and addition of 2 ml of media culture with 2 gm of carbon source (Paper). After the inoculation, the flasks were sterilized by using of autoclave at 121°C for 15 min. After cooling at room temperature was then added 10^6 per ml of fungi and incubated for 6 days. The supernatant was estimated for Whatman filter papers (No. 1 saccharifying activity of cellulases.

Determination of the optimum co-cultured media (*A. niger* and *T. viride*)

We use five co-cultured medium:

(A + T, 24 A + T, 24 T + A, 48 A + T and 48 T + A) in addition of monoculture for each fungi by using of Sunders culture medium at

30°C for 6 days.

Determination of optimum carbon source

To determine the optimum carbon source, 2 gm of (wood fibers, paper waste and cotton waste) act as culture medium at 30°C for 6 days.

Effect of pH

The working pH of co-cultured media (*A. niger* and *T. viride*) was from 3 to 6 using phosphate buffer. The optimum pH was evaluated.

Effect of temperature

The working temperature of the co-cultured medium was varied from 20 to 50°C (+5).

Determination of enzyme activity

The determination of enzyme activity was done by using of Mandels method. Incubation of 0.9 mL of substrate solution with enzyme extract at 45°C for 1 h, then added 1 mL of DNS solution. The mixture was then heated at water bath for 5 min. then let it to cool and then added 10 mL of distilled water. The equivalent solution was prepared by added 1 mL of DNS to 0.9 mL of substrate then added 0.1 mL of enzyme solution. The determination of reduction saccharides was done by using of Mandels method and then calculates the enzyme activity (Mandels et al., 1976).

RESULTS AND DISCUSSION

Figure 1 represent the highest enzyme activity by using of co-culture media (*A. niger* and *T. virid*) and it raised 2.37 unit per mL and it was very high as compared with another co-culture medias and that because of Antagonism between *A. niger* and *T. virid* (Haq et al., 2005).

The three carbon sources wood fibers, paper waste and cotton waste were optimized. Among them, cotton waste was proved to be the best for cellulase production and it was 3.40 unit per mL and it was batter from the other wastes (that use in this paper, Figure 2) due to high percentage of cellulose production and the easy of braking bonds of waste paper, but wood and cotton waste have other substances like lignin and that is difficult to analyzed by enzyme.

The effect of the H+ concentration on the activity and stability of the cellulase is shown in Figure 3. Hydrolysis of was confined to acid media. The enzyme activity being maximal (4.54) at pH 5. In contrast with the sharp pH optimum seen in the activity profile, the cellulase was stable in the absence of substrate, over a wide range of pH values. Initial pH has a direct effect on the uptake of mineral nutrients, which are present in the fermentation medium. So, the effect of different pH (3.0 - 6.0) of fermentation medium on the enzymes production was

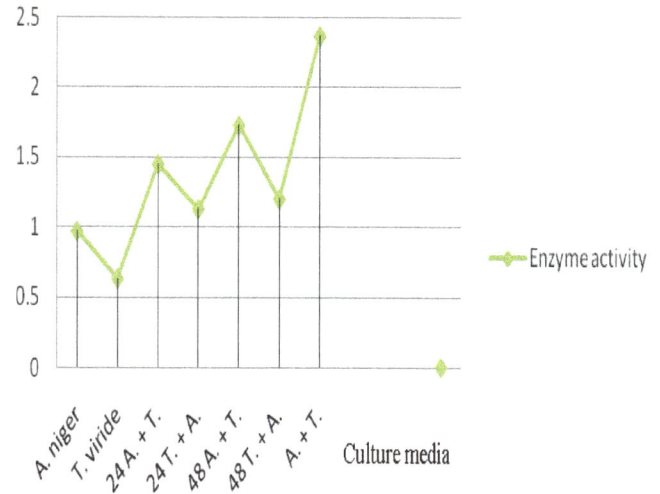

Figure 1. Enzyme activity vs culture medias.

Figure 2. Enzyme activity vs. carbon source.

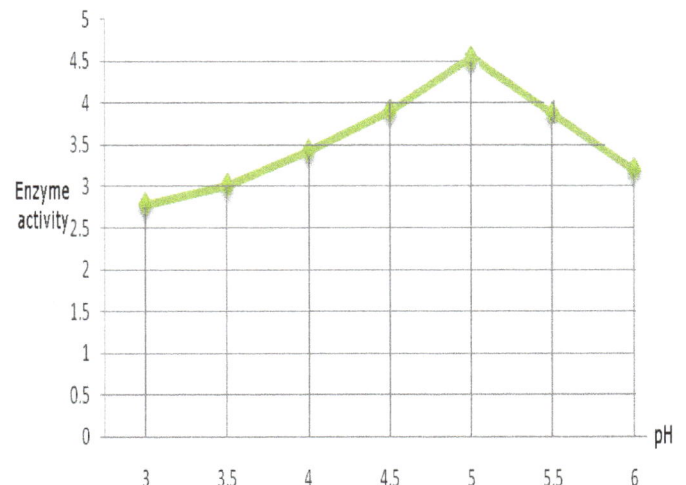

Figure 3. Enzyme activity vs pH.

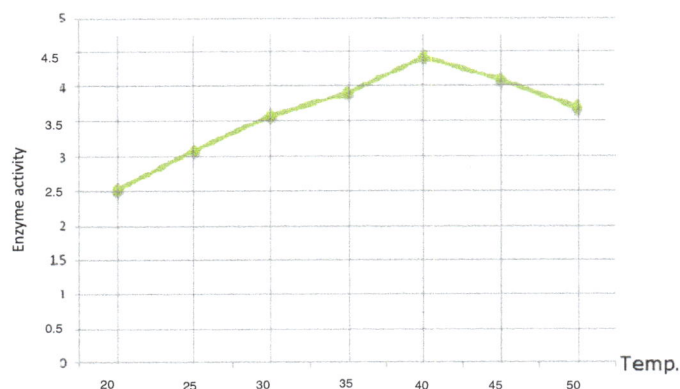

Figure 4. The enzyme activity vs temperature.

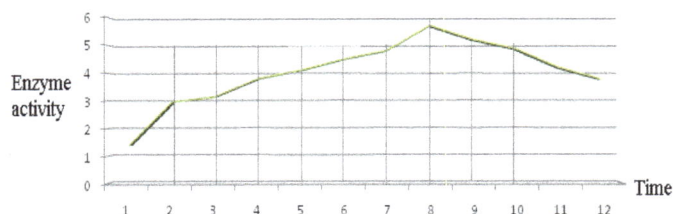

Figure 5. Enzyme activity vs time in days.

also investigated (Figure 3) with simultaneous co-culturing of *A. niger* and *T. viride*. High acidic and high basic pH, both showed negative effects, but a medium with low acidic pH 5 was ideal for enzyme fungal cultures require slightly acidic pH for their growth and enzyme biosynthesis (Haltrich et al., 1996). In the present study, a temperature of 20 - 50°C (+5°C) was used in the standard assay; however, the temperature of maximum activity under these conditions proved to be 40°C (Figure 4). Incubation temperature plays an important role in the metabolic activities of microorganism. Figure 5 represent the maximum enzyme activity at the 8th day and it was 5.70 units per mL.

REFERENCES

Acharya PB, Acharya DK, Modi AH (2008). Optimization for cellulase production by As*p*ergillus *niger* using saw dust as substrate. Afr. J. Biotechnol., 7(22) 4147-4152.

Archer DB, Peberdy JF (1997). The molecular biology of secreted enzyme production by fungi. Crit. Rev. Biotechnol., 17:273–306.

Beguin P, Anbert JP (1993). The biological degradation of cellulose. FEMS Microbiol. Rev., 13: 25-58.

Chellapandi P, Jani HM (2008). Production of endoglucanase by the native strains of Strptomyces isolates in submerged fermentation. Bra. J. Microbiol., 39: 122-127.

de Vries RP, Visser J, de Graaff LH (1999). CreA modulates the XlnF-induced expression on xylose of *Aspergillus niger* genes involved in xylan degradation. Res.

Eriksson ICE (1981). Cellulases of fungi In: Trends in the biology of fermentations for fuels and chemicals, edited by A. Hollaender, 19-32. New York: Plenum Press.

Haltrich D, Nidetzky B, Kulbe KD, Steiner W, Zupancic S (1996). Biores. Technol., 58: 137-161.

Haq UI, Javed MM, Kahn ST, Siddiq Z (2005). Cotton Saccharification activity of cellulose produced by co-cultured of *Aspergillus niger* and Trichoderma viride. Res. J. Agric. Biological Sci., 1:241-245.

Howard PJA, Howard DM, (1985). Multivariate analysis of soil physiological data (Merlewood research and development paper no. 105). Grange-over-Sands: Institute of Terrestrial Ecology. Ishii, T. Plant Sci., 1997, 127, 111–127.

Lam TBT, Iiyama K, Stone BA (1992). Cinnamic acid bridges between cell wall polymers in wheat and phalaris internodes. Phytochemistry. 31:1179–1183.

Latter PM, Howson G (1977). The use of cotton strips to indicate cellulose decomposition in the field. Pedobiologia, 17, 145-155.

Ljungdahl LG, Eriksson KE (1985). Ecology of microbial cellulose degradation. Adv. microb. Ecol., 8; 237-299.

Mandels M (1985). Applications of cellulases. Biochem. Soc. Trans. 13:414-415.

Mandels M, Andreotii RC (1976). Measurement of saccharifying cellulase. Biotechnol. Bioeng. Symp., 6:21-23.

Oosterveld A, Grabber JH, Beldman G, Ralph J, Voragen AGJ (1997) Formation of ferulic acid dehydrodimers through oxidative cross-linking of sugar beet pectin. Carbohydr. Res., 300:179-181.

Ralph J, Bunzel M, Marita JM, Hatfield RD, Lu F, Kim H, Schatz PF Grabber JH, Steinhart H (2004). Phytochem. Rev., 3, 79-96.

Ryu DDY, Mandels M (1980). Cellulase: biosynthesis and applications. Enzyme Microb. Technol., 2; 91-102.

Sagar BF (1988). Microbial cellulases and their action on cotton fibres. In: Cotton strip assay an index of decomposition in soils, edited by A.F. Harrison, P.M. Latter & D.W.H. Walton, 17-20. (ITE symposium no. 24.) Grange-over-Sands: Institute of Terrestrial Ecology.

Sharma PD (1989). Methods in microbiology and plant pathology. Rastogi and company Meerut, India. pp. 33-35.

Tarek AAM, Nagwa AT (2007). Optimization of cellulase and glucosidase induction by sugarbeet pathogen *Sclerotium rolfsii*. Afr. J. Biotechnol., 6(8): 1048-1054.

van Peij NN, Gielkens MM, de Vries BP, Visser J, de Graaff LH (1998). The transcriptional activator XlnR regulates both xylanolytic and endoglucanase gene expression in *Aspergillus niger*. Appl. Environ. Microbiol., 64:3615–3619.

van Peij NN, Visser J, de Graaff LH (1998). Isolation and analysis of xlnR, encoding a transcriptional activator co-ordinating xylanolytic expression in *Aspergillus niger*. Mol. Microbiol., 27:131–142.

Walton DWH, Allsopp D (1977). A new test cloth for soil burial trials and other studies on cellulose decomposition. Int. Biodeterior. Bull., 13, 112-115.

26

Screening of indigenous yeast isolates obtained from traditional fermented foods of Western Himalayas for probiotic attributes

Aditi Sourabh[1], Sarbjit Singh Kanwar[1]*, and Om Prakash Sharma[2]

[1]Department of Microbiology, CSK Himachal Pradesh Agricultural University, Palampur, Himachal Pradesh-176062, India.
[2]Biochemistry Laboratory, Indian Veterinary Research Institute, Regional Station Palampur, Himachal Pradesh-176061, India.

Twenty three indigenous isolates of yeast (*Saccharomyces cerevisiae*) obtained from various traditional fermented foods and traditional inocula of Western Himalayas (Himachal Pradesh) were subjected to *in vitro* probiotic tests. All the isolates were found to be intrinsically tolerant to upper gastrointestinal transit and this property was isolate dependent. Reduction in viability (in terms of log CFU/ml cells) was more in simulated gastric juice of pH 2 as compared to pH 3. These isolates were also investigated for surface hydrophobicity and autoaggregation abilities. Nine yeast isolates produced exopolysaccharide and four exhibited antioxidative activity using 1, 1-diphenyl-2-picryl-hydrazyl (DPPH) free radical scavenging assay (in methanol and buffered methanol reaction systems). Interestingly, one indigenous yeast isolate (Sc15) was found positive for siderophore production, whereas none of the isolates was positive for bile salt deconjugation activity (towards glycine and taurine conjugated bile salts) and galactosidase enzyme production.

Key words: Probiotics, yeast isolates, simulated gastric and intestinal juices, Western Himalayas, antioxidant, exopolysaccharide.

INTRODUCTION

Fermented food products are essential components of diet in a number of developing countries and are consumed either as beverage, main dish or condiment, which contribute to one third the diet of people worldwide (Campbell-Platt, 1994). A variety of indigenous fermented food products and beverages of Western Himalayas (Himachal Pradesh) have been documented (Kanwar et al., 2007). Traditionally, fermented foods are processed through naturally occurring microorganisms; however, modern conventional methods of production generally exploit the use of defined starter culture to ensure consistency and the quality of the final product (Ross et al., 2002). Cultures and species involved in fermented foods do not pose any health risk, and thus are designated as 'GRAS' (generally recognized as safe) organisms (Adams, 1999; Nout, 2001; Hansen, 2002). Therefore, some of the species of these microorganisms because of their long history of safe use in food products can be implemented as protective cultures or probiotics. Fermented foods are rich sources of these probiotics or functional microorganisms which can antagonize the growth of some of the spoilage and pathogenic microorganisms present in these foods (El-Gazzar and Marth, 1992; Vignola et al., 1993; Chiang et al., 2000). Yeasts are the most common and important microorganisms associated with the fermented foods (Yarrow, 1998).

Mostly, selected strains of bacteria belonging to *Lactobacillus* and *Bifidobacterium* genera are used as probiotics (Prasad et al., 1998), however, species belonging to the genera *Lactococcus, Enterococcus,*

*Correspondingauthor.E-mail:sskanwar1956@gmail.com.

Saccharomyces (Salminen and von Wright, 1998; Dunne et al., 1999; Sanders and in't Veld, 1999), and *Propionibacterium* (Grant and Salminen, 1998) are also considered as probiotic microorganisms. Some yeast strains such as *Saccharomyces cerevisiae* and *Saccharomyces boulardii* have also been used as probiotics in humans for many years because they exert some influence on the intestinal flora (Czerucka et al., 2007; Kumura et al., 2004). Food is the common delivery system for probiotic microorganisms. Probiotic microorganisms that are delivered through food systems have to first survive during the transit through the upper gastrointestinal tract, and then persist in the gut to provide beneficial effects for the host (Chou and Weimer, 1999). Given that fermented food products can contain probiotics, prebiotics or both, it is not surprising that their consumption has long been associated with good health.

In our laboratory, yeasts obtained from traditional fermented foods of Western Himalayas have already been characterized by using traditional and molecular tools (Pathania et al., 2010). However, these were not explored with respect to their probiotic diversity. The present study was therefore, conducted to screen these yeast isolates obtained from the traditional fermented foods of Western Himalayas, for various probiotic attributes.

MATERIALS AND METHODS

Twenty three indigenous isolates of yeast (Sc01 – Sc23) obtained from various traditional fermented foods (*chilra, babru, and bhaturu*), alcoholic beverages (*aara, chhang, chuli, faasur, lugari*) and traditional inocula (*phab, dhaeli, khameer*) of Western Himalayas, available with the Department of Microbiology, College of Basic Sciences, CSKHPKV, Palampur (H.P.) India, were used in the present investigation. These isolates were identified by employing traditional and molecular tools (Pathania et al., 2010).

Preparation of simulated gastric and small intestinal juices

Simulated gastric juice was prepared by dissolving pepsin (Merck Specialities Pvt. Ltd, Mumbai, India) in sterile saline (0.85% w/v) to a final concentration of 3 g/L. The pH of simulated gastric juice was adjusted to 2.0. Similarly, simulated intestinal juice was prepared by dissolving pancreatin from porcine pancrease, USP specifications (Sigma-Aldrich Inc. USA) to a final concentration of 1 g/L and its pH was adjusted to 8. These juices were then filter sterilized through millipore filter assembly using 0.22 µm Durapore membrane filter.

Preparation of washed cell suspension

The isolates were incubated in potato dextrose broth (PDB) at 37°C for 18 h and centrifuged at 2500 x g at 4°C for 10 min. The cell pellet obtained was washed three times in PBS buffer solution (pH 7.0). The collected cells were resuspended in sterile saline and viable count was determined by serial dilution method on PDA plates prior to the assay of transit tolerance.

Upper gastrointestinal transit tolerance assay

The tolerance of washed cell suspensions of yeast isolates to simulated gastric and small intestinal transit was determined by following the method of Charteris et al. (1998). For screening gastric transit tolerance, 0.1 ml aliquot was removed after every 30 min of intervals (upto 4 h) for determining the viable count. The small intestinal transit tolerance was evaluated by determining the viable count after 1, 4, and 8 h intervals.

Effect of ox bile on the growth rate

The effect of ox bile (HiMedia Laboratories, Pvt. Ltd., Mumbai, India) on the growth of yeast isolates was measured on the basis of time required to increase the absorbance at 620 nm by 0.3 units in PDB-thio broth with and without 0.3% ox bile (Walker and Gilliland, 1993). The difference in time (h) for attaining desirable absorbance between both culture media was considered as the lag time (LT) (Usman and Hosono, 1999).

Bile salt deconjugation activity

The ability of the yeast isolates to deconjugate bile salts was determined according to the method of Taranto et al. (1995) and Vinderola and Reinheimer (2003). Separate bile salt plates were prepared by adding 0.5% (w/v) of sodium salts (Calbiochem) of taurocholic acid (TC), taurodeoxycholic acid (TDC), glycocholic acid (GC) and glycodeoxycholic acid (GDC) to PDA, autoclaved (121°C, 15 min) and immediately used. The isolates were spot inoculated on the media and the plates were anaerobically (GasPak System-Hi Media) incubated at 37°C for 72 h. The presence of precipitated bile acid around colonies (opaque halo) was considered a positive result.

Microbial adhesion to hydrocarbon (MATH)

The test of adhesion to hydrocarbon n-hexadecane was adopted to screen yeast isolates for their cell surface hydrophobicity property. Microbial adhesion to hydrocarbon (MATH) in terms of the cell surface hydrophobicity (% H), was determined according to the method of Rosenberg et al. (1980) with slight modification as described by Vinderola and Reinheimer (2003).

Autoaggregation ability

Autoaggregation assay was performed as described by Collado et al. (2008) with minor modifications. Yeast isolates were grown at 37°C for 24 h in PDB. The cells were harvested by centrifugation and suspended in phosphate buffered saline (PBS) to 0.5 optical density (O.D.) units at 600 nm. 5 ml of this yeast suspension was incubated at 37°C for 20 h and then 3 ml of the upper suspension was carefully transferred to another tube and the O.D. was measured at 600 nm. Percent autoaggregation ability was calculated as:

1- (O.D. upper suspension/O.D. total bacterial suspension) × 100.

Production of exopolysaccharide

Exopolysaccharide (EPS) production was evaluated as reported by Mora et al. (2002). Overnight cultures were streaked on the surface of plates containing ruthenium red milk (10% w/v, skim milk powder, 1% w/v, sucrose and 0.08 g/L ruthenium red, 1.5% w/v agar). After incubation at 37C for 24 h, non-ropy strains gave red colonies due to the staining of the microbial cell wall, while ropy isolates appeared as white colonies.

Total antioxidative activity (TAA)

Preparation of whole cell extracts and intracellular cell free extracts

The yeast isolates were grown in PDB at 37°C for 24 h and harvested by centrifugation at 10000 × g at 5°C for 15 min. For the preparation of intact cells, cells were washed three times with phosphate buffered saline (PBS)/water and then were resuspended in the same. Total cell numbers were adjusted to 10^9 CFU/ml for subsequent preparation of whole cell and intracellular cell-free extracts. Ultrasonic disruption (B. Braun Biotech International, Germany) was performed for five times at 1-min intervals in an ice bath. The sonicated cell lysate was divided into two parts. In one part, cell debris was removed by centrifugation at 7800 × g for 10 min, and the resulting supernatant was the intracellular cell-free extract. In the second part, the cell debris was not removed and the resulting lysate was used as whole cell extract.

1, 1-diphenyl-2-picryl-hydrazyl (DPPH) free radical scavenging assay

1, 1-diphenyl-2-picryl-hydrazyl (DPPH) free radical scavenging assay (Sharma and Bhat, 2009) in methanolic and buffered methanol systems employing 200 μM of 1, 1-diphenyl-2-picryl-hydrazyl (DPPH) (Sigma-Aldrich Inc. USA) solution was used for screening yeast isolates for free radical scavenging (antioxidant) activity. Butylated hydroxytoluene (BHT) a standard antioxidant was used as positive control in all antioxidant assays. Protein concentration in the yeast extracts was estimated by using standard Lowry method (Lowry et al., 1951) with bovine serum albumin (BSA) as a standard. Dry weight of the yeast bacterial extracts used for antioxidant assays was also determined.

Galactosidase enzyme production

For qualitative assay of β-galactosidase, 100 mM solution of isopropyl β-D-1-thiogalactopyranoside (IPTG, Sigma) was prepared by dissolving 2.383 mg of IPTG in 100 ml of sterile double distilled water. Consequently 50 mg/ml solution of X-gal (Sigma) was also prepared in N, N' dimethyl formamide. Both the solutions were filter sterilized through 0.2 μm (Millipore) filters. For the detection of cultures producing this enzyme, 100 and 20 μl of IPTG and X-gal solutions, respectively, were spread plated on the surface of PDA plates aseptically. After that the test cultures were spot inoculated on these plates and incubated at 28°C for 24 to 48 h. Observations were then recorded as blue/white colored colonies.

Production of siderophore

Screening for siderophore production was done by the method of Schwyn and Neilands (1987). The glassware used for this experiment was acid rinsed and then three times rinsed with double distilled water. The plates prepared by adopting the aforementioned method, were spot inoculated with the test yeast cultures and incubated at 28°C for 48 h. Siderophore producing strain of Pseudomonas aeruginosa-MTCC 2581, was used as positive control for this trait. Formation of yellow-orange zone around the colony on blue colored medium plates was taken as positive result.

Statistical analysis

Statistical analysis of the obtained data was carried out by using WindoStat software version 8.0. The data were subjected to one way Analysis of Variance (ANOVA) at the significance level of 5%. The comparison of mean values was done with Duncan's Multiple Range test.

RESULTS AND DISCUSSION

Tolerance to simulated gastric and small intestinal juices

One procedure generally described concerning selection of probiotic microorganisms is the in vitro screening based on the capacity of microorganisms to survive in the simulated conditions of the digestive tract as it is an indispensable trait for a probiotic to act. The tolerance studies on 23 yeast isolates were conducted in simulated gastric juice of pH 2 (Table 1) and 3. At pH 2 of simulated gastric juice, the reduction in viability was more (2.71 to 4.12 log CFU/ml) as compared to reduction in pH 3 (0.88 to 3.06 log CFU/ml) after 240 min exposure. Lowest reduction in viability was noticed for Sc19 isolate indicating its high tolerance to gastric juice. All the yeast isolates exhibited more survival in the gastric juices as compared to bacterial isolates (Sourabh et al., 2010). This may be due to various factors like cell size, composition of cell wall etc. as reviewed by Czerucka et al. (2007). The tolerance of 23 yeast isolates in simulated intestinal juice of pH 8 is presented in Table 2. The overall reduction of 0.08 to 0.76 log CFU/ml in viability after 8 h exposure was noticed for all the indigenous yeast isolates, where lowest reduction was observed for Sc02 isolate and highest for Sc04 isolate. These food borne yeast isolates of S. cerevisiae demonstrated high tolerance to simulated human upper gastrointestinal tract juices, and thus they offer a relatively overlooked source of potential probiotics, apart from bacteria and S. boulardii. The tolerance to acidic pH has already been reported for S. cerevisiae strains (Psomas et al., 2001). Thus, indigenous yeast isolates used in the present study fulfill the preliminary in vitro selection criteria for being designated as potential probiotics.

Effect of ox bile on the growth rate of yeast isolates

The effect of ox bile on the growth rate of yeast isolates is shown in Table 3. Eight isolates showed a lag time of less than 0.5 h, indicating that they are fairly resistant to tested concentration (0.3%) of ox bile. In addition, the ability to withstand ox bile was noticed in all the indigenous yeast isolates though the lag time (except the aforementioned nine isolates) was more than 1 h. Several other workers have also reported the survival of yeast strains in the presence of bile (Psomas et al., 2001; van der Aa Kqhle et al., 2005). Gotcheva et al. (2002) similarly assessed probiotic potential of four lactic acid bacteria (LAB) and three yeast strains isolated from a traditional Bulgarian cereal-based fermented beverage

Table 1. Survival of indigenous yeast isolates in simulated gastric juice of pH 2.

S/N	Isolate	Viable count (log CFU/ml)						
		0 min	30 min	60 min	90 min	120 min	180 min	240 min
1	Sc01	10.93 (0.01)	10.87 (0.02)*	9.98 (0.01)**	9.48 (0.01)**	8.22 (0.02)**	7.67 (0.03)**	7.08 (0.02)**
2	Sc02	10.90 (0.02)	10.74 (0.02)**	9.48 (0.02)**	8.75 (0.03)**	8.21 (0.07)**	8.05 (0.04)**	7.73 (0.03)**
3	Sc03	10.79 (0.02)	9.99 (0.02)**	9.07 (0.02)**	8.83 (0.03)**	8.04 (0.03)**	7.66 (0.02)**	7.16 (0.03)**
4	Sc04	10.97(0.02)	10.23 (0.02)**	8.85 (0.04)**	8.55 (0.02)**	8.26 (0.03)**	7.86 (0.03)**	7.04 (0.03)**
5	Sc05	10.87 (0.02)	10.66 (0.02)**	9.89 (0.02)**	8.77 (0.01)**	8.25 (0.02)**	7.86 (0.03)**	7.06 (0.03)**
6	Sc06	10.88 (0.01)	10.68(0.03)**	9.82 (0.02)**	9.05 (0.01)**	8.89(0.02)**	8.04 (0.02)**	7.26 (0.02)**
7	Sc07	10.94 (0.02)	10.89 (0.01)**	10.87 (0.01)*	9.66 (0.03)**	8.8 (0.01)**	7.76 (0.03)**	6.82 (0.06)**
8	Sc08	10.95 (0.01)	10.75 (0.02)	10.24 (0.01)**	10.02 (0.02)**	9.15 (0.03)**	8.04 (0.03)**	7.66 (0.02)**
9	Sc09	10.75 (0.05)	10.59 (0.02)**	10.04 (0.05)**	9.46 (0.05)**	9.14 (0.02)**	8.45 (0.02)**	7.29 (0.05)**
10	Sc10	10.81 (0.04)	10.12 (0.02)**	9.93 (0.02)**	8.8 (0.01)**	8.14 (0.03)**	7.85 (0.04)**	7.76 (0.03)**
11	Sc11	10.88 (0.02)	10.45 (0.03)**	9.91 (0.02)**	9.49 (0.04)**	8.76 (0.02)**	8.06 (0.02)**	7.55 (0.03)**
12	Sc12	10.67 (0.03)	10.59 (0.03)	10.16 (0.04)**	9.46 (0.01)**	8.94 (0.02)**	8.45 (0.02)**	7.76 (0.03)**
13	Sc13	10.89 (0.01)	10.84 (0.01)	9.49 (0.02)**	8.88 (0.01)**	8.26 (0.03)**	7.96 (0.01)**	7.9 (0.01)**
14	Sc14	10.64 (0.04)	10.61±(0.04)	9.46 (0.030)**	9.05 (0.01)**	8.87 (0.01)**	8.15 (0.04)**	7.75 (0.02)**
15	Sc15	10.85 (0.03)	10.36 (0.02)**	9.44 (0.04)**	9.01 (0.01)**	8.27 (0.03)**	8.2 (0.03)**	7.65 (0.03)**
16	Sc16	10.8 (0.02)	10.77 (0.02)	10.7 (0.02)**	9.56 (0.02)**	9.02 (0.02)**	7.9(0.02)**	7.84 (0.02)**
17	Sc17	10.75 (0.04)	10.53 (0.03)**	9.74 (0.03)**	9.48 (0.02)**	8.89 (0.04)**	7.77 (0.04)**	7.69 (0.04)**
18	Sc18	10.74 (0.03)	10.25 (0.02)**	9.62 (0.03)**	9.55 (0.02)**	9.06 (0.04)**	8.36 (0.02)**	7.96 (0.01)**
19	Sc19	10.59 (0.04)	10.54 (0.04)	9.57 (0.04)**	9.16 (0.02) **	8.75 (0.03)**	8.16 (0.02)**	7.86 (0.03)**
20	Sc20	10.95 (0.04)	10.86 (0.04)	10.04 (0.05)**	9.75 (0.04)**	8.83 (0.05)**	8.05 (0.04)**	7.81 (0.01)**
21	Sc21	10.77 (0.02)	10.64 (0.02)**	9.56 (0.03)**	8.94 (0.03)**	8.45 (0.03)**	7.85 (0.04)**	7.31 (0.01)**
22	Sc22	10.8 (0.01)	10.71(0.01)**	10.55 (0.03)**	8.93 (0.01)**	7.81 (0.01)**	7.72 (0.01)**	6.95 (0.03)**
23	Sc23	10.8 (0.02)	10.7 (0.02)	9.56 (0.02)**	9.02 (0.02)**	8.65 (0.02)**	8.07 (0.03)**	7.76 (0.01)**

Results are shown as mean (± standard deviation), Number of replications=3; Independent sample, testing of significance of variances by F -test, significance testing of two means by using Fishers t-test and Cochran and Cox t-test ,*$P < 0.05$ but > 0.01; ** $P< 0.01$.

and reported variable tolerance to low pH (2.0 to 3.0) and high bile concentrations (0.2 to 2.0%). van der Aa Kqhle et al. (2005) investigated 18 S. cerevisiae strains isolated from various foods or beverages, and found that all yeast strains were able to withstand pH 2.5 and 0.3% oxgall which are important probiotic traits. One isolate that is Sc03 showed augmented growth in the presence of ox bile with a lag time of just -0.13 h. It is difficult to assign any particular reason for this behaviour of isolate Sc03 as authors could not find appropriate kinetic studies on yeast isolates in presence of bile salts.

Bile salt deconjugation activity

Deconjugation has been included by the World Health Organization (WHO) experts as one of the main activities of intestinal microorganisms (FAO/WHO, 2002). For qualitative assaying of the deconjugation activity, the formation of opaque or whitish halo zone around the colony due to the release of free bile acids on deconjugation of added bile salts was taken as an indication of deconjugation ability of an organism as reported by Dashkevicz and Feighner (1989). None of the 23 yeast isolates was able to show precipitation/deconjugation activity on PDA plates supplemented with 0.5% (w/v) of various glycine and taurine conjugated bile salts; however, all isolates could grow in the presence of these bile salts, a property which needs to be ascertained further.

Microbial adherence to hydrocarbon

After tolerating upper gastrointestinal transit, the next challenge for an effective probiotic is to adhere to small intestinal cells. Adhesion is a complex trait that could be a multistep process in which both non-specific and specific mechanisms play a role. Cell surface hydrophobicity is considered to be an important factor in the adhesion and proliferation of microorganisms on the intestinal epithelial cells (Del Re et al., 1998). The adsorption of microbes to hydrocarbons and their partitioning in a hydrocarbon-aqueous biphasic system,

Table 2. Survival of indigenous yeast isolates in simulated intestinal juice (pH 8).

S/N	Isolate	Viable count (log CFU/ml)			
		0 h	1 h	4 h	8 h
1	Sc01	6.91 (0.01)	6.89 (0.03)	6.81 (0.02)**	6.72 (0.01)**
2	Sc02	7.01 (0.02)	6.99 (0.02)	6.97 (0.02)*	6.93 (0.02)**
3	Sc03	7.28 (0.03)	7.15 (0.03)**	7.02 (0.02)**	6.94 (0.01)**
4	Sc04	5.81 (0.02)	5.39 (0.05)**	5.15 (0.03)**	5.05 (0.02)**
5	Sc05	6.87 (0.01)	6.83 (0.02)	6.81 (0.01)**	6.72 (0.01)**
6	Sc06	7.06 (0.02)	7.05 (0.02)	7.01 (0.02)*	6.98 (0.01)*
7	Sc07	6.65 (0.07)	6.60 (0.04)	6.54 (0.02)	6.48 (0.04)*
8	Sc08	6.52 (0.02)	6.50 (0.01)	6.37 (0.02)**	6.21 (0.02)**
9	Sc09	6.50 (0.02)	5.45 (0.01)*	6.37 (0.02)	6.26 (0.02)*
10	Sc10	6.65 (0.07)	6.65 (0.07)	6.57 (0.02)	6.17 (0.04)**
11	Sc11	6.86 (0.07)	6.78 (0.07)*	6.69 (0.02)**	6.66 (0.04)**
12	Sc12	7.60 (0.02)	7.58 (0.02)*	7.44 (0.02)**	7.39 (0.01)**
13	Sc13	7.82 (0.01)	7.41 (0.01)**	7.40 (0.02)**	7.35 (0.01)**
14	Sc14	5.81 (0.02)	5.66 (0.02)**	5.48 (0.02)**	5.11 (0.04)**
15	Sc15	7.01 (0.02)	7.01 (0.02)	6.97 (0.01)*	6.89 (0.02)**
16	Sc16	7.22 (0.02)	7.06 (0.02)**	6.87 (0.02)**	6.78 (0.02)**
17	Sc17	5.48 (0.01)	5.44 (0.02)	5.11 (0.01)**	5.04 (0.01)**
18	Sc18	6.76 (0.01)	6.55 (0.02)**	6.34 (0.02)**	6.15 (0.03)
19	Sc19	5.89 (0.02)	5.79 (0.02)**	5.75 (0.03)**	5.62 (0.03)**
20	Sc20	5.75 (0.02)	5.66 (0.03)	5.62 (0.01)	5.46 (0.01)**
21	Sc21	5.98 (0.02)	5.92 (0.08)*	5.82 (0.04)**	5.68 (0.01)**
22	Sc22	5.82 (0.01)	5.73 (0.02)*	5.65 (0.02)**	5.53 (0.03)**
23	Sc23	5.84 (0.02)	5.78 (0.03)	5.66 (0.03)**	5.51 (0.03)**

Results are shown as mean (± standard deviation), Number of replications=3; Independent sample, testing of significance of variances by F -test, significance testing of two means by using Fishers t-test and Cochran and Cox t-test ,*P < 0.05 but > 0.01; ** P< 0.01.

has been suggested to be a good method for measuring cell surface hydrophobicity (Rosenberg et al., 1980). Strains possessing high hydrophobicity exhibit good adhesion property to intestinal cell lines (Pan et al., 2006; Marin et al., 1997; Wadstrom et al., 1987). Therefore, hydrophobicity was used to ascertain the adhesive potential of the indigenous isolates. Hydrophobicity was determined with n-hexadecane because it has been reported to give more reliable results without any cell lysis as compared to other hydrocarbons for evaluation of adhesion ability of probiotics (Vanhaecke and Pijck, 1988; Richard et al., 1999). A great variability in hydrophobicity values was observed in 23 indigenous yeast isolates (Table 4).

For only nine isolates, the hydrophobicity values were above 50%. Highest (59.65%) hydrophobicity was recorded for isolate Sc01 and lowest (13.46%) for isolate Sc18. In general, the hydrophobicity values for indigenous yeast isolates were overall lower than the values as high as 79.69% reported for bacteria (Sourabh et al., 2010; Del Re et al., 2000; Collado et al., 2008).

Isolates possessing low hydrophobicity values may have low adhesion ability, which needs to be confirmed using in vivo experiments/cell line studies. It was reported that usually, yeast strains show variable adherence property in comparison to bacteria, that is why these microorganisms are to be administered repeatedly to achieve steady-state concentrations in the colon (Kumura et al., 2004). Since these yeast indigenous isolates are an integral part of traditional fermented foods being consumed regularly by people of Western Himalayas, therefore role of these isolates as probiotics seems to be important in this population.

Autoaggregation ability

Autoaggregation ability of probiotics is another trait which is associated with the adhesion ability of microorganisms (Del Re et al., 1998; Perez et al., 1998). Autoaggregation (%) ability values were found to be in the range of 18.47 to 67.65% (Table 5) for all the tested indigenous yeast

Table 3. Effect of ox bile (0.3 % w/v) on the growth rate of indigenous yeast isolates.

S/N	Isolate	Time required to increase A_{620} nm by 0.3 units [Lag time (in hours)]
1	Sc01	0.13^j
2	Sc02	1.26^{bc}
3	Sc03	-0.13^k
4	Sc04	0.72^e
5	Sc05	1.37^b
6	Sc06	0.17^{lj}
7	Sc07	0.13^j
8	Sc08	0.67^{efi}
9	Sc09	0.64^{ef}
10	Sc10	0.99^d
11	Sc11	0.73^e
12	Sc12	0.51^{fg}
13	Sc13	1.10^{cd}
14	Sc14	0.96^d
15	Sc15	1.08^{cd}
16	Sc16	0.70^{ef}
17	Sc17	0.61^{ef}
18	Sc18	0.31^{hj}
19	Sc19	1.83^a
20	Sc20	0.29^{hlj}
21	Sc21	0.25^{hlj}
22	Sc22	0.35^{ghi}
23	Sc23	0.38^{gh}

Number of replications=3. Mean values within the same column followed by different superscript letters differ significantly when compared by Duncan's Multiple Range Test.

Table 4. Microbial adhesion (expressed in terms of % hydrophobicity) to n-hexadecane exhibited by indigenous yeast isolates.

S/N	Isolate	% Hydrophobicity
1	Sc01	$59.65 (0.58)^{abcde}$
2	Sc02	$41.55 (2.78)^{mn}$
3	Sc03	$54.52 (1.26)^{gh}$
4	Sc04	$34.55 (1.92)^{opq}$
5	Sc05	$21.53 (0.94)^{stu}$
6	Sc06	$51.92 (1.55)^i$
7	Sc07	$20.91(1.14)^{tu}$
8	Sc08	$57.47 (1.25)^{abcdefg}$
9	Sc09	$30.80 (2.51)^{qr}$
10	Sc10	$42.38 (1.59)^{lm}$
11	Sc11	$49.22 (0.39)^k$
12	Sc12	$57.39 (1.67)^{abcdefg}$
13	Sc13	$49.55 (1.51)^{jk}$
14	Sc14	$28.40 (3.43)^r$
15	Sc15	$58.28 (0.95)^{abcdef}$
16	Sc16	$58.79 (0.46)^{abcde}$
17	Sc17	$55.21 (0.77)^{efgh}$
18	Sc18	$13.46 (2.63)^{wx}$
19	Sc19	$21.37 (0.52)^{tuv}$
20	Sc20	$56.12 (1.40)^{cdefgh}$
21	Sc21	$35.63 (1.83)^{no}$
22	Sc22	$32.66 (1.39)^{pqr}$
23	Sc23	$13.81(0.70)^v$

Results are shown as mean (± standard deviation); Number of replications=3. Mean values within the same column followed by different superscript letters differ significantly when compared by Duncan's Multiple Range Test.

isolates. Although hydrophobicity values for most of the isolates were low, isolates having good autoaggregation ability in conjunction with the good hydrophobicity values can strongly be related to the adhesion ability of these microorganisms. Though these traits are independent of each other; they are still related to adhesion property of a particular microbe (Rahman et al., 2008). Autoaggregation ability has been more strongly associated to adhesion as compared to hydrophobicity (Del Re et al., 2000) therefore, good amount of autoaggregation ability in spite of low hydrophobicity values may account for adherence property of these indigenous isolates.

Production of exopolysaccharide

For microbial cells, EPS are thought to play a role in protection against desiccation, toxic compounds, bacteriophages, osmotic stress, and to permit adhesion to solid surfaces and biofilm formation (De Vuyst and Degeest, 1999). Therefore, the native yeast isolates were screened for the production of EPS by using ruthenium red dye in the medium where the positive isolates gave white colonies (ropy colonies) whereas, the negative isolates gave red or pinkish colonies (non-ropy colonies). Out of 23 yeast isolates, nine yeast isolates namely, Sc03, Sc11, Sc12, Sc13, Sc15, Sc19, Sc20, Sc21 and Sc22 were positive for production of EPS (Figure 1A). Isolates Sc09 and Sc17 showed no exopolysaccharide production but they exhibited proteolysis on the medium used for detection of exopolysaccharide (Figure 1B). EPS has immunostimulatory (Hosono et al., 1997) and anti-tumoral (Ebina et al., 1995) activities, and phosphate groups in EPS play an important role in the activation of macrophages and lymphocytes (Kitazawa et al., 2000; Uemura et al., 2003). Therefore, with this background, the indigenous yeast isolates can be used as immunostimulatory adjuvants and further studies in our laboratory are going on.

Antioxidant activity

Out of 23 yeast isolates, only four isolates that is Sc01,

Table 5. Autoaggregation (%) ability of indigenous yeast isolates.

S/N	Isolate	% Autoaggregation ability
1	Sc01	64.59 (0.19)^{cd}
2	Sc02	67.59 (0.27)^{ab}
3	Sc03	63.90 (0.42)^{efg}
4	Sc04	42.43 (0.27)^{r}
5	Sc05	57.37 (0.16)^{jkl}
6	Sc06	32.84 (0.22)^{t}
7	Sc07	60.52 (0.25)^{g}
8	Sc08	31.15 (0.10)^{u}
9	Sc09	49.50 (0.32)^{n}
10	Sc10	18.47 (0.43)^{w}
11	Sc11	63.52 (0.26)^{fg}
12	Sc12	58.28 (0.22)^{h}
13	Sc13	64.38 (0.24)^{def}
14	Sc14	47.90 (0.25)^{pq}
15	Sc15	67.65 (0.33)^{ab}
16	Sc16	54.64 (0.11)^{m}
17	Sc17	57.16 (0.41)^{jkl}
18	Sc18	47.87 (0.27)^{q}
19	Sc19	35.18 (0.21)^{s}
20	Sc20	55.63 (0.34)^{l}
21	Sc21	48.57 (0.26)^{o}
22	Sc22	57.63 (0.37)^{ijk}
23	Sc23	27.74 (0.53)^{v}

Results are shown as mean (± standard deviation), Number of replications=3. Mean values within the same column followed by different superscript letters differ significantly when compared by Duncan's Multiple Range Test.

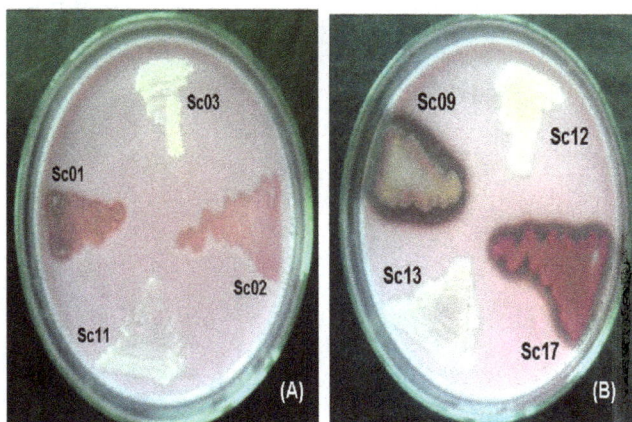

Figure 1. Exopolysaccharide production by yeast isolates.

Sc02, Sc03 and Sc16 showed scavenging of DPPH free radical in both methanol and buffered methanol reaction systems (Table 6). The intracellular cell free extract of isolate Sc16 showed maximum antioxidant activity (22.07%) while isolate Sc01 showed minimum activity (11.43%) with DPPH assay (methanol). With whole cell extract, the highest scavenging activity was shown by isolate Sc02 (25.47%). In case of DPPH assay employing buffered methanol, isolate Sc01 and Sc02 showed maximum antioxidant activity with intracellular cell free (33.30%) and whole cell extract (42.51%) respectively.

The results of the present study indicate that the radical scavenging ability of both intracellular cell-free extracts and whole cell extract of native isolates contribute towards the antioxidative effect. Thus, making indigenous yeast isolates as potential candidate for the production of functional food supplements. Various workers have reported the exhibition of antioxidative activity by bacteria like *Lactobacillus* species (Kullisaar et al., 2002; Saide and Gilliland, 2005; Jarvenpaa et al., 2007) whereas, in yeasts it is less reported. Therefore, presence of this activity in yeast isolates is an additional protective trait which may be useful as natural antioxidant.

Galactosidase enzyme production

An important way in which probiotics beneficially affects the health of the host is by providing enzymatic activities that improve the utilization of nutrients within the intestine (Rowland, 1992; Mustapha et al., 1997). The primary mechanism which is involved in this case is through the intraintestinal hydrolysis of the lactose by β - galactosidase (EC 3.2.1.23) enzyme therefore, the indigenous yeast isolates were assayed for β - galactosidase enzyme. None of the yeast isolates was found positive for this enzyme.

Production of siderophore

These indigenous yeast isolates were also subjected to screening for siderophore production, a iron chelating metabolite known to exert antagonistic effect (Kline et al., 2000). Only one yeast isolate that is Sc15, was found positive for siderophore production as judged by the formation of orange halo zone on CAS agar medium (Figure 2). Most bacteria and fungi synthesize, secrete, and take up at least one type of siderophore and yet also have the capacity to take up siderophores secreted by other organisms (Byers and Arceneaux, 1998). Although *S. cerevisiae* do not secrete siderophores, it is capable of taking up siderophore-bound iron (Neilands, 1995; Lesuisse et al., 1987), and a specific transport system for the hydroxamate-type siderophore ferrioxamine B(FOB) has been described (Yun et al., 2000). In this study, the production of siderophore by indigenous yeast isolate (*S. cerevisiae*) is an interesting observation.

The results of the present study confirm that indigenous yeast isolates showed better survivability in simulated gastrointestinal juices and in the presence of ox bile. Some of these isolates also exhibited good hydrophobicity

Table 6. Antioxidant activity of indigenous yeast isolates using scavenging of DPPH free radical.

S/N	Isolate	Protein concentration (µg/ml) of sample	Dry weight (mg) per 500 µl of sample		DPPH (methanol) in 100 µl		DPPH (buffered methanol) in 100 µl	
			Intracellular cell free extract	Whole cell extract	Intracellular cell free extract	Whole cell extract	Intracellular cell free extract	Whole cell extract
1	Sc01	179.08	0.8	1.8	11.43 (±0.40)[c]	19.47 (±0.06)[b]	33.30 (±0.30)[a]	36.27 (±0.15)[b]
2	Sc02	205.32	0.8	0.9	17.27 (±0.31)[b]	25.47 (±0.21)[a]	25.27 (±0.31)[b]	40.40 (±0.20)[a]
3	Sc03	327.45	0.9	1.4	17.10 (±0.10)[b]	16.40 (±0.10)[c]	18.43 (±0.21)[c]	32.77 (±0.25)[c]
4	Sc16	323.52	1.7	6.6	22.07 (±0.93)[a]	15.40 (±0.36)[d]	25.67 (±0.25)[b]	27.67 (±0.95)[d]

Results are shown as mean, mean (± standard deviation); Number of replications=3. Mean values within the same column followed by different superscript letters differ significantly when compared by Duncan's Multiple Range Test.

and autoaggregation abilities which are one of the essential markers for the selection of adherence potential of probiotics. In addition, the exhibition of some other traits such as antioxidant activity, exopolysaccharide and siderophore production

makes them appropriate candidate for the development of functional foods.

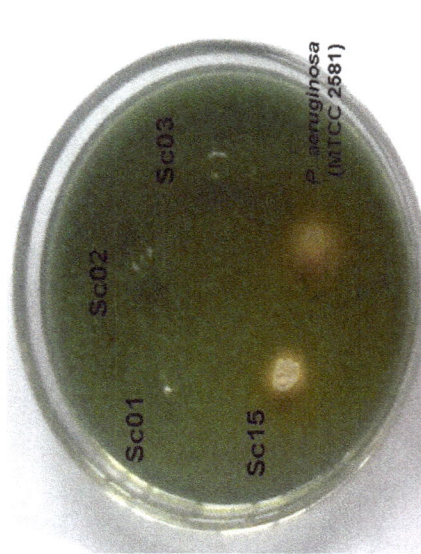

Figure 2. Siderophore production by yeast isolate Sc15 and *Pseudomonas aeruginosa*-MTCC 2581 (positive control).

REFERENCES

Adams MR (1999). Safety of industrial lactic acid bacteria. *J. Biotechnol.*, 68: 171-178.

Byers BR, Arceneaux JEL (1998). Iron Transport and Storage in Microorganisms, Plants, and Animals, Vol. 35 (Sigel A and Sigel H, eds). Marcel Dekker, Inc., New York, pp. 37-66.

Campbell-Platt G (1994). Fermented foods – A world perspective. *Food Res. Int.*, 27: 253-264.

Charteris WP, Kelly PM, Morelli L, Collins JK (1998). Development and application of an *in vitro* methodology to determine the transit tolerance of potentially probiotics *Lactobacilli* and *Bifidobacterium* species in the upper human gastrointestinal tract. *J. Appl. Microbiol.*, 84(5): 759-768.

Chiang BL, Sheih YB, Wang LH, Lino CK, Gill HS (2000). Enhancing immunity by dietary consumption of a probiotic lactic acid bacterium (*Bifidobacterium lactis*, HN019): optimization and definition of cellular immune responses. *Euro. J. Clin. Nutr.*, 54: 849-855.

Chou L, Weimer B (1999). Isolation and characterization of acid- and bile-tolerant isolates from strains of *Lactobacillus acidophilus*. *J. Dairy Sci.*, 82: 23-31.

Collado, MC, Meriluoto J, Salminen S (2008). Adhesion and aggregation properties of probiotic and pathogen strains. *Eur. Food Res. Technol.*, 226: 1065-1073.

Czerucka D, Piche T, Rampal P (2007). Review article: yeast as probiotics – *Saccharomyces boulardii*. Aliment Pharmacol. Ther., 26: 767-778.

Dashkevicz MP, Feighner SD (1989). Development of a deferential medium for bile salt hydrolase-active *Lactobacillus* spp. Appl. Environ. Microbiol., 55:11-16.

De Vuyst L, Degeest B (1999). Heteropolysaccharides from lactic acid bacteria, FEMS Microbiol. Rev., 23: 153-177.

Del Re B, Busetto A, Vignola G, Sgorbati B, Palenzona D (1998). Autoaggregation and adhesion ability in a *Bifidobacterium suis* strain. Lett. Appl. Microbiol., 27: 307-310.

Del Re B, Sgorbati B, Miglioli M, Palenzona D (2000). Adhesion, autoaggregation and hydrophobicity of 13 strains of *Bifidobacterium longum*. Lett. Appl. Microbiol., 31: 438-442.

Dunne C, Murphy L, Flynn S, O'Mahony L, O'Halloran S, Feeney M, Morrissey D, Thornton G, Fitzgerald G, Daly C, Kiely B, Quigley E.M.M, O'Sullivan GC, Shanahan F, Kevin J (1999). Probiotics: from myth to reality. Demonstration of functionality in animal models of disease and in human clinical trials. Antonie Van Leeuwenhoek, 76: 279-292.

Ebina T, Ogata N, Murata K (1995). Antitumor effect of *Lactobacillus bulgaricus* 878R. Biotherapy, 9: 65-70.

El-Gazzar RE, Marth EH (1992). Food borne-disease investigating procedures and economic assessment. J. Environ. Health, 55: 24-27.

FAO/WHO.2002. Drafting guidelines for the evaluation of probiotics in foods. Report of a Joint FAO/WHO Working Group, London Ontario, Canada, April 30 and May 1. Retrieved April 30 and May 1 2000, from http://www.who.int/foodsafety/fs_management/en/probiotic_guideline s.pdf.

Gotcheva V, Hristozova E, Hristozova T, Guo M, Roshkova Z, Angelov A (2002). Assessment of potential probiotic properties of lactic acid bacteria and yeast strains. Food Biotechnol., 16(3): 211 – 225.

Grant C, Salminen S (1998). The potential of Propionibacterium spp. as probiotics. In: Lactic acid bacteria: microbiological and functional aspects. Marcel Dekker, Inc., New York, pp. 588-603.

Hansen EB (2002). Commercial bacterial starter cultures for fermented foods of the future. Int. J. Food Microbiol., 78: 119-131.

Hosono J, Lee A, Amenati M, Natsume M, Hirayama T, Adachi S, Kaminogawa (1997). Characterization of a water-soluble polysaccharide fraction with immunopotentiating activity from Bifidobacterium adolescentis M101-4. Biosci. Biotechnol. Biochem., 61: 312-316.

Jarvenpaa S, Tahvonen RL, Ouwehand AC, Sandell M, Ja¨rvenpaa E, Salminen S (2007). A Probiotic, Lactobacillus fermentum ME-3, has antioxidative capacity in soft cheese spreads with different fats. Am. Dairy Sci. Assoc., 90: 3171-3177.

Kanwar SS, Gupta MK, Katoch C, Kumar R, Kanwar P (2007). Traditional fermented foods of Lahaul and Spiti area of Himachal Pradesh. Indian J. Trad. Knowl., 6(1): 42-45.

Kitazawa H, Ishii Y, Uemura J, Kawai Y, Saito T, Kaneko T, Noda K, Itoh T. (2000). Augmentation of macrophage functions by an extracellular phosphopolysaccharide from Lactobacillus delbrueckii subsp. bulgaricus. Food Microbiol., 17: 109-118.

Kline T, Fromhold M, McKennon TE, Cai S, Treiberg J, Ihle N, Sherman D, Schwan W, Hickey MJ, Warrener P, Witte PR, Brody LL, Goltry L, Barker LM, Anderson SU, Tanaka SK, Shawar RM, Nguyen LY, Langhorne M, Bigelow A, Embuscado L, Naeemi E (2000). Antimicrobial effects of novel siderophores linked to β-lactam antibiotics Bioorg. Med. Chem., 8(1): 73-93.25.

Kullisaar T, Zilmer M, Mikelsaar M, Vihalemm T, Annuk H, Kairane C, Kilk A (2002). Two antioxidative Lactobacilli strains as promising probiotics. Int. J. Food Microbiol., 72: 215-224.

Kumura H, Tanoue Y, Tsukahara M, Tanaka T, Shimazaki K (2004). Screening of Dairy Yeast Strains for Probiotic Applications. J. Dairy Sci., 87: 4050-4056.

Lesuisse E, Raguzzi F, Crichton RR (1987). Iron uptake by the yeast Saccharomyces cerevisiae: Involvement of a reduction step. J. Gen. Microbiol., 133: 3229-3236.

Lowry OH, Rosebrough N J, Farr A L, Randall RJ (1951). Protein measurement with the folin phenol reagent. J. Biol. Chem., 193: 265-275.

Marin ML, Benito Y, Pin C, Fernandez MF, Garcia ML, Selgas MD, Casas C (1997). Lactic acid bacteria: hydrophobicity and strength of attachment to meat surfaces. Lett. Appl. Microbiol., 24: 14-18.

Mora D, Fortina MG, Parini C, Ricci G, Gatti M, Giraffa G, Manachini PL (2002). Genetic diversity and technological properties of Streptococcus thermophilus strains isolated from dairy products. J. Appl. Microbiol., 93: 278-287.

Mustapha A, Jiang T, Savaiano D (1997). Improvement of lactose digestion by humans following ingestion of unfermented acidophilus milk: influence of bile sensitivity, lactose transport, and acid tolerance of Lactobacillus acidophilus. J. Dairy Sci., 80: 1537-1545.

Neilands JB (1995). Siderophores: structure and function of microbial iron transport compounds. J. Biol. Chem., 270: 26723-26726.

Nout MJR (2001). Fermented foods and their production. In: Fermentation and Food Safety (Adams MR & Nout MJR, eds). Aspen Publishers, Inc., Gaithersburg, Md., USA, pp. 1-30.

Pan WH, Li PL, Liu Z (2006). The correlation between surface hydrophobicity and adherence of Bifidobacterium strains from centenarian's faeces. Anaerobe, 12: 148-152.

Pathania N, Kanwar SS, Jhang T, Koundal KR, Sharma TR (2010). Application of different molecular techniques for deciphering genetic diversity among yeast isolates of traditional fermented food products of Western Himalayas. World J. Microbiol. Biotechnol., 26(9): 1539-1547.

Perez PF, Minnaard Y, Disalvo EA, De Antoni GL (1998). Surface properties of bifidobacterial strains of human origin. Appl. Environ. Microbiol., 64: 21-26.

Prasad J, Gill H, Smart J, Gopal PK (1998). Selection and

characterization of Lactobacillus and Bifidobacterium strains for use as probiotics. Int. Dairy J., 8: 993-1002.

Psomas E, Andrighetto C, Litopoulou-Tzanetaki E, Lombardi A, Tzanetakis N (2001). Some probiotic properties of yeast isolates from infant faeces and feta cheese. Int. J. Food Microbiol., 69: 125-133.

Rahman MM, Kim WS, Kumura H, Shimazaki K (2008). Autoaggregation and surface hydrophobicity. World J. Microbiol. Biotechnol., 24: 1593-1598.

Richard S, Pembrey K, Marshall C, Schneider RP (1999). Cell Surface Analysis Techniques: What Do Cell Preparation Protocols Do to Cell Surface Properties?. Appl. Environ. Microbiol., 65: 2877-2894.

Rosenberg M, Gutnick D, Rosenberg E (1980). Adherence of bacteria to hydrocarbons: a simple method for measuring cell-surface hydrophobicity. FEMS Microbiol. Lett., 9: 29-33.

Ross RP, Morgan S, Hill C (2002). Preservation and fermentation: past, present and future. Int. J. Food Microbiol., 79: 1-16.

Rowland IR (1992). Metabolic interactions in the gut. In: Probiotics the scientific basis, (Fuller R, ed.). Chapman and Hall, London, pp. 29-53.

Saide JAO, Gilliland SE (2005). Antioxidative activity of Lactobacilli measured by oxygen radical absorbance capacity. J. Dairy Sci., 88: 1352-1357.

Salminen S, von Wright A (1998). Current probiotics—safety assured?. Microb. Ecol. Health Dis., 10: 68-77.

Sanders ME, in't Veld JH (1999). Bringing a probiotic-containing functional food to the market: microbiological, product, regulatory and labeling issues. Antonie van Leeuwenhoek, 76: 293-315.

Sazawal S, Hiremath G, Dhingra U, Malik P, Deb S, Black RE (2006). Efficacy of probiotics in prevention of acute diarrhoea: a meta-analysis of masked, randomized, placebo-controlled trials. Lancet Infect. Dis., 6: 374-382.

Schwyn B, Neilands JB (1987). Universal chemical assay for the detection and determination of siderophores. Anal. Biochem., 160 (1): 47-56.

Sharma OP, Bhat TK (2009). DPPH antioxidant assay revisited. Food Chem., 113: 1202-1205.

Sourabh A, Kanwar SS, Sharma PN (2010). Diversity of bacterial probiotics in traditional fermented foods of Western Himalayas. Int. J. Probio. Prebio., 5(4): 193-202.

Taranto MP, de Ruiz Holgado AP, de Valdez GF (1995). Bile salt hydrolase activity in Enterococcus faecium strains. Microbiol. Aliments Nutr., 13: 375-379.

Uemura JN, Kitazawa H, Kawai Y, Itoh T, Oda M, Saito T (2003). Functional alteration of murine macrophages stimulated with extracellular polysaccharides from Lactobacillus delbrueckii subsp. bulgaricus OLL1073R-1. Food Microbiol., 20: 267-273.

Usman, Hosono A (1999). Bile tolerance taurocholate deconjugation and binding of cholesterol by Lactobacillus gasseri strains. J Dairy Sci., 82: 243-248.

van der Aa Kqhle A, Skovgaard K, Jespersena L (2005). In vitro screening of probiotic properties of Saccharomyces cerevisiae var. boulardii and food-borne Saccharomyces cerevisiae strains. Int. J. Food Microbiol., 101: 29-39.

Vanhaecke E, Pijck J (1988). Bioluminescence assay for measuring the number of bacteria adhering to the hydrocarbon phase in the BATH test? Appl. Environ. Microbiol., 54: 1436-1439.

Vignolo GM, Suriani F, de Ruiz Holgado AP, Oliver G (1993). Antibacterial activity of Lactobacillus strains isolated from fermented sausages. J. Appl. Bacteriol., 75: 344-349.

Vinderola CG, Reinheimer JA (2003). Lactic acid starter and probiotic bacteria: a comparative "in vitro" study of probiotic characteristics and biological barrier resistance. Food Res. Int., 36: 895-904.

Wadstrom T, Andersson K, Sydow M, Axelsson L, Lindgren S, Gullmar B (1987). Surface properties of lactobacilli isolated from the small intestine of pigs. J. Appl. Bacteriol., 62: 513-520.

Walker DR, Gilliland SE (1993). Relationship among bile tolerance, bile salt deconjugation, and assimilation of cholesterol by Lactobacillus acidophilus. J. Dairy Sci., 76: 956–961.

Yarrow D (1998). Methods for the isolation, maintenance and identification of yeasts. The Yeasts, a Taxonomic Study, 4th edn (Kurtzman CP & Fell JW, eds). Elsevier, Amsterdam, the Netherlands, pp. 77-100.

Yun CY, Ferea T, Rashford J, Ardon O, Brown PO, Botstein D, Kaplan J, Philpott CC (2000). Desferrioxamine-mediated Iron Uptake in *Saccharomyces cerevisiae*. Evidence for two pathways of iron uptake. J. Biol. Chem., 275: 10709-10715.

27

Isolation and characterization of yeast strains from local food crops

Abosede Margaret Ebabhi[1] , Adedotun Adeyinka Adekunle[2], Wahab Oluwanisola Okunowo[3] and Akinniyi Adediran Osuntoki[3]

[1]Department of Biological Sciences, McPherson University, Seriki Sotayo, Ogun State, Nigeria.
[2]Department of Botany, Faculty of Science, University of Lagos, Lagos State, Nigeria.
[3]Department of Biochemistry, College of Medicine, University of Lagos, Lagos State, Nigeria.

Isolation and identification of yeast from *Manihot esculenta*, *Zea mays*, *Cola acuminata* and *Sorghum bicolor* was done using the spread plate technique. Morphological, cultural, physiological and molecular characterizations were carried out resulting in determination of the species. Four isolates belonging to different genera which include *Pichia*, *Kluyveromyces*, *Candida* and *Saccharomyces* were identified. This present study showed that the yeast isolates have the potential to ferment both hexose and pentose sugars.

Key words: Food crops, yeast, isolation, identification, fermentation.

INTRODUCTION

Yeasts, the chemoorganotrophs obtain carbon mostly from hexose sugars, such as glucose and fructose or disaccharides such as sucrose and maltose (Barnett, 1975). Some species can also metabolize pentose sugars like xylose (Chaudhary and Qazi, 2006), alcohols and organic acids. Yeast species either require oxygen for aerobic cellular respiration (obligate aerobes), or are anaerobic but also have aerobic methods of energy production (facultative anaerobes). The useful physiological properties of yeast have led to their use in the field of biotechnology. Fermentation of sugar by yeast is the oldest and largest application of this technology. Some isolates have the ability to carry out an alcoholic fermentation while others lack this property. Thus, many types of yeasts are used for making many foods; yeast in wine fermentation (Martini, 1992; Oelofse et al., 2008) and for xylitol production (Sreenivas Rao et al., 2004).

They can also find application as biological control agents (Qing and Shiping, 2000; Chanchaichaovivat et al., 2007) and they include some of the most widely used model organisms for genetics and cell biology. Recently, the ability of yeast to convert sugar into ethanol has been harnessed by the biotechnology industry for ethanol fuel production (Oyeleke and Jibrin, 2009; Ocloo and Ayernor, 2010; Mohd et al., 2011). Thus, there is an exploration of diverse sources of yeast to seek potent species that can utilize a wide range of substrates. Kurtzman and Piskur (2006) reported that yeasts do not form a specific taxonomic or phylogenetic group and at present it is estimated that only 1% of all yeast species have been described.

In the present study, some locally available food crops were analyzed for isolation and subsequent characterization of yeast isolates which may further be

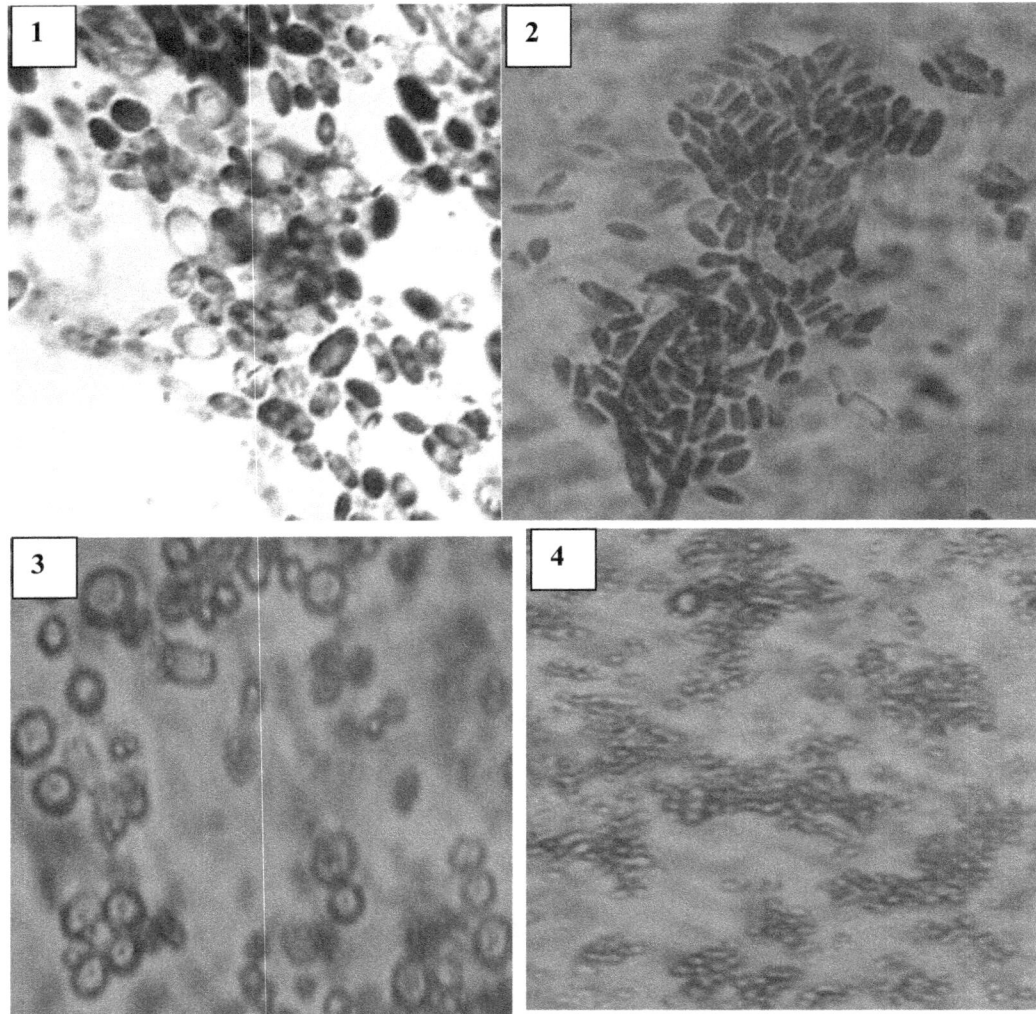

Figure 1. Microscopic characteristics of yeast strains were cultivated on MEA agar after incubation at 48 h and 28 ± 2°C 1. *K. marxianus*; 2. *C. tropicalis*; 3. *S. cerevisiae*; 4. *P. caribbica*

utilized in alcohol production.

MATERIALS AND METHODS

Collection of samples

Cassava tubers (*Manihot esculenta 92/0057*) were collected from the International Institute of Tropical Agriculture (IITA) Ibadan, Oyo State. Maize, guinea corn and kola nut were randomly bought from local markets within Lagos Metropolis, Nigeria. The samples were transported to the laboratory, dirt were removed, the cassava tubers were peeled. All samples were washed using sterile water and blended to powder form using a blender (Binatone). These particles were then sieved to obtain average particle sizes of 300 μm in diameter.

Isolation and identification of microorganisms

One gram of each sample was soaked in distilled water in 250 ml

conical flask for 72 h at 28 ± 2°C. Serial dilution of the steep of each was carried out up to 10^{-5}. An aliquot of 0.1 ml of each dilution was plated on Malt Extract Agar (MEA) plates (5 g/100 ml) using spread plate technique. The inoculated plates were incubated for 48 h at 28 ± 2°C. Chloramphenicol at the rate of 30 μg/ml was added as an antimicrobial agent to inhibit all bacteria growth.

Subculture technique

Isolates were subculture on MEA to check for purity and incubated at 28 ± 2°C for 48 h (Figure 1). Purified cultures were routinely maintained on MEA slants and kept at 4°C. The strains were stained using methylene blue and viewed under a high power microscope (100× magnification). Colour, texture and other features were observed on the colonies. Biochemical tests of the selected yeast isolates were carried out by the means of fermentation of different carbon sources using the modified method of Olutiola et al. (2000). Photomicrographs of the isolates were taken with motic camera 2.0. The identities of the isolates were confirmed by comparing the characteristics with those of known taxa using the

Table 1. Cultural characteristics of yeast isolates.

Isolate code	EAM 2	EAM 3	EAM 5	EAM 6
Name of organism	*Kluyveromyces marxianus*	*Pichia caribbica*	*Candida tropicalis*	*Saccharomyces cerevisiae*
Pigmentation	Creamy; white	Creamy; white	Creamy	White; Creamy
Colony morphology	Raised,smooth clustered	Flat, ovoid	Flat, smooth	Oblong/Eclipse
Cell length (µm)	5.0 to 10.5	1.2 to 10.8	1.5 to 2.0	2. 0 to 8.0
Cell breadth (µm)	2.0 to 5.0	1.5 to 3.5	5.0 to 6.0	0.5 to 3.0

schemes of Rhode and Hartmann (1980) and Ellis et al. (2007). Isolates were genetically identified by growing them on MEA slants in 5.0 ml McCartney bottles and sent to the Centre for Agriculture and Bioscience International (CABI) identification service, Royal Botanical Garden Kew, England. These isolates were sequenced according to CABI standard protocols. Procedure– terms and conditions was applied (Centre for Agriculture and Bioscience International, Royal Botanical Garden Kew, England).

Determination of enzyme synthesizing ability

Urea hydrolysis test was done using the method of Seeliger (1956). The cultures were grown on Sabouraud agar and were transferred to fresh slants before the urea test was done. Fresh cultures from the slants were transferred with a loop to the surface of Christensen urea medium consisting of the following: 0.1 g peptone; 0.1 g glucose; 0.5 g NaCl; 0.2 g KH_2PO_4; 1.5 g agar and 0.012 g phenol red per 1000 ml of distilled water. The ingredients were mixed and melted in a water bath. After adjusting the pH to 6.8, the medium was dispensed into test tubes in 4.5 ml amounts and autoclaved for 10 min at 121°C. The tubes were allowed to cool to 50°C. To every tube of the autoclaved medium, 0.5 ml of a 20% Seitz-filtered solution of urea was added aseptically. After mixing with the base, the contents of the tubes were allowed to solidify with a long slant and a deep butt. The inoculated tubes were incubated at the optimal temperature (28 to 30°C) of the organism for 72 h. Urea hydrolysis was indicated by a distinct colour change of the indicator from a deep pinkish red to an orange-yellow colour starting at the slanted part of the medium and progressing rapidly to the deep part of the butt.

RESULTS

Four yeast strains were isolated, purified and further identified from different food crops produced in Nigeria. Differential tests were applied including morphological, physiological and molecular which facilitated the opportunity for identification of the yeasts. These tests allowed information gathering for the studies objectives and the determination of the systemic status of the yeast. Distinct yeast was isolated from each of the food crop with each belonging to a separate genus which included *Candida*, *Kluyveromyces*, *Pichia* and *Saccharomyces* (Table 1). Only one of the isolates (EAM 5) showed imperfect state and also showed a pinkish colony appearance while the others were whitish creamy. Two isolates (EAM 5 and 6) showed pseudohyphae formation. Budding was observed in isolate EAM 6 which also had butryous colony texture. All isolates showed elliptical to

round spores. A single species belong to the sub-division Deuteromycotina and the rest to the sub-division Ascomycotina.

The physiological and biochemical tests of the yeasts carried out showed all isolates to ferment glucose for their growth. *Pichia caribbica* was able to ferment xylose of the four isolates. The assimilation and fermentation of most sugars by the isolates was variable. Urea hydrolysis was weak by three of the isolates with *Candida tropicalis* showing no trace of hydrolysis after 72 h, while hydrolysis was very vigorously registered by *P. caribbica* (EAM 3). Results of physiological and biochemical tests are presented in Table 2.

FASTA sequences of the studied yeast genomic DNA

Image 1 was obtained for the isolation of DNA from the four yeast isolates. The DNA sequence data obtained from some of the isolates (EAM 2, EAM 3 and EAM 5) have been deposited in the database library of the Royal Botanical Garden Kew. The blast sequence query showed that *Kluyveromyces marxianus* (IMI 398399) and *P. caribbica* (IMI 398400) had maximum identity (100%) with the genomic DNA sequence of EAM 2 and EAM 3, respectively at both ITS. Isolate EAM 5 was 100% homologous to *C. tropicalis* 18S rRNA gene sequence. The blast sequence query showed that *Saccharomyces cerevisiae* (GU 931323.1) has the maximum identity (88%) with the genomic DNA sequence of EAM 6 at ITS 1 and ITS 4 sequence with that in the Genbank Library Database (Table 3).

DISCUSSION

In this study, four yeast species from four genera were isolated from *Cola acuminata*, *M. esculenta*, *Zea mays*, and *Sorghum bicolor*. Respectively, they include *C. tropicalis*, *K. marxianus*, *P. caribbica* and *S. cerevisiae*. The yeasts were found to be fermentative in the breakdown of hexose and pentose sugars. Hitherto, several workers such as Oyeleke and Jibrin (2009), Mohd et al. (2011) have reported the activities of some of the yeast strains in bioethanol production. Species of yeast like *Candida* have not been extensively reported as fermentative yeast for industrial utilization such as the

Table 2. Phenotypic characterization of the isolates.

Biochemical characteristic	Isolate			
	EAM 5	EAM 2	EAM 3	EAM 6
	C. tropicalis	*K. marxianus*	*Pichia caribbica*	*S. cerevisiae*
Gram reaction	+	+	+	+
Sucrose	-	+	+	+
Glucose	+	+	+	+
Lactose	+	V	+	+
Galactose	+	+	-	+
Maltose	+	--	+	-
Xylose	-	-	+	-
Fructose	-	+	+	+
Dextrose	-	+	-	-
Urease	-	-	+	-

+ = Present; - = absent; V = variable.

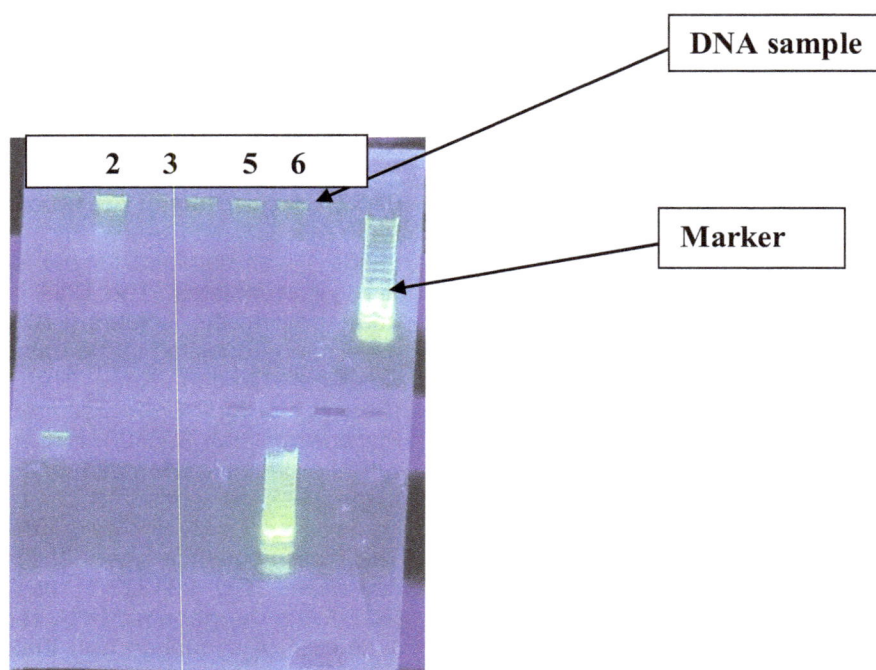

Image 1. Agarose gel electrophoresis of RAPD/PCR products of the four DNA bands (EAM 2, EAM 3, EAM 5 and EAM 6) viewed under the ultra-violet light.

production of bioethanol nor in the production of other useful organic compounds except as causal agents of human diseases. Ellis et al. (2007) reported *C. tropicalis* as the causal agent of candidiasis in man; they are opportunistic fungi which live in most human organs. However, recent reports by Kathiresan and Saravanakumar (2011) and Senthilraja et al. (2011) have shown that species of *Candida* are not just pathogens but can be useful tools for bioethanol production, as they were able to use *C. tropicalis and Candida albicans*

isolated from marine environment to produce bioethanol. There had also been reports of their isolation from dairy products such as yoghurt (Rohm et al., 1992) and milk (Gadaga et al., 2000). *P. caribbica* and *C. tropicalis* were isolated from *Z. mays* and *Cola acuminate*, respectively. This is probably the first report of isolation and characterization of yeasts from these substrates that can be used in fermentation for the production of bioethanol in Nigeria. Most workers had reported the use of *S. cerevisiae* for fermentation in the production of bioethanol

Table 3. Molecular identity of yeast species isolated by the rDNA sequencing and morphological characteristics.

Isolate code	Organisms	Origin	Identification (%)	IMI (Kew, UK) identification number/accession number	rDNA[a]
EAM 2	*Kluyveromyces marxianus*	Cassava	100	IMI 398399	ITS1/ITS4
EAM 3	*Pichia caribbica*	Maize	100	IMI 398400	ITS1/ITS4
EAM 5	*Candida tropicalis*	Kolanut	100	IMI 398401	ITS1/ITS4
EAM 6	*Saccharomyces cerevisiae*	Guinea corn	88	GU931323.1	ITS1/ITS4

[a] Region of the rDNA gene used for identification.

(Abouzied and Reddy, 1986; Adesanya et al., 2008; Oyeleke and Jibrin, 2009). This report therefore gives an array of prospective fermentative species of yeast from locally available substrates which can be of industrial benefits. The organisms were able to degrade the carbon sources because they contain the enzymes necessary for the conversion of sugars to other products. Organism like *K. marxianus* was also able to ferment galactose which is an indication of the presence of β-galactosidase, while Rajoka et al. (2003) had earlier extracted the enzyme from *K. marxianus*. Contrary to literature reports that species of *Saccharomyces* cannot ferment lactose as they lack the enzyme lactase, the *S. cerevisiae* strain (from guinea corn) isolated in the course of this study was able to ferment the lactose used in the fermentation test. This strain could be used for the purpose of fermentation to produce alcohol and other derivatives. In yeast (fungi) taxonomy, conventional methods such as physiological and morphological analysis are not enough to adequately identify yeast especially with the emergence of new strains. Molecular identification is known to provide a more objective separation of genera and species than phenotypic analysis. DNA sequence analyses was achieved through DNA extraction using CTAB procedure, amplification of regions of rDNA/Internal Transcribed Spacer sequence (ITS) and purification of the PCR products. The amplified region was done using ITS1 and ITS4 which are recommended universal primers for fungi identifications (Trost et al., 2004). Then, the data obtained was compared with known sequence in the database of the Genbank and CABI, Royal Botanical Garden, Kew. The comparison of rDNA gene sequence is an important instrument in determining the phylogenetic and evolutionary relationship of many organisms. Ellepola et al. (2003) used this method to separate a species of *Candida* which had earlier been wrongly identified as *C. albicans* into the correct species of *Candida dubliniensis*. This research shows the genome rDNA/ITS sequences of the yeast samples to be more accurate and reliable in phylogenetic typing and identification than the conventional means.

Conclusion

This study has dealt with the isolation and characterization of four yeast strains using standard microbiological procedures. They showed interesting features such as extra cellular enzyme and fermentation capability which facilitate the opportunity for identification of the yeasts. The result of this study indicated that these indigenous yeasts, isolated from food crops showed good fermentation attributes, which could enhance ethanol yield that would contribute to the cost effective role in the production of bioethanol and enzymes of industrial importance; hence, increasing the varieties of yeast and decreasing its importation.

REFERENCES

Abouzied M, Reddy CA (1986). Direct fermentation of potato starch to ethanol by coculture of *Aspergillus niger* and *Saccharomyces cerevisiae*. Appl. Environ. Microbiol. 52(5):1055-1057.

Adesanya OA, Oluyemi KA, Josiah SJ, Adesanya RA, Ofusori DA, Bankole MA, Babalola GB (2008). Ethanol production by *Saccharomyces cerevisiae* from cassava peels hydrolysate. Internet J. Microbiol. 5:1.

Barnett JA (1975). The entry of D-ribose into some yeasts of the genus *Pichia*. J. Gen. Microbiol. 90(1):1-12.

Chanchaichaovivat A, Ruenwongsa P, Panijpan B (2007). Screening and identification of yeast strains from fruits and vegetables: Potential for biological control of postharvest chilli anthracnose (*Colletotrichum capsici*). Biol. Contr. 42:326–335.

Ellepola ANB, Hurst SF, Elie CM, Morrison CJ (2003). Rapid and unequivocal differentiation of *Candida dubliniensis* from other *Candida* species using species-specific DNA probes: Comparison with phenotypic identification methods. Oral Microbiol. Immunol. J. 18(6):379-388.

Ellis D, Davis S, Alexiou H, Handke R, Bartley R (2007). *Description of Medical Fungi*. Mycological Unit, Women's and Children's Hospital, North Adelaide, p. 198.

Gadaga T, Mutukumira HAN, Narvhus JA (2000). Enumeration and identification of yeasts isolated from Zimbabwean traditional fermented milk. Int. Dairy J. 10:459-466.

Kathiresan K, Saravanakumar SK (2011). Bio-ethanol production by marine yeasts isolated from coastal mangrove sediment. Int. Multidiscipl. Res. J. 1(1):19-24.

Kurtzman CP, Piskur J (2006). Taxonomy and phylogenetic diversity among the yeasts. In: Sunnerhagen, P. and Piskur, J. (Eds). Comparative Genomics: Using Fungi as Models. Berlin: Springer-Verlag, Berlin, pp. 29-46.

Martini A (1992). Biodiversity and conversation of yeasts. Biodiver. Conserv. 1(4):324-333.

Mohd AK, Loh SK, Nasrin A, Astimar A, Rosnah MS (2011). Bioethanol production from empty fruit bunches hydrolysate using *Saccharomyces cerevisiae*. Res. J. Environ. Sci. 5(6):573-586.

Oelofse A, Pretorius IS, du Toit M (2008). Significance of *Brettanomyces* and *Dekkera* during winemaking: A synoptic review.

S. Afr. J. Enol. Viticult. 29(2):128–144.

Ocloo FCK, Ayernor GS (2010). Production of alcohol from cassava flour hydrolysate. J. Brew. Distill. 1(2):15-21.

Olutiola PO, Famurewa O, Sonntag HG (2000). Introduction to General Microbiology A Practical Approach. Bolaby Publication, Lagos, p. 267.

Oyeleke SB, Jibrin NM (2009). Production of bioethanol from guinea corn husk and millet husk. Afr. J. Microbiol. Res. 3(4):147-152.

Qing F, Shiping T (2000). Postharvest biological control of Rhizopus rot of nectarine fruits by *Pichia membranefaciens*. Plant Dis. 84(11):1212–1216.

Rhode B, Hartman G (1980). Introductory Mycology by examples. Schering Aktiengellshaft, Hamburg, p. 140.

Rohm H, Eliskases-Lechner F, Brauer M (1992). Diversity of yeasts in selected dairy products. J. Appl. Bacteriol. 72:370-376.

Rajoka MI, Khan S, Shahid R (2003). Kinetics and regulation studies on the production of β-galactosidase from *Kluyveromyces marxianus* grown on different substrates. Food Tech. Biotech. 41:315-320.

Seeliger HPR (1956). Use of a urease test for the screening and identification of *Cryptococci*. J. Bacteriol. 72(2):127–131.

Senthilraja P, Kathiresan K, Saravanakumar K (2011). Comparative analysis of bioethanol production by different strains of immobilized marine yeast. J. Yeast Fungal Res. 2(8):113–116.

Sreenivas Rao RS, Prakasham K, Krishna Prasad S, Rajesham PN, Sarma L, VenkateswarRao (2004). Xylitol production by *Candida* sp : Parameter optimization using Taguchi approach. Proc. Biochem. 39:951-956.

Trost A, Graf B, Eucker J, Sezer O, Possinger K, Göbel UB, Adam T (2004). Identification of clinically relevant yeasts by PCR/RFLP. J. Microbiol. Meth. 56(2):201-211.

28

A new ethanol-based macrochemical test combined with a cultural character in the process of identification of the cosmopolitan wood-decayer, *Ganoderma resinaceum* Boud. (Basidiomycota)

Dominique Claude MOSSEBO[1,2], Rose T. AMBIT[1], Marie-Claire MACHOUART[3] and Germain KANSCI[4]

[1]Mycological Laboratory, University of Yaoundé 1, B.P. 1456 Yaoundé, Cameroon.
[2]Mycology Section, Jodrell Laboratory, Royal Botanic Gardens, Kew, Richmond, Surrey, TW9 3AB, England.
[3]Laboratoire Stress Immunité Pathogène – EA 7300 – Parasitologie-Mycologie – Université de Lorraine, 9 avenue de la forêt de Have, 54511 Vandoeuvre les Nancy, France.
[4]Department of Biochemistry, Faculty of Science, University of Yaoundé 1, B.P. 812 Yaoundé, Cameroon.

A new macrochemical test using ethanol drops was set up and described here as a safer, quicker and more reliable substitute for the previously used match flame to reveal yellow resin on the pileus of *Ganoderma resinaceum*, no matter its geographical origin. Four concentrations (30, 70, 90 and 99%) of ethanol (CH_3-CH_2OH) were tested with distilled water as negative control, as a substitute to the old match flame test in the process of identification of this species. The positive control test was performed on 18 other species of *Ganoderma* including *Ganoderma lucidum*. All control tests were negative, ethanol concentrations ranging between 90-99% revealed a ± bright and lasting yellowish resin oozing from the pileus of *G. resinaceum*. Observations from laboratory cultures showed that in this genus, only mycelium of *G. resinaceum* so far turns yellowish as earlier established in other studies on strains of the species identified at molecular (ITS-rDNA) level. Therefore, in this very wide genus where the boundaries between numerous species are still poorly circumscribed, the new positive ethanol test combined with the occurrence of yellowish zones in mycelial cultures bring more accuracy in the identification process of *G. resinaceum,* prior to confirmation by additional taxonomic investigations.

Key words: *Ganoderma resinaceum*, identification process, ethanol, pileus, mycelial culture, yellow resin.

INTRODUCTION

Ganoderma Karst with over 250 species (Chang and Buswell, 1999; Ryvarden, 1992) is a cosmopolitan genus recorded worldwide; in tropical as well as in temperate climates. It belongs to the family Ganodermataceae with key characters, the shape and size of basidiospores and the texture of the pileus (Furtardo, 1965; Ryvarden, 2000). Several species are found in tropical Africa and numerous studies (Bresadola, 1890; Futardo, 1965; Hjortstam et al., 1993; Kengni Ayissi and Mossebo, 2014a; Kengni Ayissi et al., 2014b; Kinge and Mih, 2014; Kinge, 2012; Moncalvo and Ryvarden, 1997; Mossebo et al., 2014; Roberts and Ryvarden, 2006; Ryvarden and Johansen, 1980; Steyaert, 1967, 1972, 1980; Zoberi, 1972) have so far been carried out in this genus due to its importance in various scientific domains such as agriculture, forestry pathology and medicine.

In spite of other numerous studies in the taxonomy and molecular phylogeny carried out in the genus *Ganoderma* (Adaskaveg and Giltbertson, 1986; Furtardo, 1967; Kinge and Mih, 2011; Mohanty et al., 2011; Moncalvo et al., 1995; Ryvarden, 2004) also known as wood decay fungi causing white rot, the boundaries between the over 250 taxa of *Ganoderma* so far described are still poorly circumscribed and not universally accepted due to variations and inconsistencies in macro- and micromorphological characters of several species including *Ganoderma resinaceum* which was termed by Steyaert (1980) with *Ganoderma parvelum* as a complex due to these variations and inconsistencies. With regards to *G. resinaceum* in particular, an extensive study carried out by Kengni Ayissi and Mossebo (2014), Kengni Ayissi et al. (2014b) and Mossebo et al. (2014) on a large number of specimens collected in several countries of tropical Africa, showed that specimens of this species are highly variable in their macro- and micro-morphology, size and colour. The above mentioned studies brought more light on these variations and inconsistencies and most importantly clues for a better identification of potential collections of this species which is cosmopolitan considering that it grows as well in the cold northern hemisphere and the generally hot tropical climates of the southern hemisphere.

Etymologically and according to Boudier (1889) who first described *G. resinaceum*, the specific epithet "*resinaceum*" means "resinous" and refers to a hard-setting sticky liquid which is resin oozing mostly from fresh damaged or scratched basidiomes. According to some specific peer-reviewed URL for *Ganoderma* [(www.first-nature.com/fungi/ganoderma-resinaceum.php) in Sept. 2014], this liquid is confirmed as being yellow and oozes from the fungus when it is cut, before setting rapidly to form a shiny varnished pileus surface. Contrary to the above mentioned features, other peer reviewed URL [(www.mycocharentes.fr) in March 2011] of macrofungi from western Europe show numerous clearly identified specimens of *G. resinaceum* with a laccate pileus, however showing no shiny varnished appearance and no yellow resin oozing on the pileus, feature also rare or inconspicuous on specimens from the tropics (Kengni Ayissi and Mossebo, 2014a; Kengni Ayissi et al., 2014b; Mossebo et al., 2014). *G. resinaceum* is also

described in the above mentioned URL as showing on scratch a yellowish layer made of resin and about this resin layer, Ryvarden and Melo (2014), Ryvarden (2000, 2004), Nũnez and Ryvarden (2000) and Breitenbach and Kränzlin (1986) said that "basidiocarps are with age more reddish brown to bay and dull due to an excreted resinous layer that becomes yellowish when crushed and melts in a match flame".

The above mentioned remarks made by various authors presume that in addition to its laccate and shiny varnished pileus, *G. resinaceum* could be readily identified even on the field of collection just by cutting or scratching the pileus looking for a yellow layer underneath or rather a yellow liquid setting on fresh basidiocarps and reported as melting in a match flame. However, trials carried out on pileus of numerous specimens of *G. resinaceum* (Ambit, 2011; Kengni Ayissi and Mossebo, 2014a; Mossebo et al., 2014) collected in several tropical countries of central Africa (Table 1) showed that the above mentioned taxonomic test (cutting or scratching pileus to let resin flow over the pileus or using a match flame to melt it on the pileus) used to reveal resin hardly work on these tropical specimens. In fact, the monitoring of several tropical specimens - identified by authors (Kengni Ayissi and Mossebo, 2014a; Mossebo et al., 2014) and cross-checked by Ryvarden (University of Oslo, Norway, personal communication) as being *G. resinaceum* - shows that they barely or not at all excrete a resinous layer as it rarely appears on the pileus of very few strains of the specimens collected. This leads to the conclusion that resin and its above mentioned features used as an indicator in the process of identification of *G. resinaceum* very likely apply mostly on specimens growing in climates of the northern hemisphere where most of the descriptions are done, and rather rarely or inconspicuously on specimens growing in hot climates of tropical Africa and the southern hemisphere in general. Therefore, considering all the above mentioned difficulties in conspicuously observing yellow resin on the pileus of *G. resinaceum* and the inconveniences in using the "match flame" in order to reveal resin in the process of identification of *G. resinaceum* which is so far the only species in the genus *Ganoderma* known to produce yellow resin on the pileus, this study aims to develop other methods in order to clearly and unequivocally reveal resin on basidiomes of *G. resinaceum* from tropical Africa (south of Sahara) and more generally from all regions of the world, irrespective of the climate and environment in which these basidiomes appear.

About the cultural characters, Mohanty et al. (2011) described the first record of *G. resinaceum* and *G. weberianum* from north India based on ITS-rDNA molecular phylogeny and reported that only *G. resinaceum* showed yellowish zones on mycelium in artificial cultures, whereas this yellowish colour was reported by the authors to be totally absent in cultures of *G. weberianum* and *G. lucidum*, the latter being the reference species in the genus *Ganoderma*. Therefore, inspired by Mohanty et al. (2011),

Table 1. Geographical coordinates (GPS) of collection areas, collection date and growth substrates of specimens of *G. resinaceum* tested (Source: Kengni Ayissi and Mossebo, 2014a).

S/N	Herbarium number	Date of collection	Area of collection *(i)*	GPS coordinates	Substrate
1	HUY1-DM 45B	28/06/1996	Yaoundé/Cameroon	N 03° 52' 21" E 11° 31' 03"	Stump of angiosperm
2	HUY1-DM 47D	15/07/1996	Yaoundé/Cameroon	N 03° 52' 21" E 11° 31' 03"	Stump of oil palm tree (*Elaeis guineensis*)
3	HUY1-DM 72(ART)	3/10/2009	Mbengwi/Cameroon	N 05° 55' 18" E 10° 08' 32"	Stump of angiosperm
4	HUY1-DM 85	26/08/1996	Campus University of Yaoundé 1 (UY1)/Cameroon	N 03° 51' 31,6" E 11° 29' 59"	Stump of angiosperm
5	HUY1-DM 104	13/07/1997	UY1/Cameroon	N 03° 51' 31,6" E 11° 29' 59	Stump of angiosperm
6	HUY1-DM 105B	23/05/2000	Yaoundé/Cameroon	N 03° 52' 26" E 11° 32' 06	Stump of angiosperm
7	HUY1-DM 418	18/11/2004	Yaoundé/Cameroon	N 03° 52' 26" E 11° 32' 06"	Tree trunk in decay
8	HUY1-DM 506	07/04/2007	Dja Biosphere reserve/Cameroon	N 03° 23' 39" E 12° 43' 25"	Tree trunk in decay
9	HUY1-DM 509	08/04/2007	Dja Biosphere reserve/Cameroon	N 03° 23' 39" E 12° 43' 25"	Tree trunk in decay
10	HUY1-DM 524	12/05/2011	Yaoundé/Cameroon	N 03° 51' 38" E 11° 30' 07"	Stump of angiosperm
11	HUY1-DM 538	10/04/2007	Dja Biosphere reserve/Cameroon	N 03° 23' 39" E 12° 43' 25"	Tree trunk in decay
12	HUY1-DM 612A	12/07/2009	Yaoundé/Cameroon	N 03° 52' 28" E 11° 31' 04"	Stump of angiosperm
13	HUY1-DM 612B	June 2008	Yaoundé/Cameroon	N 03° 52' 28" E 11° 31' 04"	Tree trunk in decay
14	HUY1-DM 612 C	12/07/2009	Yaoundé/Cameroon	N 03° 52' 28" E 11° 31' 04"	Stump of angiosperm
15	HUY1-DM 617	13/06/2008	UY1/Cameroon	N 03° 51' 31,6" E 11° 29' 59	Stump of angiosperm
16	HUY1-DM 619A	12/06/2008	Yaoundé/Cameroon	N 03° 51' 23" E 11° 31' 08"	Tree trunk in decay
17	HUY1-DM 619B	13/06/2008	Yaoundé/Cameroon	N 03° 51' 28" E 11° 30' 09"	Stump of angiosperm
18	HUY1-DM 619C	13/06/2008	Yaoundé/Cameroon	N 03° 51' 28" E 11° 30' 09"	Stump of angiosperm
19	HUY1-DM 619 E	08/03/2013 21/07/2013	UY1/Cameroon	N 03° 51' 31,6" E 11° 29' 59	Stump of angiosperm
20	HUY1-DM 622	15/06/2008	Lobaye Forestry domain/ RCA	N 04° 22' 20" E 19° 27' 18"	Tree trunk in decay
21	HUY1-DM 629	09/06/2008	Yaoundé/Cameroon	N 03° 52' 21" E 11° 31' 03"	Tree trunk in decay
22	HUY1-DM 660	13/06/2008	Yaoundé/Cameroon	N 03° 52' 21" E 11° 31' 03"	Stump of angiosperm
23	HUY1-DM 696	20/08/2009	Yaoundé/Cameroon	N 03° 51' 33" E 11° 31' 02"	Trunk of oil palm tree (*Elaeis guineensis*) in decay
24	HUY1-DM 709	23/08/2011	Kpangbala Ndeke in RCA	N 04° 27' 17" E 19° 32' 28"	Tree trunk in degradation

Table 1. Contd

25	HUY1-DM 711	21/08/2008	Lobaye Forestry domain in RCA	N 04° 22' 20" E 19° 27' 18"	Roots of an angiosperm stump in decay
26	HUY1- DM 784	23/03/2013	Ipassa-Makokou in Gabon	N 0° 30' 05" E 12° 47' 42"	Tree trunk in dacay
27	HUY1-DM 785	26/03/2013	Ipassa-Makokou in Gabon	N 0° 30' 11" E 12° 47' 40.5"	Tree stump
28	HUY1 -DM 786	27/04/2013	Kisangani in Congo (Kinshasa)	N 0° 31' 59" E 25° 1146.14"	Stump of *Elaeis guineensis*
29	HUY1-DM 731	30/01/2012	Yaoundé/Cameroon	N 03° 52' 21" E 11° 31' 03"	trunk of a oil palm tree (*Elaeis guineensis*) in decay
30	HUY1-DM 732	22/11/2011	Yaoundé/Cameroon	N 03° 52' 21" E 11° 31' 03"	Stump of angiosperm
31	HUY1-DM 750	23/04/1998	UY1/Cameroon	N 03° 51'31,6" E 11° 29' 59	Stump of angiosperm
32	HUY1-DM 760A	26/05/2012	Yaoundé/Cameroon	N 03° 52' 21" E 11° 31' 03"	Stump of angiosperm
33	HUY1-DM 763	03/05/2011	Yaoundé/Cameroon	N 03° 52' 21" E 11° 31' 03"	Stump of angiosperm
34	HUY1-DM 764	06/10/2012	UY1/Cameroon	N 03° 51'31,6" E 11° 29' 59	Trunk of a living angiosperm
35	HUY1-DM 769	11/11/2009	Mbengwi/Cameroon	N 05° 55' 18" E 10° 08' 32"	trunk of angiosperm
36	HUY1-DM 777	08/11/2012	Nkoabang (suburbs of Yaoundé/Cameroon)	N 03° 53' 18" E 11° 32' 06"	Stump of angiosperm

(i) Collection areas were forestry domains and reserves as well as savannahs, urban centres and their outskirts showing various types of vegetations in several countries of central Africa.

mycelial culture tests on several species of *Ganoderma* from our collections were also set up in addition to the ethanol test as a taxonomic indicator in the process of identification of *G. resinaceum*.

MATERIALS AND METHODS

Morphological identification

Before carrying out the macrochemical test properly said, each specimen of *Ganoderma* collected was first thoroughly scrutinized in taxonomy according to the features described by Kengni Ayissi and Mossebo (2014a) in order to determine whether it actually belongs to *G. resinaceum* or to other species that could be used for the positive control test. However, some dry specimens found in the Herbarium of the Mycology Section of the Royal Botanic Gardens in Kew had been previously identified after collection in UK as *G. resinaceum* by Kew taxonomists and therefore used as such in the test as strains from temperate (cold) countries. For tropical specimens collected mostly in central Africa, the macro- and micro-morphological features were first described according to various protocols used for polypores description (Mossebo and Ryvarden, 1997, 2003; Mossebo, 2005; Mossebo et al., 2007) and more specific protocols for *Ganoderma* (Kengni Ayissi and Mossebo, 2014a; Kinge and Mih, 2014; Kinge and Mih, 2011; Kinge, 2012; Mohanty et al., 2011;

Mossebo et al., 2014; Gilbertson and Ryvarden, 1986; Ryvarden and Johansen, 1980; Ryvarden, 2000, 2004; Nûnez and Ryvarden, 2000). These features were thereafter compared with the summary presented in Table 3 and to those of existing taxa of *Ganoderma* described in the most reliable taxonomic studies (Kengni Ayissi and Mossebo, 2014a;Kinge, 2012; Mohanty et al.,2011; Mossebo et al., 2014; Nûnez and Ryvarden, 2000; Ryvarden and Johansen,1980; Ryvarden, 2000, 2004), in order to determine whether the specimen belonged to *G. resinaceum*. Double of our specimens were sent to Ryvarden (University of Oslo, Norway, personal communication) to cross-check our preliminary identifications. They were thereafter registered in the Herbarium of the Department of Botany of the University of Oslo with voucher material conserved in the mycological herbarium of the University of Yaoundé 1 in Cameroon under the HUY1-DMx herbarium number (Table 1).

Macrochemical test using ethanol to reveal resin on basidiomes of *Ganoderma resinaceum*

Freshly collected basidiocarps or exsicatta of tropical strains of *G. resinaceum* described by Kengni Ayissi and Mossebo (2014a) (Table 1) were used in the test inspired from Charbonnel (1995) who reported ethanol test on the pileus of some Agaricales species for taxonomic purposes. Some other collections [K(M) 21513-7; K(M) 21513-18; K(M) 21513-47] of *G. resinaceum* from UK found at the herbarium of the Mycology Section of the Royal Botanic

Gardens in Kew-UK were also tested as strains from temperate (cold) countries. The pileus as test surface was first cleaned whenever necessary, using a clean piece of fabric or sponge in order to get rid of dust, spore prints or any dirty remains (soil, grass, insects etc) from the collection area. Thereafter, four concentrations (30, 70, 90 and 99%) of ethanol (CH_3-CH_2OH) were prepared and filled in plastic droppers of 30 ml each. The 0% concentration was simple distilled water used as negative control. The positive control consisted of tests of the same concentrations of ethanol carried out on eighteen (18) species of Ganoderma collected over the study period. They were gradually collected mostly in the tropics in the same collection areas as mentioned in Table 1. Seven (7) were not clearly identified at species level whereas the 11 others were clearly identified as Ganoderma australe, Ganoderma applanatum, Ganoderma baudonii, Ganoderma carocalacreus, Ganoderma colossum, Ganoderma hildebrandii, Ganoderma lobenense, Ganoderma ryvardense, Ganoderma weberianum, Ganoderma zonatum and Ganoderma lucidum which is the reference species in this genus. It must be underscored that the 7 unidentified species were included in the positive control test because in the process of their identification, although their specific names were not clearly determined, most of their macroscopic and microscopic features as well as the preliminary BLAST SEARCH of their ITS-rDNA sequences (not shown here) were first entirely different from each other, and all different from those of G. resinaceum and the 11 other species tested as positive control. In order to perform the test, the test surface was first virtually subdivided in two equal parts using a special design on a piece of white paper on which two windows were cut opened as presented on Figures 1A, B, F and 1G, one designed for the ethanol test and the other for water control. The paper design was thereafter attached to the pileus using a tailor needle so as to show the two test windows ready to receive drops of the reagent. Twenty to thirty drops of a given concentration (30, 70, 90 and 99%) of ethanol were thereafter poured on the "test area" (test surface) of the pileus and the same number of drops of distilled water (0% ethanol) on the "water control area" (Figures 1A, B, F and 1G) of the same pileus. For each collection, three (3) replicates of each test were performed using 3 different basidiomes of the collection and the test was gradually upon collection, extended to all specimens presented in Table 1 and other more recent collections used as positive control. The basidiomes tested were thereafter closely monitored. In the case of positive reaction, resin was revealed on fresh basidiomes either as a yellowish to yellow spot (Figures 1B and 1G) clearly occurring 5 to 15 min after pouring the drops on the "test area" of the pileus. On exsiccata, the test surface either gradually changed to show just a ± conspicuously shiny varnished sticky liquid (Figure 1A), or a shiny varnished spot oozing resin as a yellow sticky liquid (Figure 1C). For each test carried out, the resin colour (Figures 1B, C and 1G) recorded was coded according to the colour chart of Kornerup and Wansher (1978) and for the positive control, the basidiomes of G. lucidum (Figure 1E and F) and other 17 fresh and dry basidiomes of other species were handled in the same manner as described above for G. resinaceum.

These series of tests were launched in Cameroon in 2005, continued in 2006-2007 at the Mycology Section of the Royal Botanic Gardens in Kew (UK), and thereafter extended, cross-checked and finalized from 2008 to 2014 at the University of Yaoundé 1 in Cameroon as specimens (Table 1) were gradually collected or received from colleagues from other countries.

Test of resin production in mycelial culture of *G. resinaceum*

Along the ethanol test on basidiocarps of G. resinaceum, mycelium raised from tiny pieces of context-tissue from fresh basidiocarps plated on malt extract agar (MEA) according to Mossebo (2002) and Mohanty et al. (2011) was monitored (Figure 1D) for 7 to 28 days in order to find out whether mycelium in artificial culture produces or exudes resin in comparison with results of ethanol test on the same basidiocarps. As control test here, pieces of context-tissue of fresh basidiocarps of G. lucidum and most of the other species (Table 2) tested as positive control were also plated on MEA and monitored for the same time period.

RESULTS

The following key remarks could be drawn from the detailed results presented in Table 2:

1. Ethanol at concentrations ranging from 90 to 99% reveals a ± conspicuous bright and lasting yellowish to yellow resin on pileus of fresh basidiocarps of G. resinaceum.
2. On dry basidiocarps (exsiccata), the same concentrations of ethanol most often just brighten ± conspicuously the pileus and sometimes also ooze resin.
3. The bigger the concentration of ethanol between 90 to 99% concentrations, the faster and more conspicuous is the basidiocarp response to the test.
4. At a concentration of 70%, a faint and evanescent yellowish resin appears on the pileus of fresh basidiocarps of G. resinaceum.
5. Concentrations of ethanol inferior (<) to 70% do not react or barely exude a very faint and evanescent yellowish resin on pileus.
6. The above mentioned reactions are recorded in average 5 to 15 min after pileus receives drops of ethanol, but in some cases, up to 30 min could be necessary in order to observe a conspicuous and clear reaction regardless of the ethanol concentration.
7. Distilled water as negative control does not exude resin from the pileus of G. resinaceum.
8. Ganoderma lucidum, G. applanatum, G. australe, G. baudonii, G. carocalacreus, G. colossum, G. hildebrandii, G. lobenense, G. ryvardense, G. weberianum, G. zonatum and 7 other unidentified species of Ganoderma, all tested as positive control do not exude resin with ethanol at the above mentioned concentrations which on some species rather ± fade the brightness of the pileus depending on the level of concentration used.
9. Already tested spots on the pileus of a basidiocarp should not be retested and eventual additional tests on the same basidiocarp must be carried out on different spots.
10. In addition to a positive ethanol test, the occurrence of yellowish zones on parts of mycelium in culture of the presumed specimen of G. resinaceum constitutes another major step forward since it brings more accuracy in the identification process of this species, considering that, so far, only specimens that produced these yellowish zones (Figure 1D) in mycelial cultures - colour earlier described by Mohanty et al. (2011) as occuring only in G. resinaceum - also responded positively to the ethanol test, indicating a positive relationship between these two taxonomic parameters in G. resinaceum.

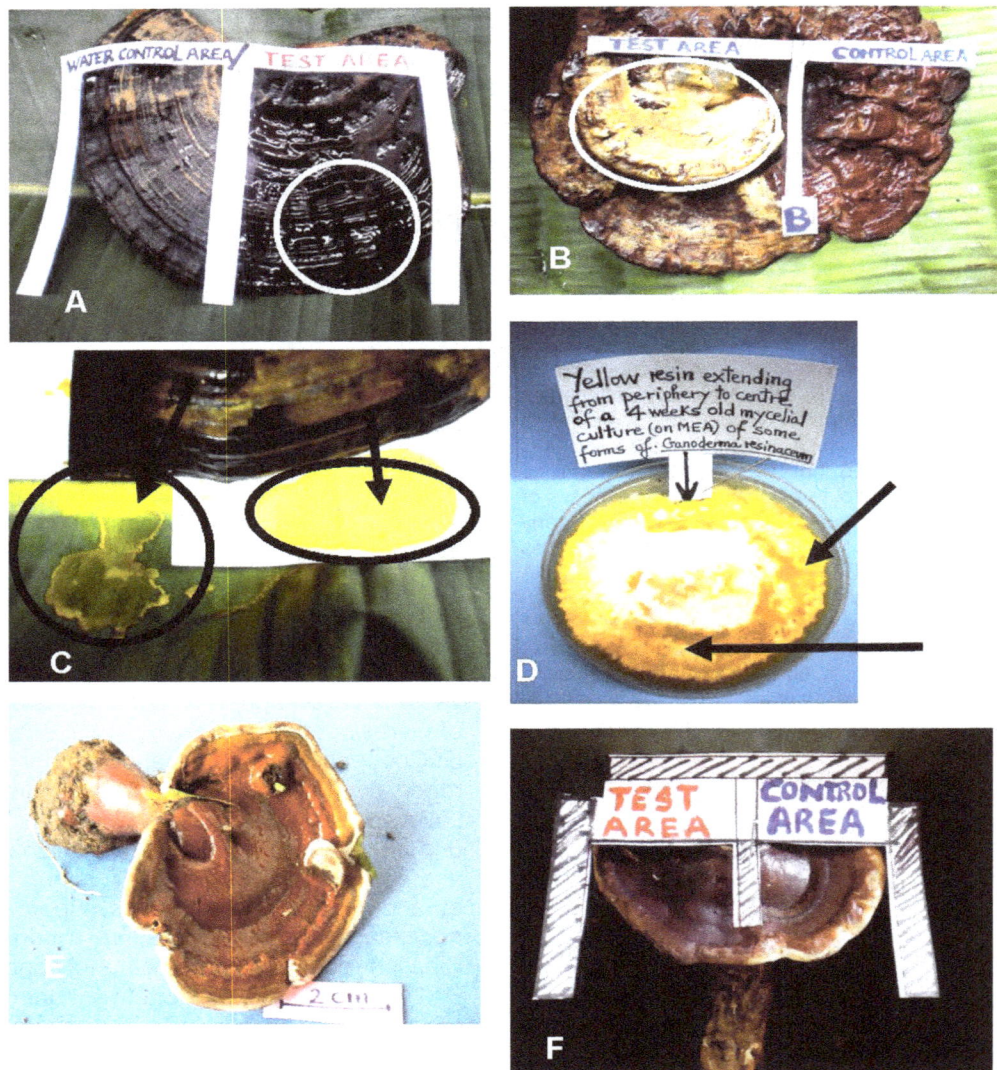

Figure 1. A: Ethanol test exuding a ± bright shiny varnished and sticky resin ("test area") on pileus of a dry basidiocarp (exsicattum) of *G. resinaceum*; **B:** ethanol test showing a spot of yellowish resin ("test area") on fresh basidiocarp of *G. resinaceum;* **C:** ethanol test exuding yellowish to yellow resin on exsicattum of *G. resinaceum;* **D:** yellowish zones occurring in mycelial cultures as a key cultural feature of *G. resinaceum*; **E:** a strain of *G. lucidum* from central Africa tested; F: a different strain of *G. lucidum* (exsicattum) tested as positive control.

The above mentioned results clearly show that ethanol could henceforth be considered as a safer, quicker and more reliable substitute to the "match flame" test previously described by several authors including Ryvarden and Melo (2014), Ryvarden (2000, 2004), Nũnez and Ryvarden (2000) and Breitenbach and Kränzlin (1986), as a reagent to reveal yellowish resin on the pileus of *G. resinaceum*. Therefore, a positive ethanol test combined with the above mentioned mycelial culture features stand as outstanding in the process of identification of *G. resinaceum*, but however still need additional macro- and micromorphological investigations of the specimens tested in order to confirm identification.

DISCUSSION

Results of this study show that positive ethanol test at concentrations ranging from 90 to 99% that reveal yellowish resin on the pileus of fresh and dry basidiocarps combined with the occurrence of yellowish zones (Mohanty et al., 2011) in mycelial cultures could be considered as an outstanding step in the process of identification of *G. resinaceum* since it brings more accuracy in this process. Both tests considered together therefore constitute a reliable taxonomic orientation test and a major progress in the taxonomy of *Ganoderma*, a genus in which the boundaries of several of the over 250

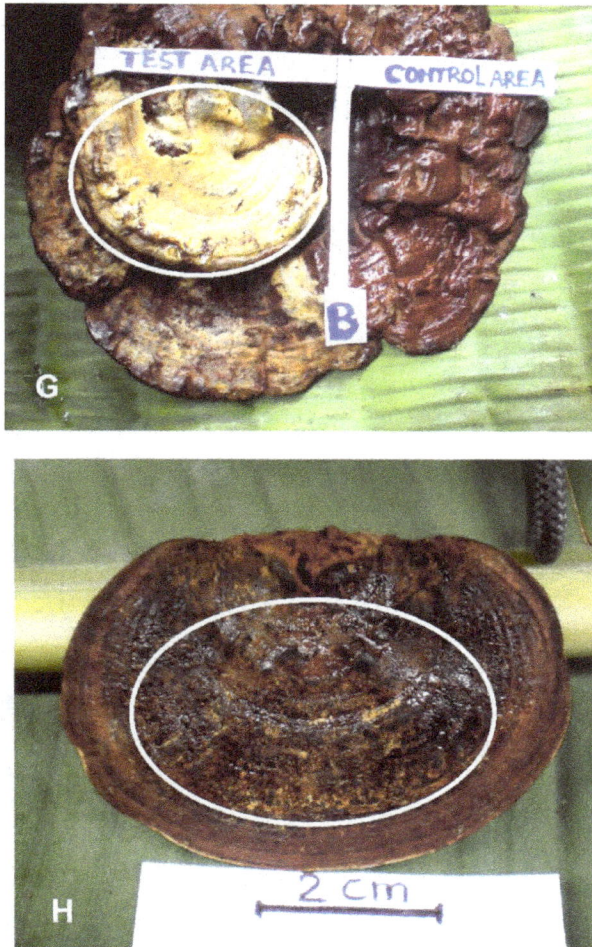

Figure 1. Contd. G: (zoom on Figure 1B) Pileus of a fresh specimen of *G. resinaceum* showing on the "control area" of the pileus surface with no trace of yellowish resin revealed rather on the "test area" by drops of ethanol at 90-99%; **H:** a different specimen of *G. resinaceum* naturally exuding on pileus and on scratch traces of coagulated yellowish resin as a key character conspicuously appearing on some strains of the species.

taxa are still poorly circumscribed in spite of numerous studies carried out by various authors (Gilbertson and Ryvarden, 1986; Kengni Ayissi and Mossebo, 2014a; Kinge and Mih, 2011, 2014; Kinge, 2012; Mohanty et al., 2011; Nŭnez and Ryvarden, 2000; Ryvarden and Johansen, 1980; Ryvarden, 2000, 2004; Steyaert 1967, 1980) in the taxonomy and molecular phylogeny of this genus. In fact, these tests save time and brings more accuracy in the process of identification of *G. resinaceum*, in that when for instance for one reason or another we are in search of *G. resinaceum* in particular among numerous specimens of *Ganoderma* collected on the field, it is only when the test reveals a ± conspicuous spot of yellowish to yellow resin on fresh specimens (Figures 1B & 1G) or a lasting shiny varnished and ±

sticky spot [Figure 1A "Test Area")] on dry sporocarps, all these combined to the occurrence of yellowish zones in mycelial cultures - the latter feature being linked to the genus only to *G. resinaceum* identified at molecular level (Mohanty et al., 2011)- that taxonomic investigations matching those described by various authors (Giltbertson and Ryvarden, 1986; Kengni Ayissi and Mossebo, 2014a; Kinge and Mih, 2014; Kinge and Mih, 2011; Kinge, 2012; Mohanty et al., 2011; Mossebo et al., 2014; Mossebo et al., 2007; Mossebo, 2005; Mossebo and Ryvarden, 1997, 2003; Nŭnez and 304 Ryvarden, 2000; Ryvarden, 2000, 2004; Ryvarden and Johansen, 1980) and summarized in Table 3 could be used to confirm identification. Otherwise, further steps in the identification process should be directed rather towards other species of *Ganoderma* different from *G. resinaceum*. The test also constitutes a major progress in taxonomic mycology because it is easier, quicker and safer to perform, and the result easier to read and interpret than the match flame so far used and not always successfully, to melt yellowish resin on the pileus as previously described by several authors including Ryvarden and Melo (2014), Ryvarden (2000, 2004), Nŭnez and Ryvarden (2000), Ryvarden and Giltbertson (1993) and Breitenbach and Kränzlin (1986). Additional tests having been carried out successfully on specimens of *G. resinaceum* from the northern hemisphere and precisely at the Herbarium of the Department of Mycology at the Royal Botanic Gardens in Kew as mentioned earlier, this new test can also be considered as fully efficient in detecting yellowish resin on potential specimens of *G. resinaceum*, irrespective of their geographical origin.

With regards to interactions between ethanol and resin, for specimens of *G. resinaceum* naturally oozing yellowish resin (Figure 1H) readily visible on parts of the pileus surface or on scratch, as for those showing no trace of yellow resin on the pileus [Figure 1A, B and 1G ("control area")], it is obvious that ethanol at 90 to 99% concentrations essentially acts as a solvent that dissolves resin either coagulated on the pileus surface (Figure 1H) or rather embedded in the cuticle cells of *G. resinaceum* and oozes it over the pileus surface [(Figures 1A, B and 1G ("test area") and Figure 1C] where it becomes conspicuously visible. Ethanol used as a solvent for resin was already earlier mentioned in some peer reviewed URL [Ref. (www.nature-energie-vitalite.com/tag/sante) in September 2014] where it is said that solvents as alcohol in general including ethanol are used to dilute "propolis" which is a kind of yellow honey-like resin produced by bees, alcohol being used here as a solvent to dilute it and easily get rid of wax and other impurities in order to obtain pure honey after filtration. Ethanol at 90% was also used by Charbonnel (1995) as a differential test between some species of *Agaricus* including *Agaricus* of the group *xanthoderma*, *Agaricus campestris*, *Agaricus arvensis*, *Agaricus silvicola* and *Agaricus romanegsii*, whereby the pileus of some species either reacted as

Table 2. Ethanol test to reveal resin on pileus and resin production in mycelial cultures of *G. resinaceum*.

Parameter		Tests results on *G. resinaceum* and other species of *Ganoderma* as positive control		
		Fresh basidiocarp of *G. resinaceum*	Dry basidiocarp (exsiccata) of *G. resinaceum*	Positive control: *G. lucidum*, *G. applanatum*, *G. australe*, *G. baudonii*, *G. carocalacreus*, *G. colossum*, *G. hildebrandii*, *G. lobenense*, *G. ryvardense*, *G. weberianum*, *G. zonatum*, *G. sp1*, *G. sp2*, *G. sp3*, *G. sp. 4*, *G. sp. 5*, *G. sp. 6*, *G. sp. 7*
Natural colour of the pileus		Variable, reddish-brown (8F5-8 and 7F7-8 to 10F8) (Figure 1B : Control Area) to bluish black (18F6-8) on parts of the lower side around the stipe, or shiny varnished brown (Figure 1H) (5E4-8, 5F6-8, 6E5-7, 6F7-8) on other specimens	Dark-grayish (10F1-2, 11F1-2, 12F1-2) to dark violaceous (16F6-8) or blackish violaceous (18F6-8) or darker (Figure 1A: Water Control Area and Figure 1C)	Bright reddish-brown (7F7-8 to 8F5-8) (Figure 1E/1F: Control Area) for dry basidiocarps of *G. lucidum*; different colours for other species
Results of ethanol tests	30%	Natural colour of pileus unchanged	Natural colour of pileus unchanged	Natural colour of pileus unchanged
	70%	Pileus slightly brightening and most often showing exudates of faint and evanescent yellowish (2A2-4) resin	Pileus brightening just slightly and occasionally showing exudates of faint and evanescent yellowish (2A2-4) resin	Natural colour of pileus unchanged or ethanol slightly degrades pileus colour or brightness without revealing yellow resin
	90%	Pileus ± conspicuously brightening (Figure 1A: Test Area) and most often also oozing exudates of bright and lasting yellowish to yellow (3B6-8 to 3B7-8) resin (Figure 1B & G: Test Area)	Pileus ± conspicuously brightening, and sometimes oozing ± sticky exudates (Figure 1A: Test Area) of yellowish to yellow resin (2B6-8) (Figure 1C)	No exudates on pileus rather absorbing ethanol without colour change or ethanol slightly degrades pileus colour or brightness without revealing yellow resin, therefore absent in *G. lucidum* (Figure 1E/1F: Test Area) and other 17 species tested as positive control
	99%	Natural colour of pileus unchanged	Natural colour of pileus unchanged	Natural colour of pileus unchanged
Water control test	Distilled water	Natural colour of pileus Unchanged	Natural colour of pileus unchanged	Natural colour of the pileus Unchanged
Mycelial culture		Mycelium exudes yellowish to yellow (3A7-8) resin in 2 to 4 weeks old cultures (Figure 1D)	///	No exudates of yellow resin. Mycelium remains mat white to cream in 2 to 4 weeks old cultures.

Figures combined with alphabets and dashes in brackets refer to the colour code in Kornerup and Wansher (1978).

yellow, yellowish or red to ethanol drops and that of other species did not react at all. Furthermore and according to the biochemist of the team of co-authors and to some specialists (Matthews and Mumford, personal communication at Silwood Park station of the Imperial College of Science and Medicine in London) in organic chemistry question on this issue in the UK during our research stay at Kew, resin is said to contain numerous resin acids which are acid components of resin with high molecular weight and which are insoluble in water and soluble in organic solvents

Table 3. Summary of the key macroscopic and microscopic features of *G. resinaceum* (Source: Kengni Ayissi and Mossebo, 2014a).

I- Key macroscopic features

-Basidioma (6)8 – 40 × 4 – 30 × 0,8 – 7 cm, most often semi-circular, sometimes dimidiate, rarely fan-shaped to spathuliform with a ± extended stalked base, or exceptionally imbricate around a central stalk.

-Pileus colour appearing as a succession of: whitish (5A1 to 6A1) (± thin or absent)/yellowish to shiny orange-yellow (5A3-4 to 6C4-6) (± thin or absent) / hiny orange-reddish-brown to violaceous-brown (8F8 to 11F8), from margins towards the ± attachment point to the substrate on young basidioma, sometimes covered by a brownish to chocolate-brown (8F5-7) spore powder and most often turning to a fading brownish to dark brown colour (7E6-8 to 7F7-8) with blackish tints on ageing basidioma; or the pileus sometimes rather shows a ± uniform brown (7F7-8 to 8F8 or darker), brownish or reddish brown (8F7-8 to 11F8) colour sometimes appearing rather as a succession of : brownish to brown (16F4-6) (± thin or absent) / dark-violaceous (16F6-8 or darker) or blackish-violaceous (18F6-8 or darker), from margins towards the attachment point to the substrate; sometimes showing ± conspicuously traces of coagulated yellowish resin over the pileus surface of rare specimens, or rather yellowish resin layer underneath revealed on scratch . Pileus sometimes rather uniformly coloured dark-violaceous (16F6-8 or darker) to blackish-violaceous (18F6-8 or darker), or rather showing on young basidioma an additional ± thin whitish margins; concentric zones ± conspicuous, sometimes crossed by radial ridges. Pileus consistency of mature sporocarps relatively hard under finger press, not soft, neither spongy as in some species of *Ganoderma*.

-Pore surface most often whitish (2A1-2) (or brownish on finger touch or when bruised) on fresh basidioma (sometimes brownish or yellowish on dried basidioma), pores circular, 4 – 5 (6) per/mm.

-Junction between pore surface and pileus margins either: rounded to applanate on mature basidioma, generally not clearly limited as well on young as on mature basidioma, and showing either an inconspicuous single line bordering both parts, or most often rather two merging structures, or: clearly limited on most mature basidiocarps by a single or several (2 to 8) ± conspicuous lines bordering both parts.

-Context built either in a distinct two layers profile with an upper 1 – 3 cm light brown (5E8 to 7E7-8) layer and a 0.5 – 1.3 cm darker, wood to chocolate-brown (7F6-8) lower layer, both layers however showing ± vertically oriented mycelial strands, or rather built in a single (0.5 – 1.8 cm thick) light brown to brown (8E4-5 to 9F5-6) layer, but sometimes showing underneath a very thin (≈ 1 mm) whitish brown (6C3 to 8C2 or 8D2-3) layer corresponding to the pore surface whenever the colour of the latter differs from that of the tube layer

-Tube layer generally a pale to greyish brown (7F4-6 on fresh basidioma and 7D5-6 on exsicatta) upper layer (0.2 – 1.3 cm thick) and a very thin (≈ 1 mm) whitish brown (6C-3) lower layer corresponding to the pore surface

II- Key microscopic features

-Basidiospores (5) 8 – *11.15* – 12 (18 – 20) × (4.5) 6 – 7 – 8 (11 – 12) µm, ± densely echinulate, ellipsoid or narrowly to broadly ellipsoid (Qm = 1.46 – *1.6* – 1.68) with truncated and non-truncated apices , some showing a ± conspicuous apiculus with oil drop(s) in others.

-Cuticle cells (14) 15 – 50 (60) × (5) 7 – 10 (15) µm, polymorphic, mostly clavate or subcylindrical to cylindrical, apically lobed in some specimens and showing ± numerous lateral and/or apical outgrowths and protuberances in others, thick-, thin-, or thin- and thick-walled with large and narrow lumen, sometimes dichotomously branched from the base.

-Hyphal system dimitic, generative hyphae hyaline, thin-walled, 1.6 – 5.2 µm diam. with clamp connections at septa level; skeletal hyphae abundant, ± branched, thick-walled, clampless

Figures combined with alphabet letters and dashes in brackets refer to the colour code in Kornerup and Wansher (1978).

such as ethanol (CH₃-CH₂OH). Their affirmation matches our results, since it clearly explains the negative reactions observed in all control tests carried out with distilled water and the positive reactions obtained with ethanol (Table 2). It is however worth mentioning that resinous-like pigments or secondary metabolites in some specimens different from *G. resinaceum* might eventually also react with ethanol at the above mentioned concentrations. However, in this case and whatever is the concentration of ethanol used, these reactions are generally faint to very faint and/or fugacious and evanescent, disappearing almost entirely on the pileus shortly after the reaction, be it with fresh or dry basidiocarps and should therefore not be confused with the conspicuous positive reactions presuming *G. resinaceum* as shown in Figures 1A, 1B, 1G ("test area") and 1C. This accounts for the necessity to first obtain both positive tests, that is, a positive ethanol test on pileus and yellowish zones in mycelial cultures (Mohanty et al., 2011), before confronting the taxonomic descriptions to existing literature in order to confirm identification of *G. resinaceum*.

Conflict of interests

The authors did not declare any conflict of interest.

ACKNOWLEDGEMENTS

The authors are very grateful to emeritus Professor Leif Ryvarden of the University of Oslo in Norway for cross-checking the preliminary identifications of *G. resinaceum* and other species of *Ganoderma* used in the macrochemical tests. The first author is particularly grateful to the Commonwealth Scholarship Commission (CSC) for the grant of a Commonwealth Fellowship [CMCF-2006-21] thanks to which part of this research work and others were carried out at the Mycology Section of the Royal Botanic Gardens, Kew, UK. The first author is also very grateful to Dr. Brian Spooner, the then head of the Mycology Section at Kew for all the logistical support put at the authors disposal during the research stay in Kew, UK.

REFERENCES

Adaskaveg JE, Giltbertson R (1986). Cultural studies and genetics of sexuality of Ganoderma lucidum and G. tsugae in relation to the taxonomy of the G. lucidum complex. Mycologia 78:694-705. http://dx.doi.org/10.2307/3807513

Ambit RT (2011). Contribution to the taxonomic and ethnomycological study of Macromycetes growing in Mbengwi (North-West region) and surrounding areas in Cameroon. M.Sc thesis, Faculty of Science, University of Yaoundé 1, Cameroon, 87 p.

Boudier EJL (1889). Etude descriptive d'une nouvelle espèce de Ganoderma de France: Ganoderma resinaceum sp.nov. Bull. Soc. Mycol. Fr. 5:72.

Breitenbach J, Kränzlin F (1986). Champignons de Suisse. Tome 2. Champignons sans lames : Hétérobasidiomycètes, Aphyllophorales, Gastéromycètes. Edt. Mykologia, CH-6000 Lucerne 9. 412 p.

Bresadola J (1890). Fungi kamerunenses. Bull. Soc. Mycol. Fr. 6: XXXII- XLIX.

Chang ST, Buswell JA (1999). Ganoderma lucidum (Curt.; Fr.) P. Karst. (Aphyllophoromycetideae). – A mushrooming medicinal mushroom. Int. J. Med. Mushroom 1:139-146. http://dx.doi.org/10.1615/IntJMedMushrooms.v1.i2.30

Charbonnel J (1995). Les réactifs mycologiques. Tome 1. Les réactifs macrochimiques. Edité par l'auteur. 344 p.

Furtardo JS (1965). Relation of microstructure to the taxonomy of the Ganodermataceae (Polyporaceae) with special reference to the structure of the cover of the pilear surface. Mycologia 57: 688-711.

Furtardo JS (1967). Some tropical species of Ganoderma (Polyporaceae) with pale context. Persoonia 4:379-389.

Gilbertson RL, Ryvarden L (1986). North American Polypores, Vol. 1. Abortiporus–Lindtneria. Fungiflora, Oslo, Norway. 433 pp.

Hjortstam K, Ryvarden L, Watling R (1993). Preliminary checklist of non-agaricoid macromycetes in the Korup National Park, Cameroon and surrounding area. Edinb. J. Bot. 50 (1):105-119. http://dx.doi.org/10.1017/S0960428600000743

Kengni Ayissi MB, Mossebo DC, Machouart MC, Kansci G, Tsigaing TF, Dogang LR, Metsebing BP, Djifack NM (2014b). A new method by correlation to forecast the optimal time of spore-prints production and collection on sporocarps of Ganoderma resinaceum Boud. (Basidiomycota) on natural substrate. Mycosphere 5(6): 758-767.

Kengni Ayissi MB, Mossebo DC (2014a). Some noteworthy taxonomic variations in the complex wood-decayer Ganoderma resinaceum

(Basidiomycota) with reference to collections from tropical Africa. Kew Bull. 69(4):1-14. http://dx.doi.org/10.1007/s12225-014-9542-9

Kinge TR (2012). Basal Stem Rot Disease of Oil palm and Identification of species of Ganoderma in South Western Cameroon. Ph.D Thesis, University of Buea, Cameroon. 218 p.

Kinge TR, Mih A (2011). Ganoderma ryvardense sp. nov. associated with basal stem rot (BSR) disease of oil palm in Cameroon. Mycosphere 2(2):179-188.

Kinge TR, Mih A (2014). Ganoderma lobenense (Basidiomycetes), a new species from oil palm (Elaies guineensis) in Cameroon. J. Plant Sci. 2(5): 242-245.

Kornerup A, Wansher JH (1978). Methuen handbook of colour. 3rd edn. Eyre Methuen, London. 252 p.

Mohanty PS, Harsh NSK, Pandley A (2011). First report of Ganoderma resinaceum and G. weberianum from north India based on ITS sequence analysis and micromorphology. Mycosphere 2(4): 469-474.

Moncalvo JM, Ryvarden L (1997). A nomenclatural Study of the Ganodermataceae Donk. Synopsis fungorum 11. 114 p.

Moncalvo JM, Wang H, Hseu RS (1995). Phylogenetic relationships in Ganoderma inferred from the internal transcribed spacers and 25S ribosomal DNA sequences. Mycologia 87(2): 223-233. http://dx.doi.org/10.2307/3760908

Mossebo DC (2002). Growth of wood-inhabiting Lentinus species from Cameroon in laboratory culture. Mycologist 16(4):168-171. http://dx.doi.org/10.1017/S0269915X02004068

Mossebo DC (2005). Contribution à la connaissance de la flore mycologique tropicale : Inventaire, taxonomie et systématique des collections de Basidiomycètes (Macromycètes) du Cameroun et d'Afrique centrale. Mémoire d'Habilitation à Diriger des Recherches (HDR), Université de Lille 2, France. p. 1-127.

Mossebo DC, Ryvarden L (1997). Fomitopsis africana sp. nov. (Polyporaceae, Basidiomycotina). Sydowia 49 (2): 147-149.

Mossebo DC, Ryvarden L (2003). The Genus Mycorrhaphium in Africa. Mycotaxon 88:229-232.

Mossebo DC, Njouonkou AL, Courtecuisse R, Amougou A (2007). Enzymatic activities and decay characteristics in some wood-rotting Basidiomycetes from Cameroon and determination of the time-dependant activity of syringaldazine in spot tests. Cryptogamie-Mycologie 28(2):107-121.

Mossebo DC, Kengni Ayissi MB, Ambit RT (2014). New taxa and potential pharmacological properties of some Ganodermataceae (Basidiomycota) from Cameroon and central Africa. In: abstracts book (Scripta Botanica Belgica 52, 291) of the Proceedings of the XXth AETFAT Congress, Stellenbosch, South Africa, 13th-17th January 2014.

Nũnez M, Ryvarden L (2000). East Asian Polypores. Volume 1. Ganodermataceae and Hymenochataceae. Synopsis Fungorum 13. Fungiflora, Oslo, Norway. 168 p.

Roberts P, Ryvarden L (2006). Poroid fungi from Korup National Park, Cameroon. Kew Bull. 61: 55-78.

Ryvarden L (1992). Genera of Polypores; nomenclature and taxonomy. Synopsis Fungorum 5. 292 p.

Ryvarden L (2000). Studies in neotropical polypores 2: A preliminary key to neotropical species of Ganoderma with a laccate pileus. Mycologia 92(1):180-191. http://dx.doi.org/10.2307/3761462

Ryvarden L (2004). Neotropical polypores, Part 1. Introduction. Ganodermataceae & Hymenochaetaceae. Synopsis Fungorum 19. 227 p.

Ryvarden L, Melo I (2014). Poroid fungi of Europe. Synopsis Fungorum 31. 455 p. Fungiflora, Oslo, Norway.

Ryvarden L, Johansen I, (1980). A Preliminary Polypore Flora of East Africa. Oslo, Fungiflora. 636 p.

Steyaert RL (1967). Les Ganoderma palmicoles. Bull. J. Bot. Nat. Belg 37 (4):465-492. http://dx.doi.org/10.2307/3667472

Steyaert RL (1972). Species of Ganoderma and related Genera, mainly of the Bogor and Leiden Herbaria. Persoonia 7: 55-118.

Steyaert RL (1980). Study of some Ganoderma species. Bull. J. Bot. Nat. Belg. 50:135-186. http://dx.doi.org/10.2307/3667780

Zoberi MH (1972). Tropical macrofungi. Some common species. The MacMillan Press Ltd, London and Basingstoke.

29

Effect of co-culturing of cellulolytic fungal isolates for degradation of lignocellulosic material

K. Mohanan[1], R.R. Ratnayake[1], K. Mathaniga[2], C. L. Abayasekara[3] and N. Gnanavelrajah[2]

[1]Institute of Fundamental Studies, Hantana Road, Kandy, Sri Lanka.
[2]Faculty of Agriculture, University of Jaffna, Sri Lanka.
[3]Department of Botany, University of Peradeniya, Peradeniya, Sri Lanka.

This study intended to compare the efficiency of fungal monocultures and co-cultures in the simultaneous delignification and saccharification of kitchen waste and *Eichhornia crassipes* in order to subject the hydrolysate into biofuel production. Three fungal isolates of genus *Trichoderma, Aspergillus, Pycnoporus* and an unidentified strain (F113) were grown in mono and co-cultures and the extracted enzymes were used for the degradation. Co-culture of *Trichoderma* spp with the other fungi improved its enzyme activity while the other co-cultures did not show significantly improved enzymatic degradation compared to monocultures. The highest percentage of saccharification (over total dry weight) achieved were 11.9% with kitchen waste after seven days and 9.8% with *E. crassipes* after 4 days. The drop in degradation rate normally seen after complete digestion of amorphous cellulose was not apparent probably due to the grinding of the substrates to fine particle size.

Key words: Fungal co-culture, biofuel, cellulase, kitchen waste, invasive weeds.

INTRODUCTION

Fossil fuel resources are limited and their usage leads to environmental problems. Hence, it is imperative to utilise alternative energy sources that are renewable and eco-friendly. Biofuels are promising in this regard. Second generation biofuels are produced from non-edible biomass, through degradation and fermentation.

Enzymes degrading lignocelluloses in nature include cellulases, xylanases and lignin degrading enzymes. The activity of lignin degrading enzyme is too slow for application in biofuel production (Lu et al., 2010). Thus, thermo-chemical pre-treatment is needed to overcome the recalcitrance (Margeot et al., 2009). Pre-treatment also results in degradation of hemicelluloses. Therefore, cellulases are the major enzymes involved in subsequent enzymatic hydrolysis.

Pre-treatment requires energy input and chemicals which adds up to a major component of the cost of production. The cost of enzymes also contributes significantly (Shi et al., 2009). Among microbes, some aerobic filamentous fungi are known to secrete high amounts of cellulases. Such strains are found in the genera *Trichoderma*, *Aspergillus*, *Pencillium* and many others (Lynd et al., 2002). The factors affecting the expression and secretion of fungal cellulases include induction, repression, metal ions, inorganic nutrients, surfactants and culture conditions (Gremel et al., 2008; Mandels and Reese, 1957; Reese and Maguire, 1969; Schmoll et al., 2005; Suto and Tomita, 2001). Cellulases belong to three classes, namely endoglucanases, exoglucanases and β-glucosidases, which act in synergy (Lynd et al., 2002). The right balance of the cellulases is crucial for optimal degradation of lignocelluloses. In *Trichoderma reesei,* large proportions of β-glucosidase remain cell-wall bound through a polysaccharide and gets released when treated with cellulase from *Aspergillus niger* (Messner et al., 1990). Endoglucanases and exoglucanases of *T. reesei* are inhibited by cellobiose whereas its β-glucosidases was inhibited by glucose (Philippidis et al., 1993). In contrast, *A. niger* secretes high amounts of β-glucosidase which is tolerant to high levels of glucose (Decker et al., 2000). For these reasons, a combination of cellulases from *T. reesei* and *A. niger* is used in the biofuel industry (Reczey et al., 1998). Co-culture of *T. reesei* and *Aspergillus* sp has been shown to result in better yield of cellulases (Ahamed and Vermette, 2008; Duff, 1985). Better utilization of substrates and formation of strong inducers for *T. reesei* cellulases by β-glucosidase of *Aspergillus* are thought to be the reasons for the increased yield (Ahamed and Vermette, 2008).

Disposal of household waste is a problem in the urban areas of Sri Lanka. Current practice is to dump the waste in the suburbs, causing unpleasant odours and health problems in those areas. Composting has been attempted in an industrial scale but has been abandoned due to operational problems (Premachandra, 2006). A possible solution to this problem could be the utilization of household waste for biofuel production. Another potential substrate is the invasive aquatic weed *Eichhornea crassipes*. Invasive weeds are probable raw materials for cellulosic biofuel production. These are non-indigenous or "non-native" plants which adversely affect the habitats and bioregions they invade economically, environmentally, and ecologically (Westbrooks, 1998). *E. crassipes is* one of a common aquatic invasive weed and another potential substrate for biofuel production. The fast growing nature of these weeds can provide raw materials in abundance for biofuel industries. Since it is less lignified, its utilisation would require low pre-treatment. If effective technologies can be developed to drive commercial products from these weeds it will be beneficial both economically and ecologically. Although a large number of microorganisms (fungi, bacteria and actinomycetes) are capable of degrading cellulose, only a few of them produce significant quantities of cell-free enzyme fractions capable of complete hydrolysis of cellulose *in vitro*. Cellulases obtained from compatible

mixed cultures of fungi appear to have more enzyme activity as compared to their pure cultures and other combinations (Jayant et al., 2011).

The objective of this study was to compare the efficiency of fungal monocultures and co-cultures in the simultaneous delignification and saccharification of kitchen waste and *E. crassipes* in order to subject the hydrolysate into biofuel production.

MATERIALS AND METHODS

Isolation of cellulolytic fungi

Samples of decaying plant material, ruminant dung, decaying kitchen waste and soil were suspended in sterile normal saline serially diluted to obtain 10^{-1}, 10^{-2} and 10^{-3} dilutions. 100 μl of each dilution was plated on potato dextrose agar (with gentamicin 50 mg/L and chloramphenicol 50 mg/L) by spread plate technique and incubated at 25°C up to a week. The fungal isolates were inoculated on Czapek dox agar without sucrose and with 1% cellulose and those showing good growth were presumed to be potential cellulose degraders and added to the culture collection. The isolates in the culture collection were screened for cellulase production by growing them without replicates as described in measurement of enzyme activities part of the work. The isolates which showed at least 0.01 FPU of cellulase activity were then tested in replicates and those with the highest enzyme activities from different genera were chosen for further study.

Measurement of enzyme activities

Production of enzymes

Fungal isolates were grown on PDA slants for 7 days. Spore suspensions were made in sterile saline, spore concentrations were adjusted to 10^7-10^8/ml and 100 μl of the suspensions were inoculated into 20 ml of a minimal medium (Mandels and Reese, 1957). Co-cultures were made by inoculating 100 μl of spore suspension from each of the relevant strain. The cultures were incubated at 28°C on a rotary shaker at 100 rpm. For initial screening, the isolates were cultured without replicates with an incubation period of 3 days. At the end of incubation, the cultures were centrifuged at 4000 g for 20 min and the supernatants were used as crude enzymes for the assays.

Total cellulase assay

Total cellulase assay was carried out using Whatmann No.1 filter paper as the substrate (Mandels et al., 1976; Ghose, 1987). Reducing sugars formed were measured by using di-nitro salicylic acid reagent (Sumner, 1921; Miller, 1959), with glucose as standard. The total cellulase activity is expressed as filter paper units/ml (FPU/ml).

Xylanase assay

Xylanase activities were measured by a method modified from Gottschalk et al. (2010) using 1% (w/v) beech wood xylan (Sigma) as the substrate. Reducing sugars formed were measured using di-nitro salicylic acid (DNS) reagent (Sumner, 1921; Miller, 1959), with xylose as standard.

β-Glucosidase assay

β-glucosidase activities were measured by using cellobiose as the substrate (Ghose, 1987; Sternberg et al., 1977). Glucose formed during the assay was measured using a commercial blood glucose meter (*One Touch Ultra 2*) based on glucose oxidase, calibrated with glucose standards in 0.05 M citrate buffer (pH = 4.8). Positive results of β-glucosidase activities were verified by high-performance liquid chromatography (HPLC) as described in enzymatic degradation of lignocellulosic materials part of this work.

Laccase assay

Laccase activities of *Pycnoporus cinnabarinus* was measured with ABTS (2,2'-azino-bis(3-ethylbenzothiazoline- 6-sulphonic acid)) as the substrate (Bourbonnais et al., 1995).

Enzymatic degradation of lignocellulosic materials

Kitchen waste and mature leaves (to represent the most recalcitrant type for degradation, as they are the most lignified) of *E. crassipes* were dried in an oven at 50°C to a constant weight. The dried material was then ground in a plant grinder and sieved through 93 µm (kitchen waste) and 50 µm (*E. crassipes*) sieve. Powdered kitchen waste (500 mg) and *E. crassipes* (200 mg) were added separately into boiling tubes, to which 6 ml of 50 mM citrate buffer (pH 4.8) was added and autoclaved at 121°C for 15 min. To each tube, 3 ml of crude enzyme from fungal monoculture or co-culture was added and incubated in a water bath at 50°C with reciprocal shaking at 100 rpm for five to seven days. Samples were withdrawn daily and total sugar concentrations were determined using DNS reagent with glucose standards (Sumner et al., 1921). The sugar components of the hydrolysate of *E. crassipes* at the end of 4th day of degradation were measured by High-performance liquid chromatography (HPLC) using agilent Hi-plex H column (p/n PL1170-6830 300x7.7 mm) at 65°C with deionised water as the mobile phase (flow rate: 0.6 ml/min, injection volume: 10 µl). Sugars were detected with an RI detector. D-cellobiose, D-glucose, D-xylose and L-arabinose (Sigma Aldrich) standards were used for calibration. High-performance liquid chromatography was not performed on kitchen waste due to the highly variable nature of its content.

Data analysis

Statistical comparisons were made by ANOVA using Minitab software (version 14). $\alpha = 0.05$ unless otherwise stated.

RESULTS AND DISCUSSION

Isolation of cellulolytic fungi

A total of 145 fungal strains were isolated from different samples. During the initial screening for cellulase activity, 35 isolates were found to have greater than 0.01 FPU/ml of activity. The isolates which showed significant cellulase activities were shown in Table 1. Most of them belong to *Trichoderma* species, while strains of *Penicillium*, *Aspergillus* and an unidentified fungal strain were also present. Fungi were tentatively identified using macroscopic and microscopic morphological characteris-

tics. Slide culture technique was used to aid the fungal identifications.

Co-culture of fungi

Fungal isolates belonging to genera of *Trichoderma* (F1, F16,F F118), *Penicillium* (F24) and an unidentified isolate (F113) were selected for co-culturing. The selection was made to include different genera and strains with β-glucosidase activity. The isolates were divided into two groups (F1, F16 and F118) and (F24, F80, F113) and co-cultures were made in all possible combinations within each group. Cellulase and xylanase activities of the co-cultures and corresponding mono-cultures were measured (Figures 1 and 2).

Among the group containing F1, F16 and F118, the co-cultures showed lower cellulase activities compared to the corresponding monocultures. Among the group containing F24, F80 and F113, the co-culture F80 and F113 showed higher cellulase activity than F113, but the difference was statistically insignificant. The co-cultures F24/F113 and F24/F80/F113 showed significantly higher xylanase activities compared to the corresponding monocultures. Other co-cultures showed either no significant difference or reduced xylanase activity.

Degradation of kitchen waste

The isolates F24, F113 and F118 were selected for degradation of kitchen waste. The cumulative sugar contents measured at 1st, 2nd 3rd and 7th days of degradation of fresh kitchen waste was shown in Figure 3. At the end of seven days of degradation, the highest amount of sugars were released by enzymes from F24 (59.7 mg) followed by the co-culture F24/F113 (58 mg). Enzymes from the co-culture F113/F118 effected significantly higher degradation than F118 (*Trichoderma*) alone. This is probably due to the lack of secreted β-glucosidase activity by F118 being complemented by F113.

Degradation of *E. crassipes*

Enzymes from the fungal strains F24, F113, F118 and a woody mushroom from Sri Lanka (M21), identified as *Pycnoporus cinnabarinus*, were used for the degradation of *E. crassipes*. Screening revealed that some organisms were more efficient than the others. M21 was found to have a significant cellulase activity (0.21 FPU) and laccase activity (50 IU/ml). The cumulative sugar contents measured at 1st, 2nd, 3rd and 4th days of enzymatic degradation of *E. crassipes* was shown in Figure 4. The sugar contents at the 4th day was shown in Figure 5 for comparison. The component sugars as

Table 1. Fungal isolates with highest cellulase activities.

Isolate no.	Total cellulase (FPU/ml)	Xylanase (IU/ml)	β-glucosidase (IU/ml)	Genus
F118	0.21	4.31	Not detected	*Trichoderma*
F80	0.16	2.03	Not detected	*Trichoderma*
F1	0.15	5.48	Not detected	*Trichoderma*
F16	0.14	5.22	0.11	*Trichoderma*
F22	0.13	5.22	0.09	*Trichoderma*
F24	0.12	4.98	Not detected	*Penicillium*
F54	0.11	1.77	Not detected	*Trichoderma*
F27	0.11	5.38	Not detected	*Trichoderma*
F98	0.10	4.79	0.05	*Trichoderma*
F10	0.09	2.14	Not detected	*Trichoderma*
F56	0.08	5.62	0.05	*Trichoderma*
F88	0.07	5.25	0.13	*Trichoderma*
F40	0.06	3.35	0.07	*Aspergillus*
F113	0.06	1.02	0.15	Unidentified

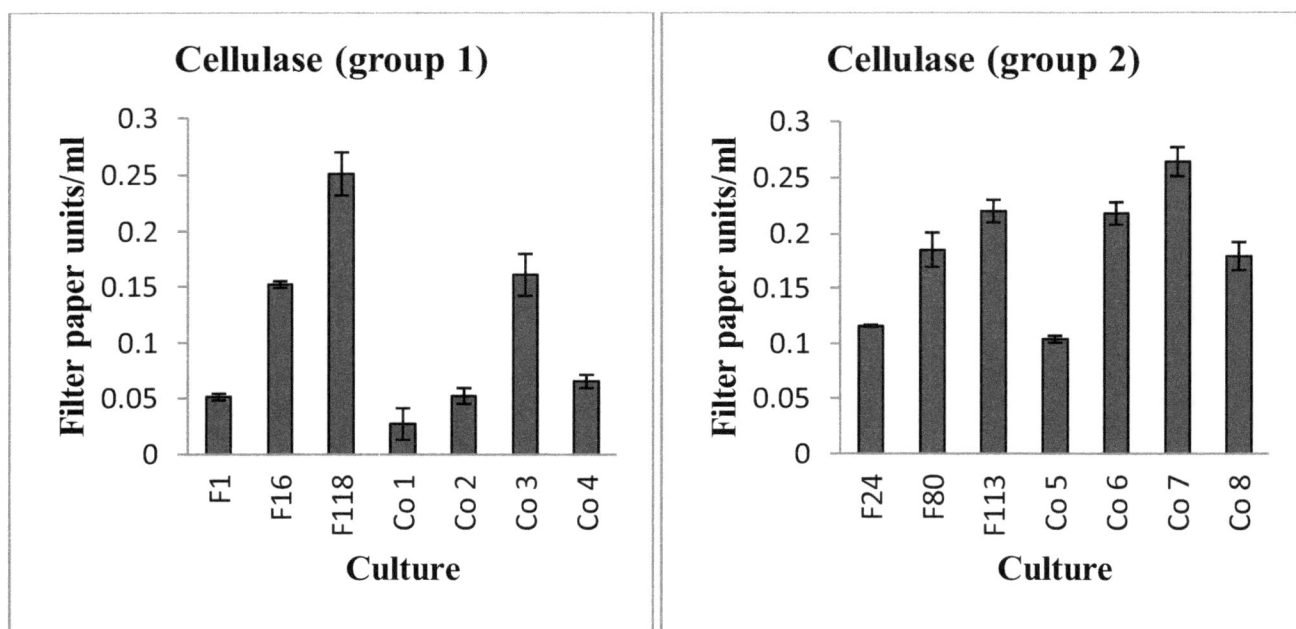

Figure 1. Cellulase activities of monocultures and co-cultures of selected fungi. Key: F1-*Trichoderma* spp., F16 - *Trichoderma* spp., F118-*Trichoderma* spp., Co1- F1 and F16, Co2-F1 and F118, Co3- F16 and F118, Co4- F1, F16 and F118, F24-*Penicillium* spp., F80-*Trichoderma* spp., F113-Unidentified, Co5- F24 and F80, Co6- F24 and F113, Co7- F80 and F113, Co8- F24, F80 and F113. Error bars indicate standard errors of the means.

measured by HPLC at the end of 4th day are shown in Figure 6. The amount of simple sugar produced and released to the medium can depend on activities of cellulase systems and physiological characteristics of particular species.

The highest quantity of total sugar content (19.6 mg) at the end of four days of degradation of *E. crassipes* was obtained with enzymes from the co-culture F113/F118 However, the mono-cultures F24, F113 and co-cultures F24/F113, F24/F118 and F24/M21 gave slightly lower quantities and the differences are not statistically significant. F118 and F113/M21 and M21 showed significantly lower degradation rate. It was noted that while the amount of xylose formed by M21 was similar to

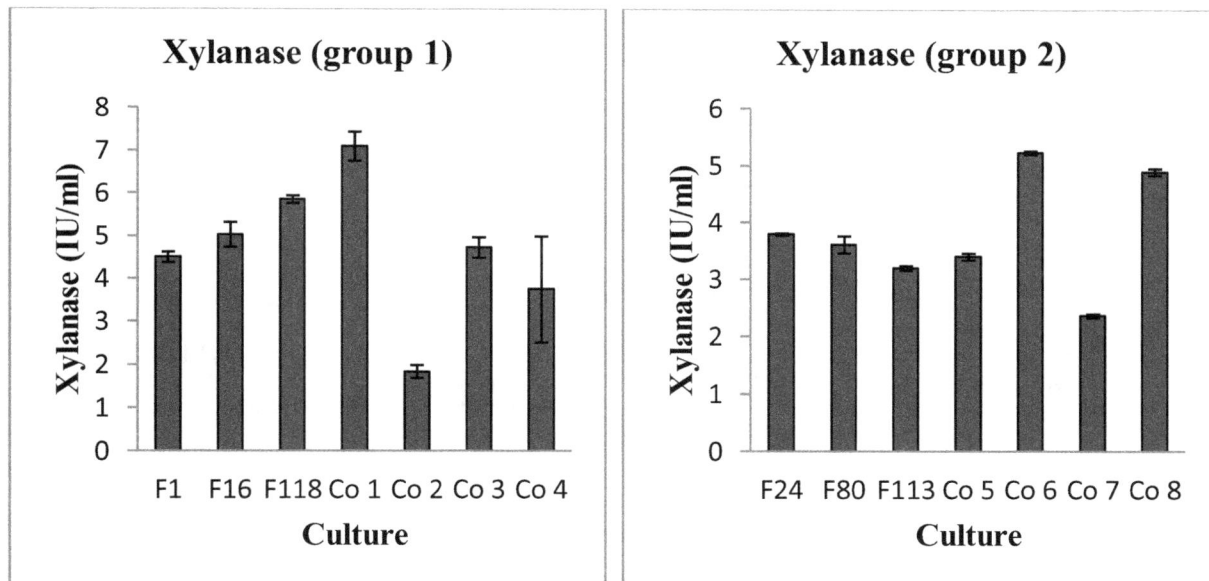

Figure 2. Xylanase activities of monocultures and co-cultures of selected fungi. Key: F1-*Trichoderma* spp., F16 - *Trichoderma* spp., F118-*Trichoderma* spp., Co1- F1 and F16, Co2-F1 and F118, Co3- F16 and F118, Co4- F1, F16 and F118, F24-*Penicillium* spp., F80-*Trichoderma* spp., F113-Unidentified, Co5- F24 and F80, Co6- F24 and F113, Co7- F80 and F113, Co8- F24, F80 and F113. Error bars indicate standard errors of the means.

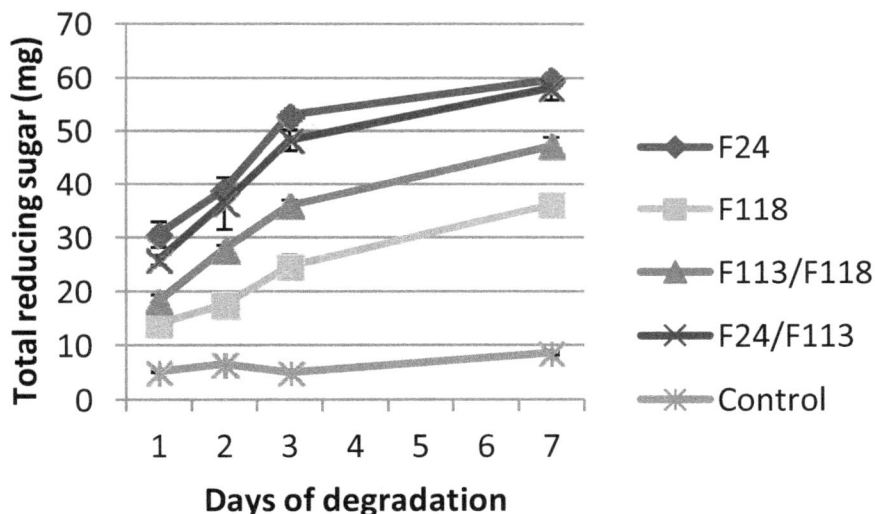

Figure 3. Total sugar content (cumulative) released from un-decomposed kitchen waste by enzymes

other isolates, the amount of arabinose formed was much lower. The percentage of maximum sugar yield over total dry weight was 11.9 and 9.8% respectively from kitchen waste and *E. crassipes*. The percentages against total polysaccharide content should be higher.

Grinding the substrate to a very fine powder (93 and 50 μm) would reduce the length of the polysaccharide chains and increase the number of free ends available for the activity of exoglucanases. It would also increase the surface area of the substrate available for enzyme activity. During the enzymatic hydrolysis of cellulose, amorphous portions are quickly degraded followed by slow degradation of crystalline regions (Mandels, 1975). Thus a change in the rate of degradation, that is, slope of the degradation curve, indicates the end of degradation of amorphous portions. This change should occur roughly

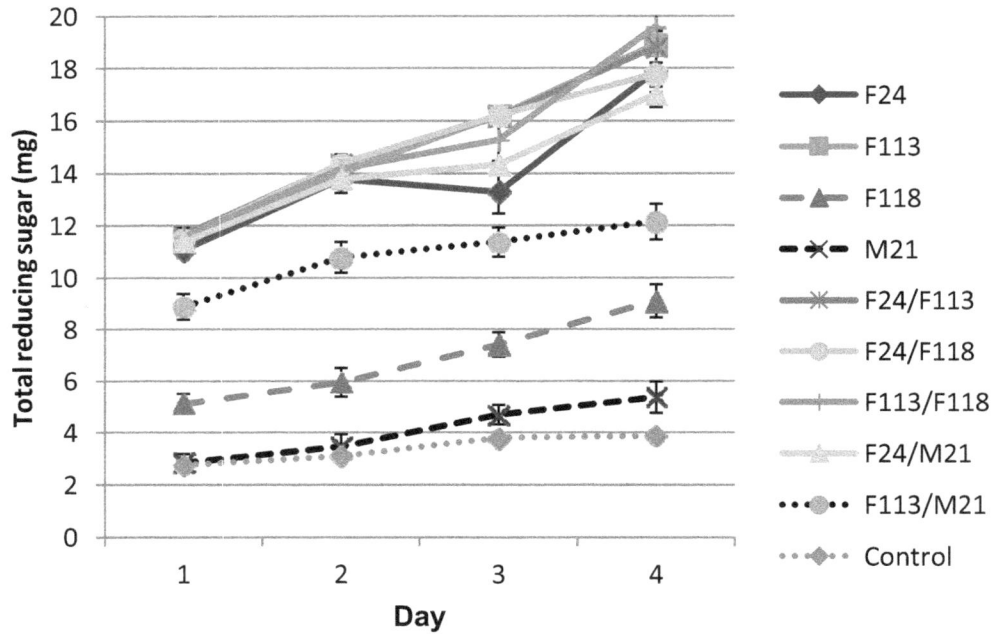

Figure 4. Total reducing sugars measured at daily intervals during enzymatic degradation of *E. crassipes*

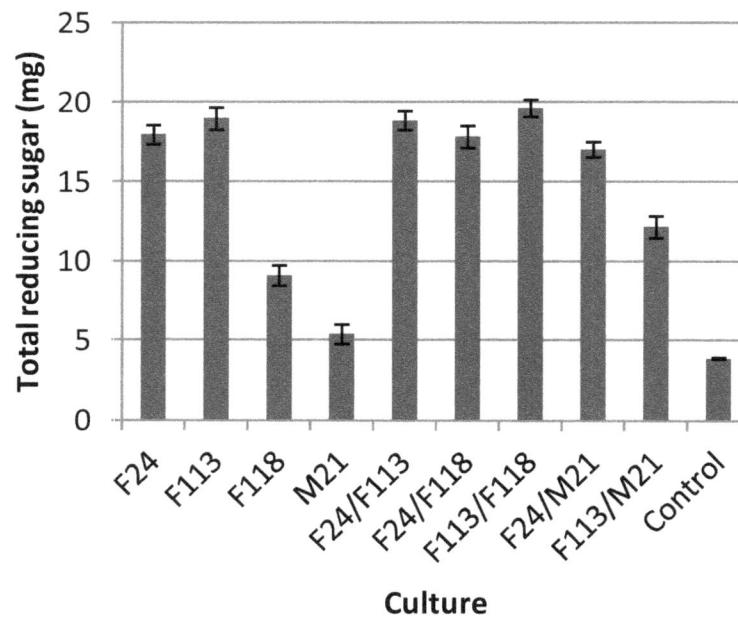

Figure 5. Total reducing sugar accumulated by degradation of *E. crassipes* by the end of 4th day of degradation

at about the same percentage of degradation for different enzymes. However, in the present study, such change of rate did not seem to occur at roughly equal percentage of degradation. This could be because at low particle sizes, the effect of enzyme loading becomes more prominent.

Conclusions

Co-culturing of *Trichoderma* with other cellulolytic fungi improved the activity of lignocellulose degrading enzymes compared to monoculture of *Trichoderma*. The co-culture

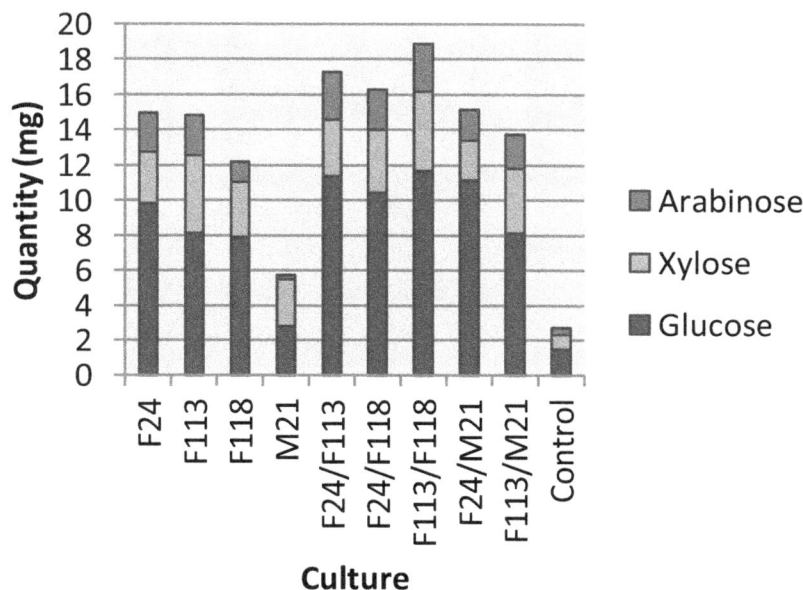

Figure 6. Quantities of glucose, xylose and arabinose of the enzyme hydrolysate of *E. crassipes* at the end of 4th day of degradation.

of other fungi did not result in significant improvement in the activity compared to corresponding monocultures. Significant percentage of degradation of kitchen waste and *E. crassipes* was achieved with the monoculture, without pre-treatment. Fine powdered material was used to eliminate the cost of pre-treatment. Enzyme from *Pycnoporus cinnabarinus* the laccase producing strain was found to be ineffective for lignocellulose degradation.

REFERENCES

Ahamed A, Vermette P (2008). Enhanced enzyme production from mixed cultures of *Trichoderma reesei* RUT-C30 and *Aspergillus niger* LMA grown as fed batch in a stirred tank bioreactor. Biochem. Eng. J. 42: 41-46.

Bourbonnais R, Paice MG, Reid ID, Lanthier P, Yaguchi M. (1995). Lignin oxidation by laccase isozymes from *Trametes versicolor* and role of the mediator 2,2'-azinobis (3-ethylbenzthiazoline-6-sulfonate) in kraft lignin depolymerization. Appl. Environ. Microbiol. 61: 1876-1880.

Decker CH, Visser J, Schreier P (2000). β-Glucosidases from five black Aspergillus species: study of their physico-chemical and biocatalytic properties. J. Agr. Food Cchem. 48: 4929-4936.

Duff SJ, Cooper DG, Fuller OM (1985). Cellulase and beta-glucosidase production by mixed culture of *Trichoderma reesei* Rut C30 and *Aspergillus phoenicis*. Biotechnol. Lett. 7:185-190.

Ghose T (1987). Measurement of cellulase activities. Pure Appl. Chem. 59:257-268.

Gottschalk LMF, Oliveira RA, Bon EPS (2010). Cellulases, xylanases, β-glucosidase and ferulic acid esterase produced by *Trichoderma* and *Aspergillus* act synergistically in the hydrolysis of sugarcane bagasse. Biochem. Eng. J. 51: 72-78.

Gremel G, Dorrer M, Schmoll M (2008). Sulphur metabolism and cellulase gene expression are connected processes in the filamentous fungus *Hypocrea jecorina* (anamorph *Trichoderma reesei*). BMC Mmicrobiol. 8: 174.

Jayant M, Rashmi J, Shailendra M, Deepesh Y (2011). Production of cellulase by different co-culture of *Aspergillus niger* and *Penicillium*

chrysogenum from waste paper, cotton waste and baggase. J. Yeast Fungal. Res. 2:24-27.

Lu C, Wang H, Luo Y, Guo L (2010). An efficient system for pre-delignification of gramineous biofuel feedstock *in vitro*: Application of a laccase from *Pycnoporus sanguineus* H275. Process Biochem. 45:1141-1147.

Lynd LR, Weimer PJ, van Zyl WH, Pretorius IS (2002). Microbial cellulose utilization: fundamentals and biotechnology. Microbiol. Mol. Biol. R. 66:506 - 577.

Mandels M. (1975). Microbial sources of cellulase. In Biotechnol. Bioeng. Symp. 5, pp. 81-105.

Mandels M, Andreotti R, Roche C (1976). Measurement of saccharifying cellulase. Biotechnology and Bioengineering Symposium (6)21 - 33.

Mandels M, Reese ET (1957). Induction of cellulase in Trichoderma viride as influenced by carbon sources and metals. J. Bacteriol. 73:269.

Margeot A, Hahn-Hagerdal B, Edlund M, Slade R, Monot F (2009)New improvements for lignocellulosic ethanol. Curr. Opin. Biotech. 20:372 - 380.

Messner R, Hagspiel K, Kubicek C P (1990). Isolation of a β-glucosidase binding and activating polysaccharide from cell walls of Trichoderma reesei. Arch. Microbiol. 154:150-155.

Miller GL (1959). Use of Dinitrosalicylic acid reagent for determination of reducing sugar. Anal. Chem. 31:426 - 428.

Philippidis GP, Smith TK, Wyman C E (1993). Study of the enzymatic hydrolysis of cellulose for production of fuel ethanol by the simultaneous saccharification and fermentation process. Biotechnol. Bioeng. 41: 846-853.

Premachandra HS (2006). Household Waste Composting & MSW Recycling in Sri Lanka. In *Asia 3R conference, Tokyo (available from: www.env.go.jp/recycle/3r/en/asia/02_03-3/08.pdf* [Accessed 14 December 2013].

Reczey K, Brumbauer A, Bollok M, Szengyel Z, Zacchi G (1998). Use of hemicellulose hydrolysate for β-glucosidase fermentation. In *Biotechnology for Fuels and Chemicals* (pp. 225-235). Humana Press.

Reese ET, Maguire A (1969). Surfactants as stimulants of enzyme production by microorganisms. Appl. Microbiol. 17:242-245.

Schmoll M, Franchi L, Kubicek C P (2005). Envoy, a PAS/LOV domain protein of Hypocrea jecorina (Anamorph Trichoderma reesei), modulates cellulase gene transcription in response to light.

Eukaryotic cell 4:1998-2007.

Shi J, Ebrik M, Yang B, Wyman CE (2009). The potential of cellulosic ethanol production from municipal solid waste: a technical and economic evaluation. (available from: http://escholarship.org/uc/item/99k818c4 [Accessed 15 December 2013].

Sternberg D, Vuayakumar P, Reese ET. (1977). β-Glucosidase: microbial production and effect on enzymatic hydrolysis of cellulose. Can. J. Microbiol. 23: 139-147.

Sumner JB (1921). Dinitrosalicylic acid: a reagent for the estimation of sugar in normal and diabetic urine. J. Biol. Chem. 47: 5-9

Suto M, Tomita F (2001). Induction and catabolite repression mechanisms of cellulase in fungi. J. Biosci. Bioeng. 92:305-311.

Westbrooks R. (1998). Invasive plants, changing the landscape of America: Fact book. Federal Interagency Committee for the Management of Noxious and Exotic Weeds (FICMNEW), Washington, D.C. 03 pp. Available at http://digitalcommons.usu.edu/cgi/viewcontent.cgi?article=1489&context=govdocs

30

Selection of cellulase-producing *Trichoderma* sp. and optimization of cultivation conditions

Tian Haijiao, Qu Chang, Wu Hao, Li Yuhua and Yang Hongyan

College of Life Sciences/Daqing Bio-tech Institute, Northeast Forestry University, Harbin 150040, China.

The aim of this study was to select the wild microorganism with the capacity to produce cellulases. Microbial sources were rotten grass, leaves and straws. Natural lignocellulose as the sole carbohydrate was used to enrich microorganisms. After isolation, the strains were inoculated into liquid media. Filter paper activity (FPA) was used as an index for selection. Sequencing of 26S rDNA D1/D2 region was used to identify the microorganism. Lactose, lactobionic acid, Tween 80 and yeast extract were added into cultivation system to improve FPA. Optimum temperature and pH of cellulase were investigated. After enrichment, isolation and selection according to FPA, one strain was selected. *Trichoderma* sp. A6 was named according to the sequencing results. When lactose (10 g/L) was added into the cultivation system, the FPA was improved from 0.16 to 0.56 IU. Lactose addition could improve FPA significantly. After addition, the FPA at two days amounted to the highest value. However, the addition of lactobionic acid did not increase FPA. Tween 80 addition only improved the FPA at four days to small extent. The 15 g/L yeast extract could increase the FPA at 2 days from 0.60 to 0.92 IU. The combination of Tween 80 and yeast extract did not improve the FPA compared with the treatment of Tween 80 addition. The optimum range of temperature and pH were 37-60 and 4.5-5.5°C, respectively.

Key words: Cellulase, *Trichoderma*, selection, cultivation condition, optimization.

INTRODUCTION

Energy security, petroleum depletion, and global warming have become the main driving forces for developing renewable fuels that can replace petroleum-derived fuels. The renewable fuel expected to be widely used around globe is bioethanol, which is largely produced by fermenting starch- or sugar-containing feedstocks. However, the supply of these crops is relatively limited and many of them can be considered human food resources. Non-food crops will be more advantageous for bioethanol production (Alriksson et al., 2009; Sticklen, 2008).

Lignocellulose is considered as one of the most impor-

tant carbon sources on Earth and a less-expensive raw material with a great potential to produce energy. Lignocellulosic biomass used for producing ethanol has been a major focus (Lynd et al., 2008; Niranjane et al., 2007). It is known that the general process for converting lignocellulosic biomass into ethanol mainly includes feedstock pretreatment, enzymatic hydrolysis, sugars fermentation, separation of lignin residue, recovery and purifying the ethanol to meet fuel specifications. Currently, there are technological and economic limitations in each step in the conversion process (Alvira et al., 2010). Cost effective pretreatment processes, cheaper hydrolytic enzymes and fermentation of pentose sugars have been always common research goals (Keshwani and Cheng, 2009).

Utility cost of enzymatic hydrolysis is low as compared to acid or alkaline hydrolysis, because enzymatic hydrolysis is usually conducted at mild conditions (pH 4.8) and temperature (45-50°C) and does not have a corrosion problem (Singh et al., 2009) However, at present, hydrolysis enzymes are expensively produced in microbial bioreactors (Sticklen, 2008). It is still important that selecting microorganisms with a high capacity to produce enzymes and improving cellulase-producing characteristics (Wang et al., 2010).

Filamentous fungi are widely used in producing cellulases currently. *Trichoderma*, *Aspergillus* and *Penicillium* are recognized with high potentials in the cellulase production (Li and Hou, 2010). In this study, we obtained one strain with cellulase-producing capacity using plate isolation and streaking method. Initial identification and optimization of culture conditions were carried out. This strain could be considered as a wild-type strain for the further physical and chemical mutagenesis. The optimization results of cultivation conditions will provide the instructions for the possible industrialization application of this strain.

MATERIALS AND METHODS

Cultivation and selection

Rotten weeds, leaves and straws were used as microbial sources. The modified Czapek medium (MCD) was used for enrichment media (Gharieb et al., 1999; Li and Hou, 2010). The medium (1 L) contained : NaNO$_3$, 2 g; KH$_2$PO$_4$, 1 g; MgSO$_4$. 7H$_2$O, 1 g; KCl, 0.5 g; FeSO$_4$, 0.01 g; agar ,15 g. The flask (250 mL) contained 100 mL medium. All media were autoclaved at 121°C for 15 min and cooled. After solidification, 1 g of sterile straws (10 cm in length) was placed on the agar, as the carbon source. Powders (0.5 g) of decaying weeds, leaves and straws were accessed to the MCD agar (100 mL), respectively. After a 14-day cultivation at 30°C, about 0.05 g straws covered with microorganisms were transferred into another MCD agar (Yang et al., 2011). After five-time repeats, the samples which could grow stably were used for further isolation. Plate dilution method was used to isolate microorganisms according the clolony morphology. The medium was PDA (Wu et al., 2012).

The strains isolated were used to inoculate into the liquid Mandels medium (Wang et al., 2011) (15 g/L microcrystalline cellulose as carbon source). After shaking for seven days, the filter paper activities (FPAs) of the cultivation system were measured. The method of enzyme assay was shown as section 1.2. The strain with the highest enzyme activity was selected for identification and optimization of culture conditions. The *Trichoderma reesei* RUT-C30 were purchased from China Center of Industrial Culture Collection (CICC).

Determination of enzyme activity

Culture supernatants obtained after centrifugation at 5800 × *g* for 10 min at 4°C were used to determine the FPA. Whatman No.1 filter paper (Whatman, England) was used as the substrate to determine enzyme activity. A filter paper strip (1.0 × 6.0 cm, ≈50 mg) was

added into 1mL 50 mM citric-phosphate buffer (pH 4.8). After pre-warming at 50°C for 5 min, 0.5 mL crude enzyme solution was added. Then the mixture was incubated at 50°C for 60 min. The release of reducing sugar was measured using the 3, 5-dinitrosalycilic acid method (Ghose, 1987). Glucose was used as the standard for FPA measurements. The color formed by cellulase bound to the substrate was subtracted from that of the standard and the sample tube. Therefore, the amount of cellulase bound to the substrate is not neglibible. One unit (IU) of enzyme activity was defined as the amount of enzyme releasing 1 μmol reducing sugar in 1 min reaction.

Microbial identification

Genomic DNA of pure culture strain was extracted using the EZNA mini DNA kit (OMEGA, USA) according to manufacture's instruction. The 50 μL PCR mixtures contained 15 ng of template DNA, 1× PCR buffer (Mg^{2+} free), 0.16 mM of each dNTP, 1.5 mM MgCl$_2$, 0.45 μM of each primer, and 1 U of Takara rTaq DNA polymerase (Takara, Japan). The primers for amplification of the D1/D2 region of the fungal 26S rRNA gene were NL1 (5'-GCATATCAATAAGCGGAGGAAAAG-3') and NL4 (5'-GGTCCG TGTTTCAAGACGG-3') (Baleiras Couto et al., 2005). The thermocycle program consisted of initial DNA denaturation at 95°C for 5 min, followed by 30 cycles of denaturation at 95°C for 1 min, annealing at 52°C for 45 s, and elongation at 72°C for 1 min 30 s, and ending with a final elongation step at 72°C for 6 min. The amplified fragment was confirmed by agarose gel electrophoresis, and then sent to Sangon Biotech (China) for sequencing. The sequences generated in this study were compared with those in GenBank (http://blast.ncbi.nlm.nih.gov/Blast.cgi). A neighbor-joining tree was constructed using the MEGA 5.0 software (Tamura et al., 2011).

Optimization of culture conditions

Mandels medium was used as the basic medium. A total amount of carbon source of 1 L culture system was 15 g. The addition of lactobionic acid and lactose was 0~15 g, respectively. Accordingly, the addition of microcrystalline cellulose was 15~0 g. The medium (100 ml) was poured into 250 mL flask. All media were autoclaved at 121°C for 15 min. All determinations are equipped with three replications. After optimization of carbon source, yeast extract and Tween 80 were used. The addition of yeast extract was 0~20 g, and Tween 80 was 0-2.0 mL. Each addition level is equipped with three replications. Each replication was measured three times. Therefore, each FPA value in this study is means of nine measurements. Enzyme activities at two, four and seven days of cultivation were measured to determine final cultivation conditions.

Data processing

Data were subjected to ANOVA using Duncan's test of the Statistic Analysis System (Version 8.2, SAS Inst. Inc., Cary, NC). Significance level was 5%.

RESULTS

Screening of strains

After a series of screening, a fungal strain was obtained due to the highest enzyme activity. The growth status of the strain cultivated on PDA medium for three days was

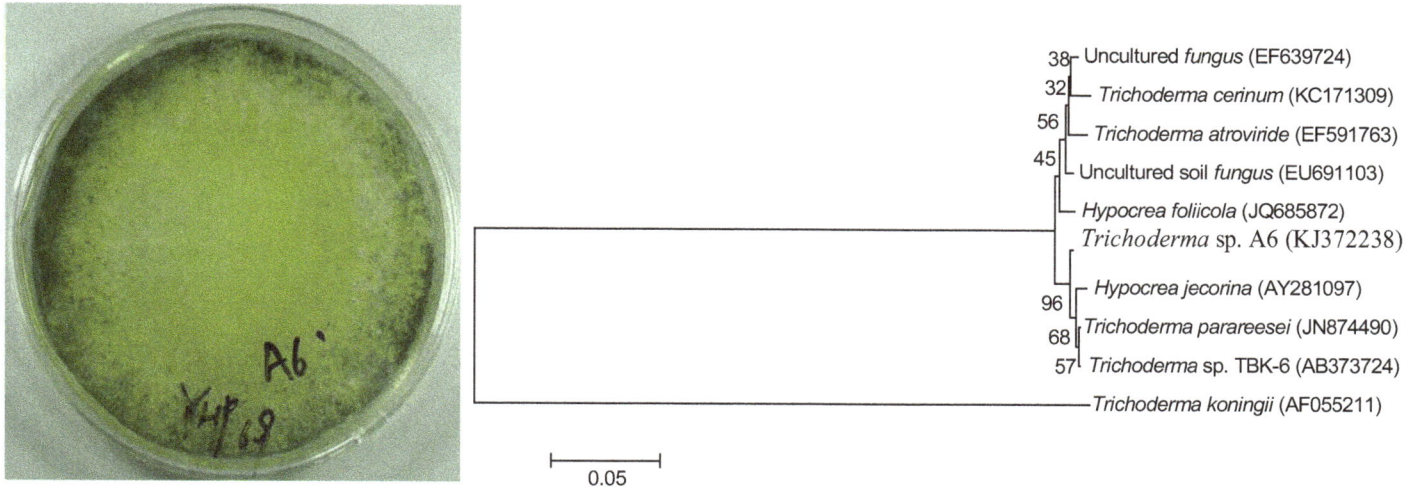

Figure 1. Isolated strain and phylogenetic analysis the fungal 26S rDNA D1/D2 region sequences by means of the neighbor-joining method. The bar represents 5% sequence divergence. The number in parentheses shows the GenBank accession number.

shown in Figure 1. After extraction of genomic DNA, PCR amplification of 26S rDNA D1/D2 region was carried out. After sequencing, phylogenetic tree was constructed and shown in Figure 1. The closest microorganism was *Trichoderma* sp. TBK-6. With 99% sequence similarity, it was identified as *Trichoderma* sp. and named as *Trichoderma* sp. A6.

Optimization of culture conditions

To further enhance A6 enzyme activity, optimization experiments of culture conditions (carbon source, nitrogen source and unknown growth factor) were carried out.

Addition of lactose and lactobionic acid

The previous report showed that the combination of cellulose and lactose as the carbon source could improve the cellulase-producing capacity of *Trichoderma reesei* (Muthuvelayudham et al., 2006). Janas reported that lactose and lactobionic acid could induce cellulase production of Trichoderma *reesei* (Janas, 2002). In this study, lactose and lactobionic acid were respectively added into the medium containing microcrystalline cellulose. The result was shown in Figures 2 and 3.

The FPAs at two days of cultivation were significantly higher with the increase of lactose addition (Figure 2). The FPAs at four days of cultivation were increased significantly in the treatments of 5 and 10 g/L lactose addition. At seven days of cultivation, the FPAs did not significantly change with 5-15 g/L addition of lactose. In all treatments of lactose addition, the FPAs expressed the downward trend from two days of cultivation. This result

showed that the improvement of FPAs due to lactose addition occurred at early stage of cultivation. The addition of 10 g/L lactose is the most optimized content.

The result of lactobionic acid addition was shown in Figure 3. The addition of lactobionic acid did not serve to improve enzyme activity. Compared the treatment without any addition, all FPAs in various addition treatments decreased. This result indicated that addition of lactobionic acid did not provide positive effect for enzyme production of *Trichoderma* sp. A6

Addition of Tween 80

Reese and Maguire found that Tween 80 can improve enzyme activity of *Trichoderma* that grew in the medium. However, the mechanism was unclear (Reese and Maguire, 1969). Domingures et al. (2000) reported that the addition of Tween 80 may increase cell permeability and affect cell morphology. As a result, the enzyme activity was improved. In this study, on the basis of adding 10 g/L lactose, the effects of Tween 80 on enzyme production were investigated. The addition gradient was set from 0 to 2 mL/L. The FPAs was shown as Figure 4. The addition of Tween 80 did not affect enzymatic activities at two and seven days of cultivation. At four days of cultivation, the FPAs increased with the addition of Tween 80 in the range of 0.5 to 1.0 mL/L. Finally, the addition concentration of Tween 80 was 1.0 mL/L for the following experiment.

Addition of yeast extract

The main components of yeast extract were peptides, amino acids, flavor nucleotides, B vitamins and trace

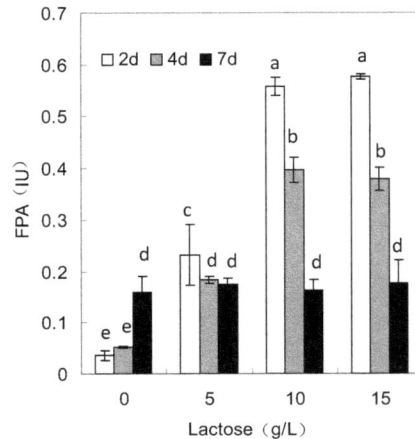

Figure 2. Effects of lactose on filter paper activity in the cultivation system. FPA: filter paper activity. The addition level of lactose was set as 0, 5, 10, and 15 g/L. The corresponding microcrystalline cellulose was 15, 10, 5 and 0 g/L. 2, 4 and 7 days expressed days of cultivation. Different letters on the column indicated significant difference (*P*<0.05).

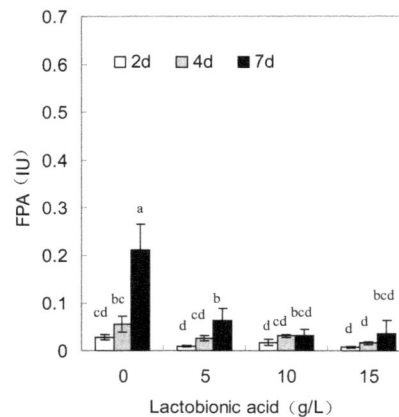

Figure 3. Effects of lactobionic acid on filter paper activity in the cultivation system. FPA: filter paper activity. The addition level of lactobionic acid was set as 0, 5, 10, and 15 g/L. The corresponding microcrystalline cellulose was 15, 10, 5 and 0 g/L. 2, 4 and 7 days expressed days of cultivation. Different letters on the column indicated significant difference (*P*<0.05).

elements. Some reports showed that yeast extract could stimulate cell growing and cellulase production of *T. reesei* (Domingues et al., 2000; Ryu and Mandels, 1980). In this study, the concentrations of yeast extract were set as 5, 10, 15 and 20 g/L. The results were shown as Figure 5.

The FPAs increased with yeast extract addition in the range of 5 - 15 g/L at 4 days of cultivation. The yeast

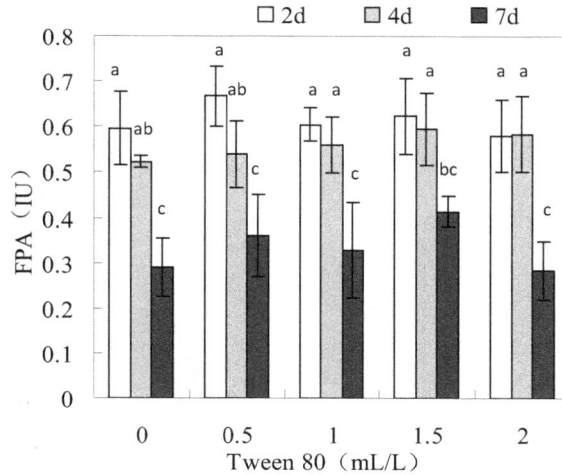

Figure 4. Effects of Tween 80 on filter paper activity in the cultivation system. Tween 80 was set as 0, 0.5, 1.0, 1.5 and 2.0 mL. FPA: filter paper activity. 2d, 4d and 7 days expressed days of cultivation. Different letters on the column indicated significant difference ($P<0.05$).

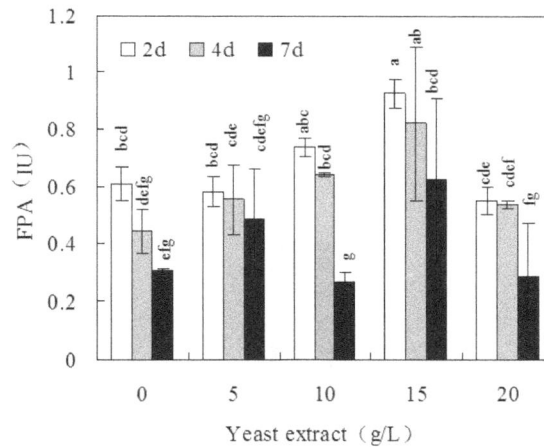

Figure 5. Effects of Yeast extract on filter paper activity in the cultivation system. The addition level of yeast extract was set as 0, 5, 10, 15 and 20 g/L. FPA: filter paper activity. 2, 4 and 7 days expressed days of cultivation. Different letters on the column indicated significant difference ($P<0.05$).

extract with 20 g/L addition had no significant effect on the increase of enzyme activity. This result indicated that addition of yeast extract could increase enzyme activity at some extent.

Combination of Tween 80 and yeast extract

Tween 80 and/or yeast extract were used to investigate whether their combination was better. As shown as

Figure 6, the treatments with yeast extract addition and combination addition had no significant differences. The enzyme activities of with Tween 80 addition were lower than those of the other treatments.

Enzyme activity of A6 and RUT-C30 during the cultivation

After optimization of cultivation media, the enzyme

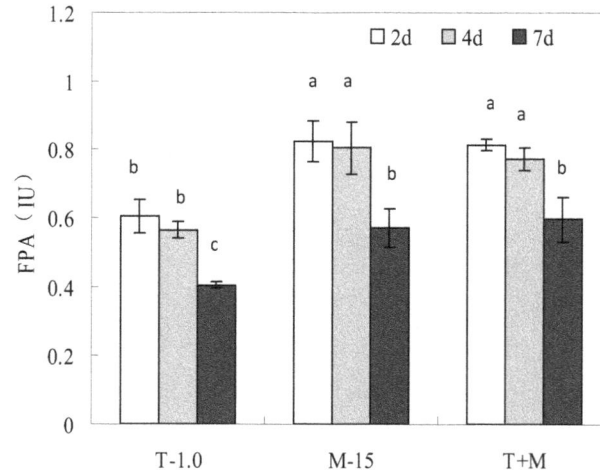

Figure 6. Effects of Tween 80 and yeast extract on filter paper activity in the cultivation system. T-1.0, Tween 80 1.0 mL/L; M-15 yeast extract 15 g/L; T+M, Tween 80 1.0 mL/L plus yeast extract 15 g/L; FPA: filter paper activity. 2, 4 and 7 days expressed days of cultivation. Different letters on the column indicated significant difference ($P<0.05$).

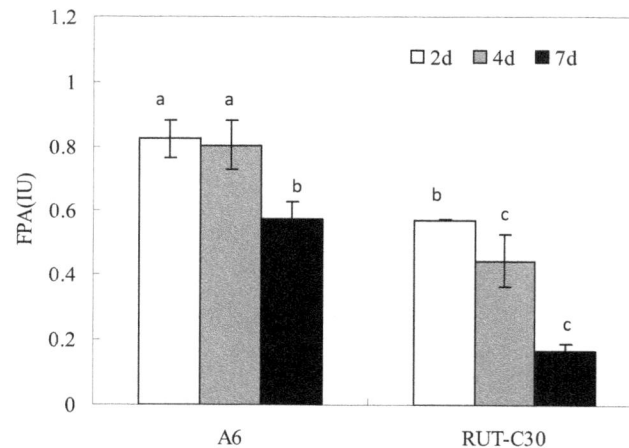

Figure 7. Enzyme activities of A6 and Rut-C30 during the cultivation. The media was Mandels medium (5 g/L microcrystalline cellulose plus 10 g/L lactose as carbon source. Yeast extract 15 g/L was added.). Different letters on the column indicated significant difference ($P<0.05$).

activities of *Trichoderma* A6 and *Trichocerma reesei* RUT-C30 during the cultivation were compared. From Figure 7, the FPAs of A6 were significantly higher than those of RUT-C30 during the cultivation.

Characteristics of crude enzyme

Fermentation liquid at 4 days of cultivation was taken to determine characteristics of crude enzyme. The tempe-

rature and pH of enzyme were tested. As shown in Figure 8, the relative activity of enzyme maintained above 85% in the range of 37 - 60°C. From 4.5 to 5.5 for pH, the relative enzyme activity maintained above 95%.

DISCUSSION

Breeding good microorganism is a crucial factor to improve the cellulase activity. At present, the microbes

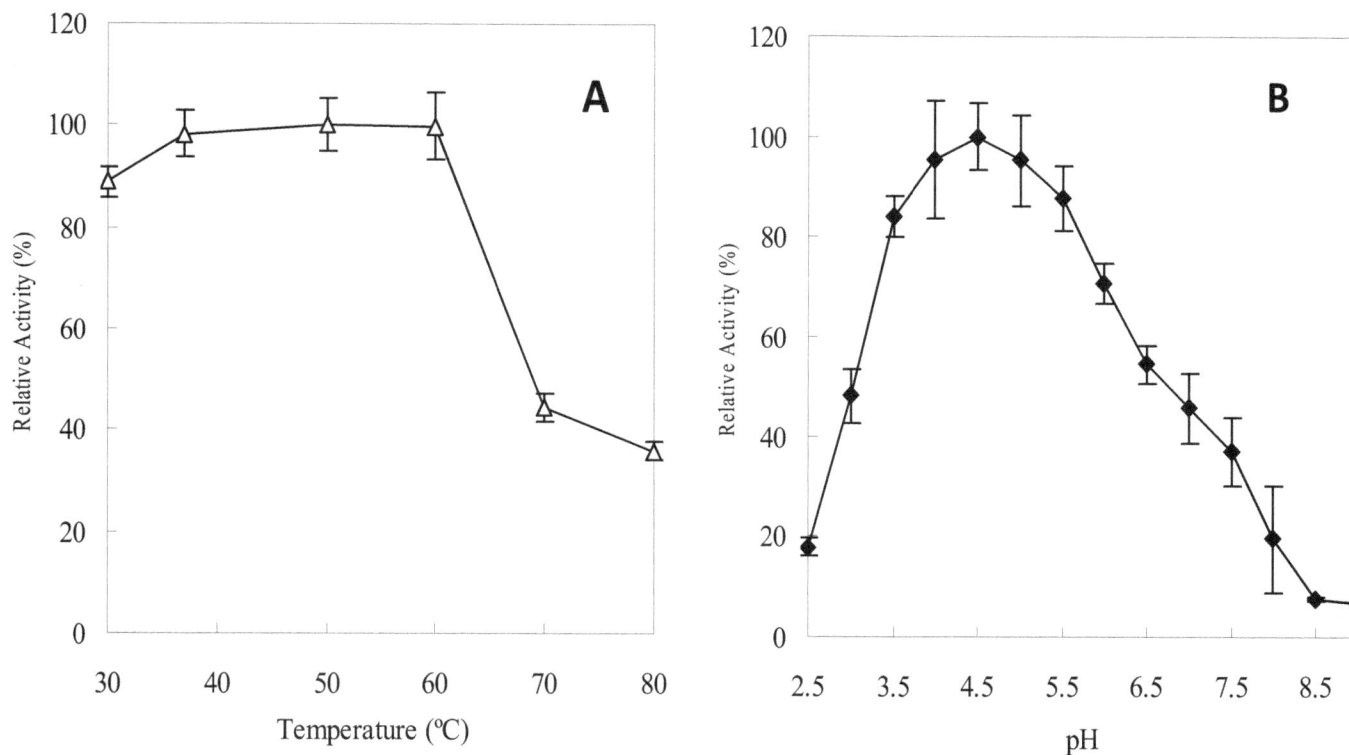

Figure 8. Relative enzyme activity at different temperatures (A) and pH levels (B).

used for cellulase production mostly belong to fungi. *Trichoderma*, *Aspergillus* and *Penicillium* are common microorganisms (Zhang et al., 2009). Currently, methods to improve the cellulase activity of strains are mainly physical, chemical mutagenesis and genetic engineering. Firstly, the parental strains should be with a high capability of cellulase production. Therefore, selection of the parental strains is important and necessary.

In this study, microorganism was derived from natural cellulose-degrading environment. During selection, natural lignocellulose was used as the sole carbon source. FPA was used as the index of cellulase production. Finally, *Trichoderma* sp. A6 was obtained. Previous reports showed that addition of lactose and lactobionic acid in the medium could increase enzyme production (Janas, 2002; Muthuvelayudham et al., 2006). In this study, lactose and lactobionic acid were added to the media, respectively. The results showed that lactose played a significant role in promoting cellulase production of the strain. Lactobionic acid was negative for enzyme production (Figure 3). A possible reason was that the addition of lactobionic acid changed the pH of cultivation system at a large extent. On the basis of lactose addition, the effects of Tween 80 and yeast extract were investigated. Results showed that addition of Tween 80 could increase cellulase, but not obvious (Figure 4). Yeast extract could significantly improve the enzyme activity of the cultivation system (Figure 5). The effect of

two-components-combined addition treatment showed no significant difference with the treatment addition of yeast extract (Figure 6). These results indicated that lactose and yeast extract play key roles in cellulase production of *Trichoderma* sp. A6. Compared with *Trichoderma reesei* RUT-C30, A6 owned higher FPAs while cultivated with the optimized media in this study.

Generally, the optimum temperature of cellulase was 45 - 65°C (Akila and Chandra, 2003). In the study, optimum temperature of the A6 strain screened was 37 ~ 60°C. As shown as Figure 7, when the temperatures were 37 and 60°C respectively, the enzyme activity reached 98.5 and 99.5% of the highest activity. In the cellulase application process, the reaction could be effectively progressed at lower temperature. The enzyme activity could be well maintained. At the same time, the energy and costs during the reaction will be reduced (Lv et al., 2007). Therefore, the enzyme from the A6 strain could be considered applied at a relatively low temperature environment.

Conclusion

A wild strain, *Trichoderma* sp. A6, was selected due to a high cellulase production capacity. Lactose and yeast extract could significantly improve the FPA in the liquid cultivation system. Addition proportions of lactose and yeast extract were 10 and 15 g/L, respectively. The proper temperature range of celluase was 37 - 60°C. The suitable pH range was 4.5 to 5.5.

ACKNOWLEDGEMENTS

This study was supported by the Fund of "Daqing Technology Innovation Program" (SCYH-2011-95) and "the Fundamental Research Funds for the Central Universities" (DL11BA11).

REFERENCES

Akila G, Chandra TS (2003). A novel cold-tolerant *Clostridium* strain PXYL1 isolated from a psychrophilic cattle manure digester that secretes thermolabile xylanase and cellulose. FEMS Microbiol. Lett. 219:63

Alriksson B, Rose SH, Van Zyl WH, Sjöde A, Nilvebrant NO, Sjöde A (2009) Cellulase production from spent lignocellulose hydrolysates by recombinant *Aspergillus niger*. Appl. Environ. Microbiol. 75 (8):2366-2374.

Alvira P, Tomás-Pejó E, Ballesteros M, Negro MJ (2010). Pretreatment technologies for an efficient bioethanol production process based on enzymatic hydrolysis: A review. Bioresour. Technol. 101:4851-4861.

Baleiras Couto MMB, Reizinho RG, Duarte FL (2005). Partial 26S rDNA restriction analysis as a tool to characterise non-*Saccharomyces* yeasts present during red wine fermentations. Int. J. Food Microbiol. 102:49-56.

Domingues FC, Queiroz JA, Cabral JMS, Fonseca LP (2000). The influence of culture conditions on mycelial structure and cellulase production by *Trichoderma reesei* Rut C-30. Enzyme Microb. Technol. 26:394-401.

Gharieb MM, Kierans M, Gadd GM (1999). Transformation and tolerance of tillurite by filamentous fungi: accumulation, reduction, and volatilization. Mycol. Res. 103 (3):299-305.

Ghose TK (1987). Measurement of cellulase activities. Pure Appl. Chem. 59 (2):257-268.

Janas P (2002) New inducers for cellulases production by *Trichoderma reesei* M-7. Food Sci Technol 5 (1):1-10.

Keshwani DR, Cheng JJ (2009). Switchgrass for bioethanol and other value-added applications: A review. Bioresour. Technol. 100:1515-1523.

Li X, Hou H (2010) Research progress in breeding of high-yield cellulase strains (in Chinese). Liquor-Making Sci. Technol. 5:92-94.

Lv M, Lv F, Fang Y, Liu S, Wang S (2007). Screening and identification of a cold-adapted cellulase-producing strain and characterization of cellulase (In Chinese). Food Sci. 28 (12):235-239.

Lynd LR, Laser MS, Bransby D, Dale BE, Davison B, Hamilton R, Himmel M, Keller M, McMillan JD, Sheehan J, Wyman CE (2008) How biotech can transform biofuels. Nat. Biotechnol. 26:169-172.

Muthuvelayudham R, Deiveegan S, Viruthagiri T (2006) Triggering of cellulase protein production using cellulose with lactose by *Trichoderma reesei*. Asian J. Microbial Biotechnol. Environ. Sci. 8 (2):33-35.

Niranjane AP, Madhou P, Stevenson TW (2007) The effect of carbohydrate carbon sources on the production of cellulase by *Phlebia gigantea*. Enzyme Microb. Technol. 40:1464-1468.

Reese ET, Maguire A (1969) Surfactants as stimulants of enzyme production by microorganisms. Appl. Microbiol. 17:242-245.

Ryu DDY, Mandels M (1980) Cellulases: biosynthesis and application. Enzyme Microb Technol 2:91-102.

Singh R, Kumar R, Bishnoi K, Bishnoi NR (2009) Optimization of synergistic parameters for thermostable cellulase activity of *Aspergillus heteromorphus* using response surface methodology. Biochem. Eng. J. 48:28-35.

Sticklen MB (2008). Plant genetic engineering for biofuel production: towards affordable cellulosic ethanol. Nat. Rev. Genet. 9:433-443.

Tamura K, Peterson D, Peterson N, Stecher G, Nei M, Kumar S (2011) MEGA5: Molecular evolutionary genetics analysis using maximum likelihood, evolutionary distance, and maximum parsimony methods. Mol. Biol. Evol. 28(10):2731-2739.

Wang W, Kang L, Lee YY (2010). Production of cellulase from kraft paper mill sludge by *Trichoderma reesei* Rut C-30. Appl. Biochem. Biotechnol. 161:382-394.

Wang X, Li P, Yuan X, Wang X, Cui Z (2011) Comparison of four microbial communities with wheat straw-degradation capability at normal temperature (in Chinese). J. China Agricultural University 16 (1):24-29.

Wu H, Yang H, You X, Li Y (2012) Isolation and characterization of saponin-producing fungal endophytes from *Aralia elata* in Northeast China. Int. J. Mol. Sci. 13 (12):16255-16266.

Yang HY, Wu H, Wang XF, Cui ZJ, Li YH (2011) Selection and characteristics of a switchgrass-colonizing microbial community to produce extracellular cellulases and xylanases. Bioresour. Technol 102:3546-3550.

Zhang X, Liu Y, Gao Y (2009) Research progress of cellulase and cellulase producing microorganisms screening (in Chinese). Feed Ind. 30 (22):14-16.

31

Effect of inoculum morphology on production of Nigerloxin by solid state fermentation

VASANTHA, K. Y., SALEEM JAVEED, CHAKRADHAR, D. and SATTUR, A. P.

Fermentation Technology and Bio Engineering Department Central Food Technological Research Institute Mysore, 570020 India.

Nigerloxin (2-amido-3-hydroxy-6-methoxy-5-methyle-4-(prop-1'enyl) benzoic acid) is a Lipoxygenase and Aldose reductase inhibitor produced by *Aspergillus niger* MTCC 5116. It is produced only under solid state fermentation (SSF) and the inoculum required for the inhibitor production is developed in submerged condition. As this is a newly discovered enzyme inhibitor with potential commercial success against diabetic complications, such as neuropathy and cataract, its complete fermentation process parametric study is not reported. In this study, the role of physical parameters (spore suspension, initial pH, incubation temperature and agitation) in spore germination and pellet size in inoculum development broth in submerged fermentation for the enhanced production of nigerloxin through SSF has been studied. It was concluded that 500 µl of spore suspension in the inoculum development broth at pH 7, incubated at 30°C and 200 rpm gave an ideal pellet size 1.23 mm resulting in 6.0 mg of nigerloxin/g dry weight of wheat bran in SSF.

Key words: Nigerloxin, spore germination, solid state fermentation, *Aspergillus niger*.

INTRODUCTION

Inoculum development is one of the major unit operations in a fermentation process, involving production of required quantity of viable desired microbial biomass in its most productive state (Hockenhull, 1980). There are successful commercial fermentations using pellets and others use dispersed forms. Much information exists on inoculum development for submerged fermentations. In fungal solid state fermentation, the thin line between success and failure of a productive fermentation process is the quality of biomass produced as inoculum. Unlike bacteria, fungal inoculum can be manipulated to required

pellet sizes or suspension forms through physical fermentation parameters to subsequently yield large quantities of the product in the main fermentation process. Further, not many reports exist on the influence of physical parameters on the fungal morphology in inoculum development broth on the production of desired metabolites through solid state fermentation.

Nigerloxin, produced by solid state fermentation of *Aspergillus niger* MTCC 5116 on wheat bran medium, is a Lipoxygenase and Aldose reductase inhibitor (Rao et al., 2002). It shows beneficial effects against diabetic

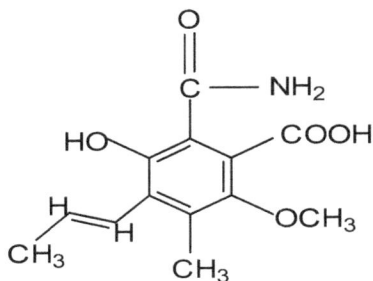

Figure 1. Structure of nigerloxin

complications such as oxidative stress and cataract formation in *vivo* (Suresh et al., 2012; Suresh et al., 2013). Production of nigerloxin is associated with the sporulation of the organism in SSF. The fermentation process is not completely reported for this potential commercially successful inhibitor against diabetic complications. While we have earlier described conditions for its optimum production (Chakradhar et al., 2009), here, we report the effect of physical parameters on inoculum by submerged fermentation and its subsequent production of nigerloxin in SSF.

MATERIALS AND METHODS

Culture

Aspergillus niger MTCC 5116 used in the present study was maintained on potato dextrose agar (Hi Media, Mumbai, India), at 4°C and subcultured once in every three weeks.

Solid state fermentation

Experiments were conducted in 500 ml Erlenmeyer flasks containing 10 g of wheat bran supplemented with 5 % (w/w) trisodium citrate with an initial moisture content of 60%. This medium served as a control wheat bran medium. After a thorough mixing, the flasks were autoclaved at 121°C for 1 h, cooled to room temperature and 2 ml of *A. niger* MTCC 5116 cell suspension was inoculated and incubated at 30°C for 6 days (Rao et al., 2005).

Extraction and determination of nigerloxin

At the end of fermentation, 100 ml of ethyl acetate was added to the fermented bran and kept on a rotary shaker at 200 rpm for 2 h. The bran was then filtered through cheesecloth followed by Whatman No 1 filter paper. The solvent was evaporated to yield a crude extract. 1.36 g of crude was suspended in 25 ml of chloroform and centrifuged at 2000 rpm for 20 min to obtain an orange precipitate. This was resuspended in 50 ml of warm ethanol to which 200 mg of activated charcoal was added. This content was filtered through Whatman No 1 filter paper and concentrated to obtain nigerloxin, which was used as standard. The nigerloxin concentration in samples was determined at 292 nm in UV-VIS spectrophotometer (Shimadzu UV 1601) (Rao et al., 2002)

Spore germination count

The spores from three day old slant were scraped and suspended in 4 ml of 2 % Tween 20 (v/v) solution for several minutes to facilitate wetting, and centrifuged for 1 min at 4000 rpm. The supernatant was then poured off and the spores resuspended in 2 ml distilled water. This suspension was then added to 5 ml of inoculum broth. Germination was determined based on the number of empty spore cases counted at 12th hour of growth using a Spencer Bright-Line Haemocytometer (American Optical Company) under magnifications of 10X and 40X and expressed as percentage spore germination (Braun, 1971).

Fungal pellet measurement

Fungal pellets grown in inoculum broth were collected and the measurements of fungal pellets were done either directly under microscope or a centimetre scale on a digital photograph projection (Pazouki and Panda, 2000).

Optimization of inoculum development parameters

To optimize the inoculum development parameters several media components and cultural conditions were altered in submerged condition. The inoculum produced was inoculated to controlled wheat bran media and observed for the enhanced production of nigerloxin: 1) The effect of various standard media like potato dextrose broth, oatmeal broth, czapeckdox broth, tryptone yeast extract broth, modified egg yolk broth, yeast extract malt extract broth, potassium tellurite broth, and glycerol asperagin broth on the production of inoculum was studied by inoculating loopful (5 X 10^5 spores) *A. niger* CFR-W-105 to each broth and pellet morphology, media pigmentation, spore germination and biomass production was studied. Nigerloxin production was evaluated by using inoculum developed by each standard medium and inoculating on to controlled wheat bran media; 2) the effect of initial pH of the inoculum development broth on biomass production, pellet size and spore germination was determined by altering the initial pH of the fermentation media with the addition of acid or alkali. Nigerloxin production was evaluated by using inoculum developed by each pH range and inoculating on to controlled wheat bran media; 3) effect of temperature was studied by incubating the organism at various temperatures ranging from 10 to 50°C in inoculum development broth.The nature of pellets, rate of spore germination, biomass production and nigerloxin production was studied; 4) The effect of agitation condition on the development of pellet size, spore germination, biomass production and nigerloxin production was studied by incubating the inoculated flasks at various agitation conditions on a rotary shaker. Nigerloxin production was evaluated by inoculating the inoculum developed under all temperature ranges.

RESULTS AND DISCUSSION

Nigerloxin (2-amino–3hydroxy-6-methoxy-5-methyl-4-(prop-1'-enyl) benzoic acid) (Figure 1) with a molecular weight of 265 and molecular formula $C_{13}H_{15}NO_5$ was discovered in our laboratory as a potent inhibitor of rat eye lens aldose reductase and lipoxygenase with a free radical scavenging property. The inhibitor is produced only by solid state fermentation and not in submerged

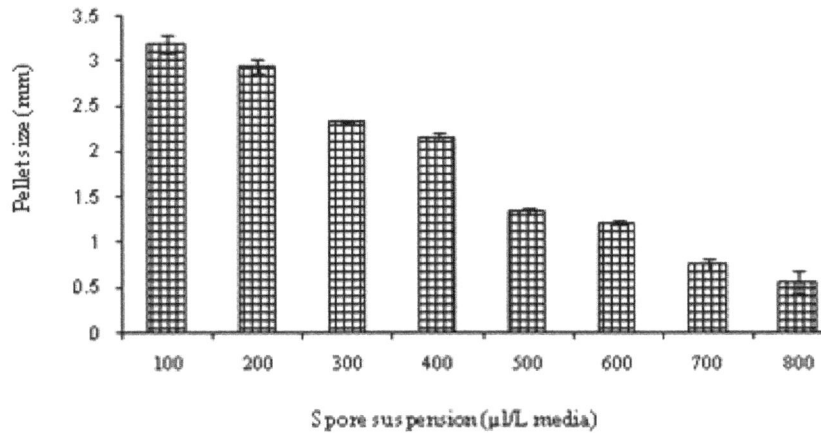

Figure 2a. Effect of spore suspension on pellet size.

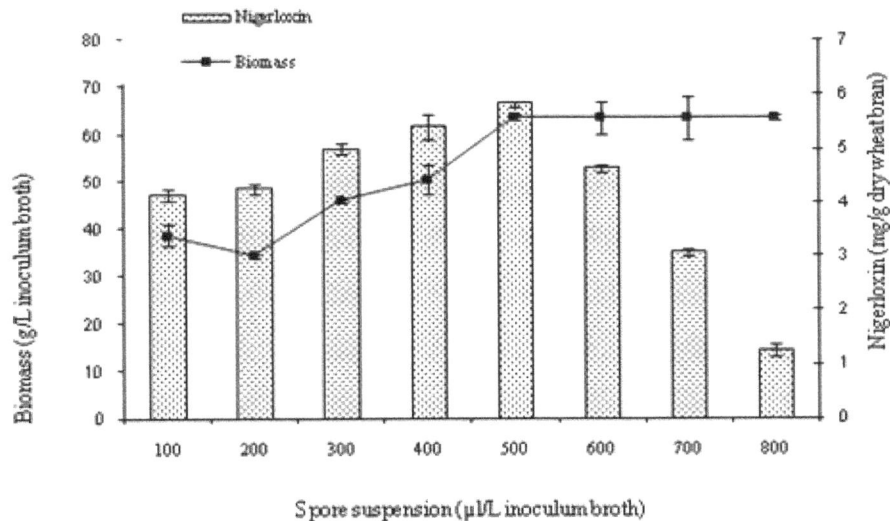

Figure 2b. Effect of spore suspension on biomass of inoculums and nigerloxin production by SSF.

fermentation conditions and the production of the inhibitor is directly related to sporulation of the culture (Chakradhar et al., 2009).

Optimization of physical parameters of inoculum

For these studies, Czapekdox broth with 6 g/L yeast extract was used as the inoculum medium since it showed good production compared to other nutritional media and solid state fermentation using wheat bran medium with 5% trisodium citrate, for the evaluation of nigerloxin production.

Effect of spore suspension

The amount of spores provided to a particular media in submerged fermentation condition has a direct impact on the morphology of pellets developed (Papagianni and Moo-Young, 2002). Increasing spore suspension from 100 µl/L inoculum broth to 800 µl/L inoculum broth showed a steady decrease in pellet size (Figure 2a). Spore suspensions below 400 µl/L media showed pellets sized above 2.1 mm which were not ideal for nigerloxin production in SSF (Figure 2b) perhaps as these yielded a biomass below 50 g/L. Spore suspensions greater than 700 µl/L produced smaller pellets perhaps due to lack of

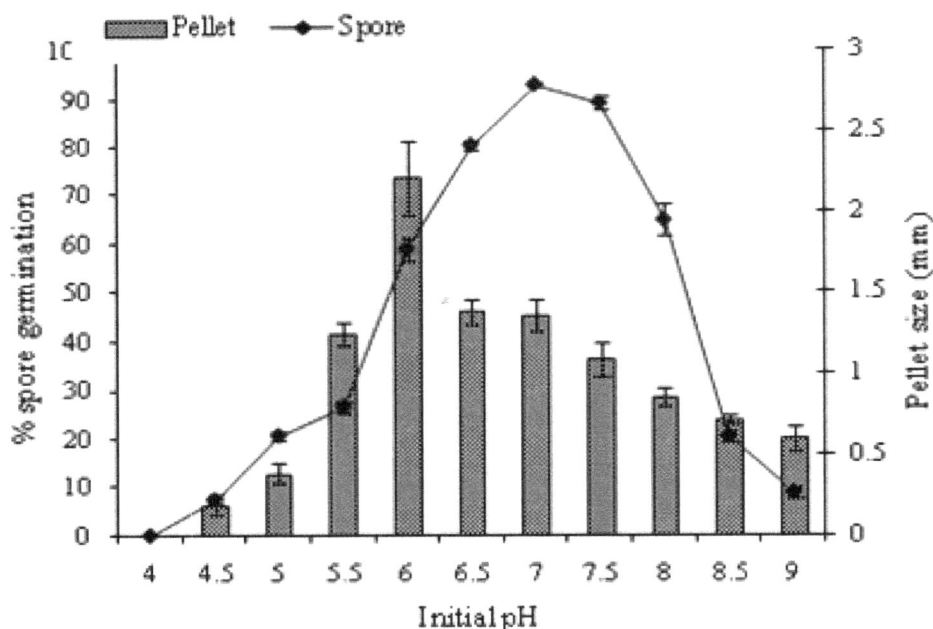

Figure 3a. Effect of initial pH of inoculum development broth on the spore germination, and pellet size.

space and did not translate into more amount of biomass and saw a significant decrease in Nigerloxin production (Figure 2b). In fact, the biomass produced above 500 µl/L, at around 50 g/L, was the same till 800 µl/L, whereas nigerloxin fell below 2 mg/g dry wheat bran. These results are in agreement with that of Van Suijdam et al. (1980) and Calam (1987) where higher concentration of spore suspension failed to produce pellets and suspension form of inoculum was produced. The highest Nigerloxin concentration was produced by using 500 µl/L inoculum broth spore suspension which yielded 5.81±0.04 mg/g dry wheat bran. Hence, 500 µl/L, from the stock of 7 x 10^9/ml spores concentration was standardized for the rest of the experiments.

Effect of pH

The effect of initial pH of inoculum medium on pellet size, spore germination and biomass production was studied by adjusting the pH of the broth from 2.0 to 9.0 (Figure 3a). It was observed that spore germination was less than 20% below pH 5 and above 8.5 with the highest germination of 92% at pH 7. Further, as pH increased from 4 to 7, there was an increase in pellet size but interestingly, the pellet size over the entire pH spectrum tested was less than 1.5 mm, except for pH 6.0 where it reached 2.4 mm. Production of biomass of 40-53 g/L was seen between pH 5.5 to 7.0 with a complete absence at

pH 4 and a slight increase between pH 4.5 to 5. There seems to be a correlation between spore germination and biomass production rather than with pellet size. The phenomenon of pellet formation being strongly influenced by pH is in agreement with the result of Galbraith and Smith and Carlsen et al. (1969, 1995). There was a complete absence of nigerloxin production by SSF when inoculum was developed below pH 5 (Figure 3b). The highest nigerloxin production of 5.7 mg/g dry wheat bran was observed at pH 6.5 after which it fell drastically.

Effect of temperature

Temperature plays an important role in the development of pellets in inoculum medium and spore germination in solid state fermentation (Estrada et al., 2000). It was seen that spore germination below 20 and above 40°C was less than 30% increasing to 91% at 30°C (Figure 4a). A similar pattern was observed with pellet size formation in the range of 1.2 to 1.3 mm between 25 and 35°C and around 0.4 mm at the extreme temperatures tested. Unlike the results seen in initial pH of the medium, biomass produced correlated to both spore germination and pellet sizes. The highest biomass produced was 58 g/L inoculum broth at 30 °C. The absence in nigerloxin production was seen in both extreme ends of temperature and the highest nigerloxin was at 30 °C which produced 5.76±0.07 mg/g dry wheat bran (Figure 4b).

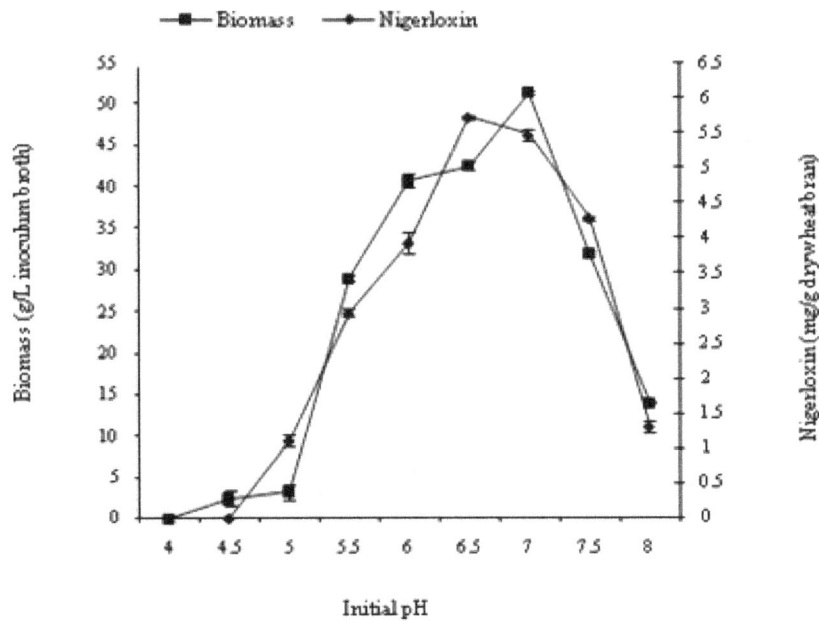

Figure 3b. Effect of initial pH of inoculum development broth on the production of biomass and nigerloxin.

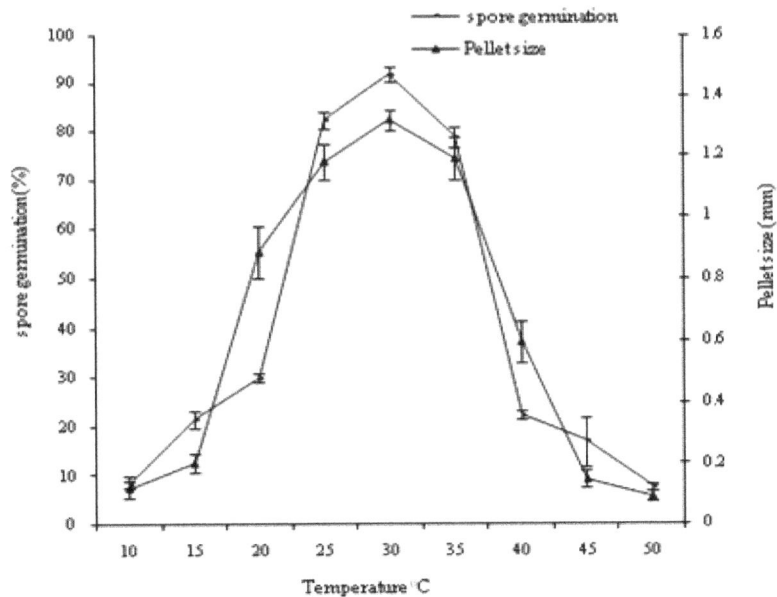

Figure 4a. Effect of incubation temperature of inoculum development broth on spore germination, and pellet size

Effect of agitation

Agitation is the by far the most important physical parameter in inoculum development for SS. Hence the effect of agitation condition on pellet size and spore germination was studied in inoculum broth and nigerloxin production in SSF.

Interestingly, the spore germination remained almost the

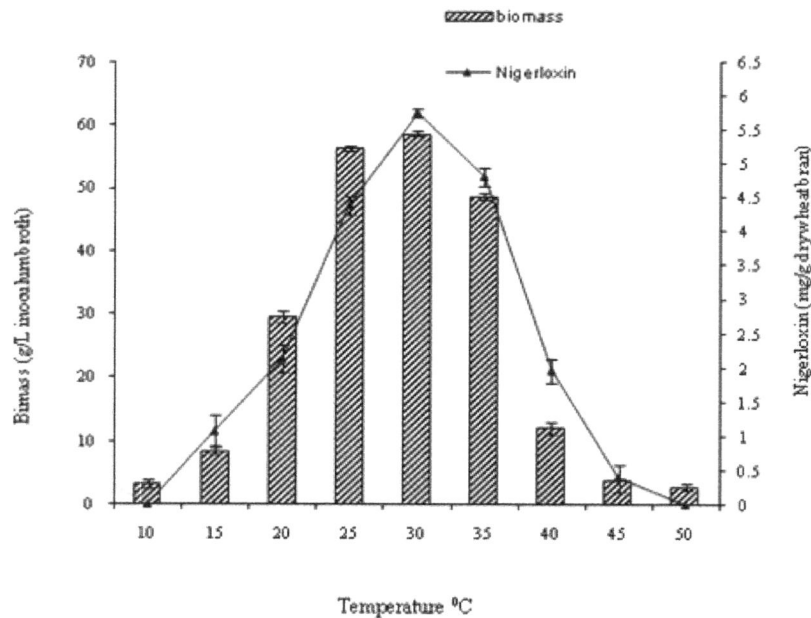

Figure 4b. Effect of incubation temperature of inoculum development broth on biomass and nigerloxin production.

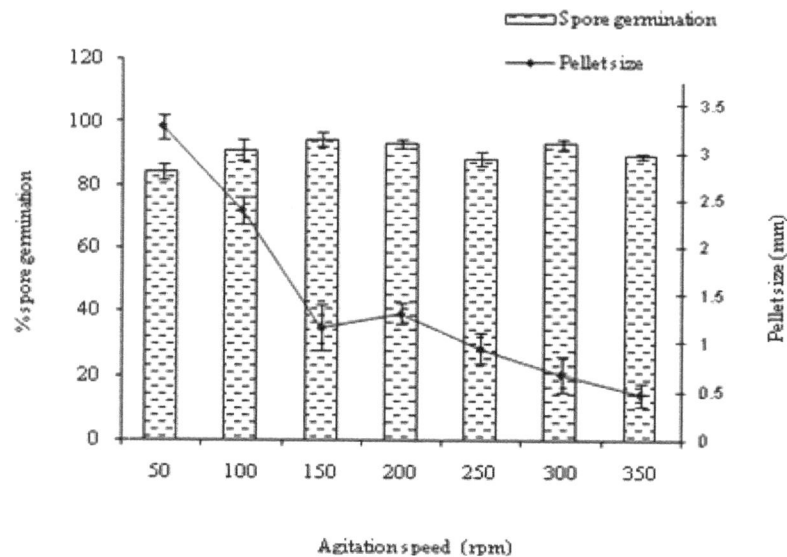

Figure 5a. Effect of agitation condition of inoculum development broth on spore germination, and pellet size

same in all agitation speeds employed with the highest at 93% at 200 rpm, whereas pellet sizes fell steadily from the highest of 3.2 mm at 50 rpm to 0.5 mm at 350 rpm (Figure 5a). Further, biomass produced was more than 40 g/L under all ranges of agitation conditions. The

highest production of nigerloxin was observed at 200 rpm which yielded 6 mg nigerloxin/g dry wheat bran (Figure 5b).

The optimized inoculum condition for the better production of nigerloxin was at 500 µl of spore

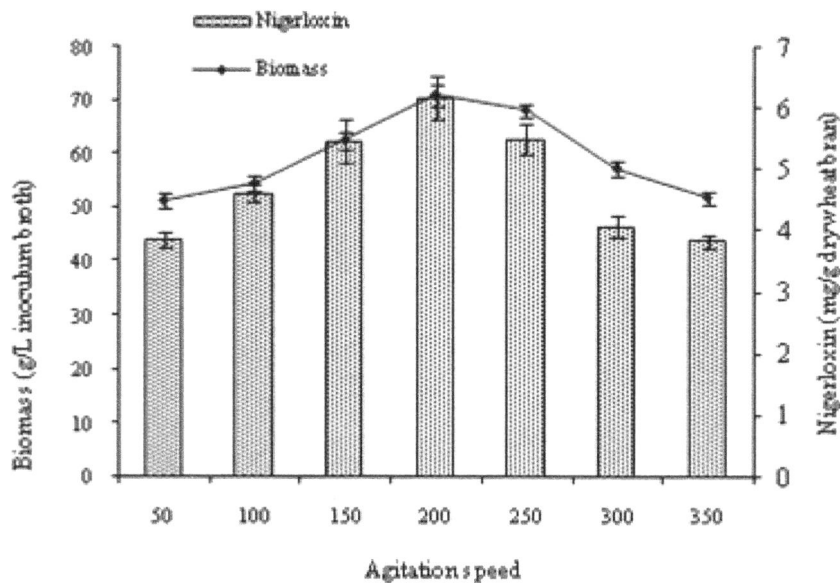

Figure 5b. Effect of agitation condition of inoculum development broth on biomass and nigerloxin production.

suspension in the inoculum development broth at pH 7, incubated at 30 °C and 200 rpm gave an ideal pellet size of 1.23 mm resulting in 6.0 mg of nigerloxin/g dry weight of wheat bran in SSF.

Production of secondary metabolite from microorganism is a complex task as they require a number of optimal conditions. Inoculum is one of such factors, which if not optimized, may lead to poor or absence of desired product yield. The successful production of a fungal metabolite requires knowledge of the growth characteristics and the physiology of the fungus in question. Thus, for each fermentation, the precise physiological condition and optimised media for inoculum development must be established. The present work was an attempt to understand the type of inoculum conditions required to enhance the production of nigerloxin in solid state fermentation using controlled wheat bran medium by *A. niger* CFTR-W-105. The data generated in this study would be helpful in producing nigerloxin on large scale and has the potential for further developing it into a powerful therapeutic molecule against diabetic complications.

Conflict of Interests

The author(s) have not declared any conflict of interests.

ACKNOWLEDGEMENT

The authors thank the Department of Biotechnology, New Delhi, India for financial support. VKY thanks CSIR for a Senior Research Fellowship award.

REFERENCES

Braun KL (1971). Spore germination time in *Fuligo septica*. The Ohio. J. Sc. 71:305-309.

Calam CT (1987). Process development in antibiotic fermentations. Cambridge Studies in Biotechnology, vol. 4. Cambridge: Cambridge Univ. Press.

Carlsen M, Spo"hr AB, Nielsen J, Villadsen J (1995). Morphology and physiology of an a-amylase producing strain of *Aspergillus oryzae* during batch cultivations. Biotechnol. Bioeng. 49:266-276.

Chakradhar D, Saleem J, Sattur AP (2009). Studies on the production of nigerloxin using agro industrial residues by solid-state fermentation. J. Ind. Microbiol. Biotechnol. 36:1179-1187.

Estrada AB, Dodd JC, Jeffries P (2000). Effect of humidity and temperature on conidial germination and appressorium development of two Philippine isolates of the mango anthracnose pathogen olletotrichum gloeosporioides. Plant Pathol. 49:608-618.

Galbraith JC, Smith JE (1969). Filamentous growth of *Aspergillus niger* in submerged shake culture. Trans. Brit. Mycol. Soc. 52:237-246.

Hockenhull DJD (1980). Inoculum development with particular reference to *Aspergillus* and *Penicillium*, in: Smith, JE, Berry DR, Kristiansen B (Eds.), Fungal biotechnology, Academic press, New York, pp. 1-24.

Papagianni M, Moo-Young M (2002). Protease secretion in glucoamylase producer *Aspergillus niger* cultures: fungal morphology and inoculum effects. Process. Biochem. 37:1271-1278.

Pazouki M, Panda T (2000). Understanding the morphology of fungi. Bioprocess. Eng. 22:127-143.

Rao KCS, Divakar S, Babu KN, Karanth NG, Sattur AP (2002). Nigerloxin, a novel inhibitor of aldose reductase and lipoxygenase with free radical scavenging activity from *Aspergillus niger* CFR-W-105. J. Antibiotic. 55:789-793.

Rao KCS, Karanth NG, Sattur AP (2005). Production of nigerloxin, an enzyme inhibitor and a free radical scavenger, by Aspergillus niger using solid state fermentation. Process Biochem. 40(7):2517-2522.

Suresh BS, Vasantha KY, Sattur AP, Srinivasan K (2013). Beneficial influence of fungal metabolite nigerloxin on diabetes-induced oxidative stress in experimental rats. Can. J. Physiol. Pharmacol. 91(2):149-156.

Suresh BS, Sattur AP, Srinivasan K (2012). Beneficial influence of fungal metabolite nigerloxin on eye lens abnormalities in experimental Diabetes. Can. J. Physiol. Pharmacol. 90(4):387-394.

Van Suijdam JC, Kossen NWF, Paul PG (1980). An inoculum technique for the production of fungal pellets. Eur. J. Appl. Microbiol. Biotechnol. 10:211-221.

Permissions

All chapters in this book were first published in JYFR, by Academic Journals; hereby published with permission under the Creative Commons Attribution License or equivalent. Every chapter published in this book has been scrutinized by our experts. Their significance has been extensively debated. The topics covered herein carry significant findings which will fuel the growth of the discipline. They may even be implemented as practical applications or may be referred to as a beginning point for another development.

The contributors of this book come from diverse backgrounds, making this book a truly international effort. This book will bring forth new frontiers with its revolutionizing research information and detailed analysis of the nascent developments around the world.

We would like to thank all the contributing authors for lending their expertise to make the book truly unique. They have played a crucial role in the development of this book. Without their invaluable contributions this book wouldn't have been possible. They have made vital efforts to compile up to date information on the varied aspects of this subject to make this book a valuable addition to the collection of many professionals and students.

This book was conceptualized with the vision of imparting up-to-date information and advanced data in this field. To ensure the same, a matchless editorial board was set up. Every individual on the board went through rigorous rounds of assessment to prove their worth. After which they invested a large part of their time researching and compiling the most relevant data for our readers.

The editorial board has been involved in producing this book since its inception. They have spent rigorous hours researching and exploring the diverse topics which have resulted in the successful publishing of this book. They have passed on their knowledge of decades through this book. To expedite this challenging task, the publisher supported the team at every step. A small team of assistant editors was also appointed to further simplify the editing procedure and attain best results for the readers.

Apart from the editorial board, the designing team has also invested a significant amount of their time in understanding the subject and creating the most relevant covers. They scrutinized every image to scout for the most suitable representation of the subject and create an appropriate cover for the book.

The publishing team has been an ardent support to the editorial, designing and production team. Their endless efforts to recruit the best for this project, has resulted in the accomplishment of this book. They are a veteran in the field of academics and their pool of knowledge is as vast as their experience in printing. Their expertise and guidance has proved useful at every step. Their uncompromising quality standards have made this book an exceptional effort. Their encouragement from time to time has been an inspiration for everyone.

The publisher and the editorial board hope that this book will prove to be a valuable piece of knowledge for researchers, students, practitioners and scholars across the globe.

List of Contributors

M. A. Abdel-Rahman Tahany
Cairo University, Faculty of Science, Cairo, Egypt

A. K. Hegazy
Cairo University, Faculty of Science, Cairo, Egypt

A. Mohsen Sayed
Cairo University, Faculty of Science, Cairo, Egypt

H. F. Kabiel
Cairo University, Faculty of Science, Cairo, Egypt

T. El-Alfy
Cairo University, Faculty of Pharmacy, Cairo, Egypt

S. M. El-Komy
Cairo University, Faculty of Pharmacy, Cairo, Egypt

R. Narayanan Krishnamoorthy
Department of Agricultural Microbiology, Centre for Plant Molecular Biology and Biotechnology, Tamil Nadu Agricultural University, Coimbatore, Tamil Nadu, India, Pin: 641 003

K. Vijila
Department of Agricultural Microbiology, Centre for Plant Molecular Biology and Biotechnology, Tamil Nadu Agricultural University, Coimbatore, Tamil Nadu, India, Pin: 641 003

K. Kumutha
Department of Agricultural Microbiology, Centre for Plant Molecular Biology and Biotechnology, Tamil Nadu Agricultural University, Coimbatore, Tamil Nadu, India, Pin: 641 003

Mukaram Shikara
Biotechnology Division, Department of Applied Sciences, University of Technology, Baghdad, Iraq

M. A. Belewu
Department of Animal Production, Microbial Biotechnology and Dairy Science Laboratory, University of Ilorin, Ilorin, Kwara State, Nigeria

R. Sam
School of Medical Sciences, University of Cape Coast, Ghana

Sideney Becker Onofre
Laboratory of Microbiology of the Paranaense University – UNIPAR – Unit Campus Francisco Beltrão, Paraná, Brasil

Paula Steilmnn
Laboratory of Microbiology of the Paranaense University – UNIPAR – Unit Campus Francisco Beltrão, Paraná, Brasil

Julia Bertolini
Laboratory of Microbiology of the Paranaense University – UNIPAR – Unit Campus Francisco Beltrão, Paraná, Brasil

Daniele Rotta
Laboratory of Microbiology of the Paranaense University – UNIPAR – Unit Campus Francisco Beltrão, Paraná, Brasi

Aline Sartori
Laboratory of Microbiology of the Paranaense University – UNIPAR – Unit Campus Francisco Beltrão, Paraná, Brasil

Francini Yumi Kagimura
Laboratory of Microbiology of the Paranaense University – UNIPAR – Unit Campus Francisco Beltrão, Paraná, Brasil

Sara Ângela Groff
Laboratory of Microbiology of the Paranaense University – UNIPAR – Unit Campus Francisco Beltrão, Paraná, Brasil

Luciana Mazzali
Laboratory of Microbiology of the Paranaense University – UNIPAR – Unit Campus Francisco Beltrão, Paraná, Brasil

Liberata Nyang'oso Mwita
Department of Molecular Biology and Biotechnology, College of Natural and Applied Sciences, University of Dar es Salaam, Tanzani

Anthony Manoni Mshandete
Department of Molecular Biology and Biotechnology, College of Natural and Applied Sciences, University of Dar es Salaam, Tanzania

Sylvester Leonard Lyantagaye
Department of Molecular Biology and Biotechnology, College of Natural and Applied Sciences, University of Dar es Salaam, Tanzania

Harpreet Sharma
Department of Biotechnology, Lovely Professional University, Phagwara-144402 Punjab, India

Leena Parihar
Department of Biotechnology, Lovely Professional University, Phagwara-144402 Punjab, India

M. M. Rashid
Bangladesh Rice Research Institute, Regional Station, Sonagazi, Feni, Bangladesh

A. B. M. Ruhul Amin
Rupali Bank Ltd., Bangladesh

F. Rahman
Student, Plant Pathology Division, Bangladesh Agricultural University, Mymensingh, Bangladesh

João V. B. de Souza
Instituto Nacional de Pesquisas da Amazônia, Manaus, AM, Brasil

Alita M. Lima
Instituto Nacional de Pesquisas da Amazônia, Manaus, AM, Brasil

Eveleise S. de J. Martins
Instituto Nacional de Pesquisas da Amazônia, Manaus, AM, Brasil

Julia I. Salem
Instituto Nacional de Pesquisas da Amazônia, Manaus, AM, Brasil

Zeng Defang
School of Resource and Environmental Engineering, Wuhan University of Technology, Wuhan, People's Republic of China
Hubei Key Laboratory of Mineral Resources Processing and Environment, Wuhan 430070, People's Republic of China

Tu Renjie
School of Resource and Environmental Engineering, Wuhan University of Technology, Wuhan, People's Republic of China
Hubei Key Laboratory of Mineral Resources Processing and Environment, Wuhan 430070, People's Republic of China

Wu Shan
School of Resource and Environmental Engineering, Wuhan University of Technology, Wuhan, People's Republic of China
Hubei Key Laboratory of Mineral Resources Processing and Environment, Wuhan 430070, People's Republic of China

G. O. Ihejirika
Department of Crop Science and Technology, Federal University of Technology, P. M. B. 1526, Owerri, Nigeria

M. I. Nwufo
Department of Crop Science and Technology, Federal University of Technology, P. M. B. 1526, Owerri, Nigeria

S. O. Anagboso
Department of Crop Science and Technology, Federal University of Technology, P. M. B. 1526, Owerri, Nigeria

O. M. David
Department of Microbiology, Ekiti State University, Ado-Ekiti, Nigeria

E. D. Fagbohun
Department of Microbiology, Ekiti State University, Ado-Ekiti, Nigeria

A. O.Oluyege
Department of Microbiology, Ekiti State University, Ado-Ekiti, Nigeria

A. Adegbuyi
Department of Pharmacy, University Teaching Hospital, Ado-Ekiti, Nigeria

Sideney B. Onofre
Departamento de Ciências Biológicas, Laboratório de Microbiologia, Universidade Paranaense, UNIPAR, Unidade-Campus Francisco Beltrão, Av. Júlio Assis Cavalheiro, 2000 – Bairro Centro - 85601-010,Francisco Beltrão – Paraná – Brasil

Cristiane R. Kasburg
Departamento de Ciências Biológicas, Laboratório de Microbiologia, Universidade Paranaense, UNIPAR, Unidade-Campus Francisco Beltrão, Av. Júlio Assis Cavalheiro, 2000 – Bairro Centro - 85601-010,Francisco Beltrão – Paraná – Brasil

Danusa de Freitas
Departamento de Ciências Biológicas, Laboratório de Microbiologia, Universidade Paranaense, UNIPAR, Unidade-Campus Francisco Beltrão, Av. Júlio Assis Cavalheiro, 2000 – Bairro Centro - 85601-010,Francisco Beltrão – Paraná – Brasil

Silvana Damin
Departamento de Ciências Biológicas, Laboratório de Microbiologia, Universidade Paranaense, UNIPAR, Unidade-Campus Francisco Beltrão, Av. Júlio Assis Cavalheiro, 2000 – Bairro Centro - 85601-010,Francisco Beltrão – Paraná – Brasil

Andréia Vilani
Departamento de Ciências Biológicas, Laboratório de Microbiologia, Universidade Paranaense, UNIPAR, Unidade-Campus Francisco Beltrão, Av. Júlio Assis Cavalheiro, 2000 – Bairro Centro - 85601-010,Francisco Beltrão – Paraná – Brasil

Jéssica A. Queiroz
Departamento de Ciências Biológicas, Laboratório de Microbiologia, Universidade Paranaense, UNIPAR, Unidade-Campus Francisco Beltrão, Av. Júlio Assis Cavalheiro, 2000 – Bairro Centro - 85601-010,Francisco Beltrão – Paraná – Brasil

Francini Y. Kagimura
Departamento de Ciências Biológicas, Laboratório de Microbiologia, Universidade Paranaense, UNIPAR, Unidade-Campus Francisco Beltrão, Av. Júlio Assis Cavalheiro, 2000 – Bairro Centro - 85601-010,Francisco Beltrão – Paraná – Brasil

S. Rehman
Indian Institute of Integrative Medicine (CSIR), Sanatnagar, Srinagar, India

Tariq Mir
Maternity and Child Hospital, Dammam, India

A. Kour
Indian Institute of Integrative Medicine (CSIR), Sanatnagar, Srinagar, India

P. H. Qazi
Indian Institute of Integrative Medicine (CSIR), Sanatnagar, Srinagar, India

P. Sultan
Indian Institute of Integrative Medicine (CSIR), Sanatnagar, Srinagar, India

A. S. Shawl
Indian Institute of Integrative Medicine (CSIR), Sanatnagar, Srinagar, India

P. Senthilraja
Department of Zoology, Faculty of Science, Annamalai University, Annamalai nagar, India

K. Kathiresan
Department of Zoology, Faculty of Science, Annamalai University, Annamalai nagar, India

K. Saravanakumar
Department of Zoology, Faculty of Science, Annamalai University, Annamalai nagar, India

M. Cristina Romero
Facultad de Ciencias Veterinarias, calle 60 y 119, s/n°, Universidad Nacional de La Plata, 1900 La Plata, Argentina

M. Inés Urrutia
Facultad de Ciencias Agrariasy Forestales, calle 60 y 119, s/n°, Universidad Nacional de La Plata, 1900 La Plata, Argentina

H. Enso Reinoso
Facultad de Ciencias Veterinarias, calle 60 y 119, s/n°, Universidad Nacional de La Plata, 1900 La Plata, Argentina

M. Moreno Kiernan
Minist. Salud. Prov. B.A., calle 60 y 119, s/n°, Universidad Nacional de La Plata, 1900 La Plata, Argentina

P. O. Bankole
Department of Botany, University of Lagos, Akoka, Lagos State

A. A. Adekunle
Department of Botany, University of Lagos, Akoka, Lagos State

Dominic Menge
Jomo Kenyatta University of Agriculture and Technology (JKUAT), P. O. Box 62000-00100 Nairobi, Kenya
Cashew Research Programme, Naliendele Agricultural Research Institute (NARI), P. O. Box 509, Mtwara, Tanzania
University of Göttingen, Grisebachstrasse 6, 37077 Göttingen, Germany

Martha Makobe
Jomo Kenyatta University of Agriculture and Technology (JKUAT), P. O. Box 62000-00100 Nairobi, Kenya

Shamte Shomari
Cashew Research Programme, Naliendele Agricultural Research Institute (NARI), P. O. Box 509, Mtwara, Tanzania

Andreas. V. Tiedemann
University of Göttingen, Grisebachstrasse 6, 37077 Göttingen, Germany

Caiying Mei
College of Life Sciences, Sun Yat-sen University, Guangzhou 510275, China
Guangdong Entomological Institute, Guangzhou 510260, China

Yi Zhang
Guangdong Entomological Institute, Guangzhou 510260, China

Xiongmin Mao
Guangdong Entomological Institute, Guangzhou 510260, China

Keqing Jiang
College of Life Sciences, Sun Yat-sen University, Guangzhou 510275, China
Guangdong Entomological Institute, Guangzhou 510260, China

Li Cao
Guangdong Entomological Institute, Guangzhou 510260, China

Xun Yan
Guangdong Entomological Institute, Guangzhou 510260, China

Richou Han
Guangdong Entomological Institute, Guangzhou 510260, China

U. N. Bhale
Research Laboratory, Department of Botany, Arts, Science and Commerce College, Naldurg, Tq. Tuljapur, Osmanabad District, 413602 (M.S.) India

P. M. Wagh
Department of Biology, S. S. and L. S. Patkar College of Arts and Science, Goregaon (W), Mumbai- 400063 (M. S.) India

J. N. Rajkonda
Department of Botany, Yeshwantrao Chavan College, Tuljapur, Osmanabad District, 413601 (M. S.) India

Kaminee Ranka
Centre for Genome Research, Department of Microbiology and Biotechnology Centre, Faculty of Science, the Maharaja Sayajirao University of Baroda, Vadodara -390002, India

Bharat B. Chattoo
Centre for Genome Research, Department of Microbiology and Biotechnology Centre, Faculty of Science, the Maharaja Sayajirao University of Baroda, Vadodara -390002, India

M. Moghtader
Department of Biodiversity, International Center for Science, High Technology and Environmental Sciences, Kerman, Iran

M. Moghtader
Department of Biodiversity, Institute of Science and High Technology and Environmental Sciences, Graduate University of Advanced Technology, Kerman, Iran

Chiadikobi Ejikeme Onyia
South-East Zonal Biotechnology Centre, University of Nigeria, Nsukka, Enugu State, Nigeria

Chukwudi Uzoma Anyanwu
Department of Microbiology, University of Nigeria, Nsukka, Enugu State, Nigeria

Ali A. Juwaied
Department of Applied Science, Biochemical Division, University of Technology, Baghdad, Iraq

Suhad Adnan
Department of Applied Science, Biochemical Division, University of Technology, Baghdad, Iraq

Ahmed Abdulamier Hussain Hussain Al-Amiery
Department of Applied Science, Biochemical Division, University of Technology, Baghdad, Iraq

Aditi Sourabh
Department of Microbiology, CSK Himachal Pradesh Agricultural University, Palampur, Himachal Pradesh-176062, India

Sarbjit Singh Kanwar
Department of Microbiology, CSK Himachal Pradesh Agricultural University, Palampur, Himachal Pradesh-176062, India

Om Prakash Sharma
Biochemistry Laboratory, Indian Veterinary Research Institute, Regional Station Palampur, Himachal Pradesh-176061, India

Abosede Margaret Ebabhi
Department of Biological Sciences, McPherson University, Seriki Sotayo, Ogun State, Nigeria

Adedotun Adeyinka Adekunle
Department of Botany, Faculty of Science, University of Lagos, Lagos State, Nigeria

Wahab Oluwanisola Okunowo
Department of Biochemistry, College of Medicine, University of Lagos, Lagos State, Nigeria

Akinniyi Adediran Osuntoki
Department of Biochemistry, College of Medicine, University of Lagos, Lagos State, Nigeria

Dominique Claude MOSSEBO
Mycological Laboratory, University of Yaoundé 1, B.P. 1456 Yaoundé, Cameroon
Mycology Section, Jodrell Laboratory, Royal Botanic Gardens, Kew, Richmond, Surrey, TW9 3AB, England

Rose T. AMBIT
Mycological Laboratory, University of Yaoundé 1, B.P. 1456 Yaoundé, Cameroon

Marie-Claire MACHOUART
Laboratoire Stress Immunité Pathogène – EA 7300 – Parasitologie-Mycologie – Université de Lorraine, 9 avenue de la forêt de Have, 54511 Vandoeuvre les Nancy, France

Germain KANSCI
Department of Biochemistry, Faculty of Science, University of Yaoundé 1, B.P. 812 Yaoundé, Cameroon

K. Mohanan
Institute of Fundamental Studies, Hantana Road, Kandy, Sri Lanka

R.R. Ratnayake
Institute of Fundamental Studies, Hantana Road, Kandy, Sri Lanka

K. Mathaniga
Faculty of Agriculture, University of Jaffna, Sri Lanka

C. L. Abayasekara
Department of Botany, University of Peradeniya, Peradeniya, Sri Lanka

N. Gnanavelrajah
Faculty of Agriculture, University of Jaffna, Sri Lanka

Tian Haijiao
College of Life Sciences/Daqing Bio-tech Institute, Northeast Forestry University, Harbin 150040, China

Qu Chang
College of Life Sciences/Daqing Bio-tech Institute, Northeast Forestry University, Harbin 150040, China

Wu Hao
College of Life Sciences/Daqing Bio-tech Institute, Northeast Forestry University, Harbin 150040, China

Li Yuhua
College of Life Sciences/Daqing Bio-tech Institute, Northeast Forestry University, Harbin 150040, China

Yang Hongyan
College of Life Sciences/Daqing Bio-tech Institute, Northeast Forestry University, Harbin 150040, China

K. Y VASANTHA
Fermentation Technology and Bio Engineering Department Central Food Technological Research Institute Mysore, 570020 India

SALEEM JAVEED
Fermentation Technology and Bio Engineering Department Central Food Technological Research Institute Mysore, 570020 India

D. CHAKRADHAR
Fermentation Technology and Bio Engineering Department Central Food Technological Research Institute Mysore, 570020 India

A. P. SATTUR
Fermentation Technology and Bio Engineering Department Central Food Technological Research Institute Mysore, 570020 India